P9-DWP-829

Statistics for Social Science and Public Policy

Advisors:
S.E. Fienberg, D. Lievesley, J. Rolph

Springer
New York
Berlin
Heidelberg
Barcelona
Budapest
Hong Kong
London
Milan
Paris
Santa Clara
Singapore
Tokyo

Statistics for Social Science and Public Policy

Devlin/Fienberg/Resnick/Roeder (Eds.): Intelligence, Genes, and Success: Scientists Respond to *The Bell Curve*.

Zeisel/Kaye: Prove It with Figures: Empirical Methods in Law and Litigation.

R

Bernie Devlin Stephen E. Fienberg
Daniel P. Resnick Kathryn Roeder
Editors

Intelligence, Genes, and Success

Scientists Respond to *The Bell Curve*

Computational Genetics Prog...
Pittsburgh, PA 15260
USA

Daniel P. Resnick
Department of History
Carnegie Mellon University
Pittsburgh, PA 15213
USA

Stephen E. Fienberg
Department of Statistics
Carnegie Mellon University
Pittsburgh, PA 15213
USA

Kathryn Roeder
Department of Statistics
Carnegie Mellon University
Pittsburgh, PA 15213
USA

Advisors:

Stephen E. Fienberg
Department of Statistics
Carnegie Mellon University
Pittsburgh, PA 15213
USA

Denise Lievesley
ESRC Data Archive
University of Essex
Colchester, Essex CO4 3SQ
UK

John Rolph
Department of Information and
Operations Management
Graduate School of Business
University of Southern California
Los Angeles, CA 90089
USA

Library of Congress Cataloging-in-Publication Data
Intelligence, genes, and success : scientists respond to the bell curve
/edited by Bernie Devlin . . . [et al.].
p. cm.
Includes bibliographical references and index.
ISBN 0-387-98234-5 (hardcover : alk paper). — ISBN 0-387-94986-0
(softcover : alk. paper)
1. Intellect. 2. Nature and nurture. 3. Intelligence levels-
-United States. 4. Intelligence levels—Social aspects—United
States. 5. Herrnstein, Richard J. Bell curve. I. Devlin, Bernie.
BF431.I527 1997
305.9′082–dc21 97-9792

Printed on acid-free paper.

Production managed by Steven Pisano; manufacturing supervised by Johanna Tschebull.
Photocomposed by Bartlett Press, Marietta, GA.
Printed and bound by R.R. Donnelley and Sons, Harrisonburg, VA.
Printed in the United States of America.

9 8 7 6 5 4 3 2 1

ISBN 0-387-98234-5 Springer-Verlag New York Berlin Heidelberg SPIN 10574687

PREFACE

The publication of Richard Herrnstein and Charles Murray's *The Bell Curve: Intelligence and Class Structure in American Life* in the fall of 1994 was accompanied by a major marketing campaign that saw the book mount all of the best-seller lists in the United States. It quickly produced an engaged public response, with reviews in a wide variety of newspapers, magazines, and other periodicals, many of which were colored by political perspectives. Critics on the left described the authors as "un-American" and "pseudo-scientific racists" and the vision of American society painted by their book as "alien and repellent." Defenders on the right characterized the authors as "brave and respectable scholars," simply explaining the overwhelming scientific evidence on the role of IQ and its large heritable component. They described the book as "lucid" and "powerfully written," and the reported analyses as "overwhelmingly convincing."

What had captured the critics' and the public's attention? It is the portion of the book that discusses the stark difference between the IQ distribution of African and Caucasian Americans. Although they never explicitly accept either a genetic or environmental explanation for the disparity, Herrnstein and Murray employ seemingly elaborate statistical analyses and cite evidence from diverse scientific studies, all of which seem to point to a genetic explanation for intelligence and the impact of intelligence on a host of different social outcomes. Their claims and work, we believe, require careful examination.

This volume began in an innocuous fashion as Daniel Resnick was asked to prepare a background paper by the Carnegie Commission Task Force on Early Primary Education regarding the scientific content of *The Bell Curve* following its publication in 1994. That background paper ultimately gathered three co-authors and, although it remained unpublished, it did lead the four of us to write a pair of reviews of the book, which appeared in *Chance* and in the *Journal of the American Statistical Association*.[1] While we worked to prepare those pieces,we realized that an in-depth examination of the scientific claims and, in particular, the reported statistical work in *The Bell Curve* would require much input from others, and considerable space of the sort not available in the usual professional journals or in standard nontechnical publications. Thus we conceived of an edited volume of response that attempted to take stock, in depth and from a variety of disciplinary perspectives, of the claims in *The Bell Curve*.

The contributors to this volume were selected both for their expertise and for their interest in *The Bell Curve*. Many have written separate reviews of the book, and the chapters in this volume often contain follow-ups and technical details. No prior publication survives here intact, but we are grateful to *The Atlantic Monthly* for permission to include the chapter by Nicholas Lemann in a form that resembles its earlier appearance in that magazine.

We were fortunate to have a series of early discussions with John Kimmel, Statistics Editor for Springer-Verlag, who encouraged us to focus on a scientific response to *The Bell Curve* and not to rush to print with what was readily available. While that advice and our other professional activities and preoccupations delayed the completion of the present volume, we believe the final product is far better than what we had originally proposed to do. John has counselled us throughout the subsequent two years and played a major role in bringing this book to print. We are deeply indebted to him for his advice and support.

By now, hundreds upon hundreds of reviews of or commentaries on *The Bell Curve* have appeared in print. We conclude this volume with a bibliography of the bulk of these reviews and commentaries. Much of the nontechnical commentary focuses on the theme of racial differences in IQ and often reads into *The Bell Curve* conclusions that Herrnstein and Murray did not draw explicitly. Many of the harshest criticisms appear to come from those who scarcely refer to statements and claims actually found in the book! Nonetheless, the diversity of previously published material has probably influenced our work, even as we tried to distance ourselves from it.

There have also been several book-length compilations of reviews and readings related to *The Bell Curve*. Several of the reviews referred to earlier have been reprinted in Fraser[2] and Jacoby and Glauberman.[3] The volume by Kincheloe et al. is far less technical and much more emotional in its response.[4] More comprehensive and technically detailed are the volumes by Schultze et al.[5] and by Fischer et al.,[6] but they are far from complete. We see the present volume as an attempt to give a fuller description of the scientific criticism, with sufficient detail that the reader can also understand its implications for public policy.

The Bell Curve has also provoked several "spinoffs." For example, there is a new best-seller by Daniel Goleman, *Emotional Intelligence*.[7] It contains only one reference to *The Bell Curve* and gives a broad review of the role of emotion in success. But the success of this book owes much to the theme trumpeted on its cover—"Why it can matter more than IQ"—clearly aimed at readers and critics of Herrnstein and Murray's book. There is also the new 1996 edition of Stephen Jay Gould's best-seller, *The Mismeasure of Man*, which on the cover claims to contain "[t]he definite refutation to the argument of *The Bell Curve*." In fact, the additions to the original consist of a new 30-page introduction and two final chapters based on his review of *The Bell Curve* for *The New Yorker* and another related essay. Gould is someone whose work we have read with interest, but his new edition of *Mismeasure* is not a replacement for the volume we have assembled.

Finally, we would like to thank our Carnegie Mellon colleagues for their interest in and encouragement of our work on this volume. We could not have produced it without the able assistance of Howard Fienberg and several staff members from the Department of Statistics at Carnegie Mellon, including Norene Mears and Margie Smykla. Heidi Sestrich took control at some point, and she succeeded in producing an integrated manuscript, with a recognizable style and structure that had earlier existed only in our minds. She corresponded with contributors, made us aware of issues, and managed an otherwise chaotic process. Her efforts were essential to bringing the volume to print and we thank her profusely.

Bernie Devlin
Stephen E. Fienberg
Daniel P. Resnick
Kathryn Roeder

References

1. Devlin, B., Fienberg, S.E., Resnick, D.P., and Roeder, K. (1995a), Wringing *The Bell Curve*: A cautionary tale about the relationships among race, genes and IQ. *Chance*, 8, 27–36. Devlin, B., Fienberg, S.E., Resnick, D.P., and Roeder, K. (1995b), Galton Redux: Eugenics, intelligence, race, and society. *Journal of the American Statistical Association*, 90, 1483–1488.
2. Fraser, S. (Ed.) (1995), *The Bell Curve Wars: Race, Intelligence, and the Future of America*, Basic Books, New York.
3. Jacoby, R., and Glauberman, N. (Eds.) (1995), *The Bell Curve Debate: History, Documents, and Opinion*, Random House, New York.
4. Kincheloe, J.L., Steinberg, S.R., and Gresson, A.D., III (Eds.) (1996), *Measured Lies: The Bell Curve Examined*, St. Martin's Press, New York.
5. Schultze, C.L., Dickens, W.T., and Kane, T.J. (1995), *Does "The Bell Curve" Ring True?*, Brookings Institution, Washington, D.C.
6. Fischer, C.S. Hout, M., Jankowski, M.S., Lucas, C.R., Swidler, A., and Voss, K. (1996), *Inequality by Design: Cracking the Bell Curve Myth*, Princeton University Press, Princeton, NJ.
7. Goleman, D. (1995), *Emotional Intelligence*, Bantam, New York.

CONTENTS

PART I

OVERVIEW

How much does the reader need to bring to the table to understand *The Bell Curve*? Not very much, say the book's authors. Its arguments are self-contained: "... everything you need to know to follow all of the discussion is contained within the book." (p. xix). But this work is neither transparent nor self-contained; its arguments push out beyond the book's confining covers. The authors, drawing selectively on the methods and findings of genetics, psychology, statistics and sociology, finesse much of the history of the eugenics movement and move beyond their own data to attack problems of public policy. So far-ranging a work calls for a response from the disciplines on which it draws.

That measured response begins with our first chapter, "Reexamining *The Bell Curve*," by Stephen Fienberg and Daniel Resnick, one a statistician and the other a historian. The two use this chapter to examine the lineage of arguments about cognitive castes, social divisions, and dysgenesis, and to preview the chapters that follow.

They note that the stance taken by Herrnstein and Murray is not without antecedents. Both authors had staked out individually in earlier publications the positions that they present in *The Bell Curve*. In 1973, Herrnstein, a psychologist, argued for the laws of biology in *IQ in the Meritocracy*. In 1984, Murray, a sociologist, declared his opposition to social legislation and the welfare state in *Losing Ground*. But like so many of today's defenders of arguments about intelligence, heredity and class or caste, Herrnsein and Murray owe their greatest debt to Francis Galton whose *Hereditary Genius: Its Laws and Consequences* appeared in 1869.

Two converging streams of the movement that Galton founded, one British and the other American, flow into *The Bell Curve*. From the British came the pioneering statistical studies and bio-social investigations of Galton, Karl Pearson, and Ronald Fisher. From the Americans came the psychology of individual differences. American psychologists developed the view of tested IQ as native intelligence. They identified "giftedness," supported school tracking, and put in place the selection systems—within school, school to college, and college to post-college—that Herrnstein and Murray find essential for producing and reproducing cognitive castes.

The Bell Curve can, however, be approached, if not fully understood, in its own terms. Terry Belke, a psychologist who trained under Richard Herrnstein at Harvard, presents a synopsis of *The Bell Curve* in Chapter 2. He does so with fidelity and nuance, reserving any critical judgment. Tracing the argument as it is presented, from the formation of cognitive elites and an "underclass" to the heritability of traits, he moves on to the demography of intelligence and race, gender, and ethnicity. He concludes with a summary of Herrnstein and Murray's position on affirmative action, compensatory education, and America's future.

chapter one

Reexamining *The Bell Curve*

STEPHEN E. FIENBERG AND DANIEL P. RESNICK

Occasionally, very occasionally, big books appear in the social sciences that make scholars and the lay public take notice. Richard Herrnstein and Charles Murray's *The Bell Curve: Intelligence and Class Structure in American Life* is one of those publishing events.[1] *The Bell Curve* is big both in size, more than 850 pages, and in scope. It draws on a large social science database and deals with themes of broad social significance. The authors have provoked a counterliterature of criticism, qualification, and confrontation that has advanced the enterprise of social research. Without the contest over methods, argument, and policy implications that are generated by the publication of such big books, public understanding would lag even further behind scholarship, and scholarship itself would lose its edge.

In *The Bell Curve*, Herrnstein and Murray ponder America's future. Growing inequality has been a major theme of the landmark social research of the last quarter-century. Other big books—such as Christopher Jencks's study of family and schooling,[2] James Coleman's examination of public and Catholic schools,[3] and William Julius Wilson's explorations of the urban underclass[4]—have dealt with the origins, expression, and consequences of inequality. We believe that *The Bell Curve* has an important place in this literature.

In a democracy, social problems seek political solutions. Wise governmental action, in turn, requires focused social research, and in our society since the Progressive period the social sciences have provided that base of knowledge. From early concerns with public health, water, and sewage to housing, transportation, and the economy, the range of issues investigated by researchers has continued to broaden. Employment, education, income, and crime are now part of the research and policy agenda, and they are the themes that Murray and Herrnstein address.

Political solutions to social problems have brought a continuing expansion of government at the federal, state, and local levels. The welfare state, introduced with limited goals in the Great Depression, has expanded rapidly since the 1960s. Charles Murray, co-author of *The Bell Curve*, argued in an earlier book, *Losing Ground*, against the moral and financial costs of this expansion. Health, public assistance, education, social insurance, and housing all ballooned. But the welfare system was not only costly, it was counterproductive, encouraging "dysfunctional values and behaviors" and producing welfare dependency.[5]

In *Losing Ground*, Murray found more poverty in the 1980s than thirty years earlier, despite the intervening social legislation, and therein lies a tale. In explaining growing inequality, Murray argued as a classical political economist, a sociologist, and a moralist. The generation of poverty was traced to bad interventionist government policies and irresponsible social behavior. He did not seek out biological factors to explain differences in the social successes of individuals or groups. The distribution or reproduction of IQ played no role in his earlier analysis of the failures of welfare policy.

As a psychologist, Richard Herrnstein, Murray's co-author in *The Bell Curve*, also found himself at war with the assumptions of interventionist government, particularly in the areas of education and employment. In his 1973 book, *I.Q. in the Meritocracy*, he wrote the following: "... [O]ur environmentalist bias has prevented us from noticing that society is willy-nilly subject to the laws of biology. There is evidence not only for the genetic ingredients in mental capacity but also in social status" (Ref. 6, p. 19). Herrnstein defended the IQ test score as a measure of real differences in human capacity and envisaged a society that would be increasingly meritocratic.

The Bell Curve's Argument

The product of Herrnstein and Murray's collaboration, *The Bell Curve*, was ten years in the making and presents a synthesis of these two themes, with an updated review of the scientific literature and new statistical analyses. It merges a genetic view of individual capacity with a libertarian critique of the welfare state.

Taylor succinctly summarizes the main themes of *The Bell Curve* via the following syllogism:

First premise: Measured intelligence (IQ) is largely genetically inherited.

Second premise: IQ is correlated positively with a variety of measures of socioeconomic success in society, such as a prestigious job, high annual income, and high educational attainment; and is inversely correlated with criminality and other measures of social failure.

Conclusion: Socioeconomic successes (and failures) are largely genetically caused.[7]

As corollaries, Herrnstein and Murray claim that because IQ is in large measure genetically determined, it is therefore resistant to educational and environmental interventions. They argue that American society is becoming increasingly stratified into a cognitive caste system, with the cognitively disadvantaged trapped at the bottom of society. This lower caste includes a large portion of the African-American population. They believe money spent and laws aimed to ameliorate this inequity will be wasted because IQ is largely genetically determined. Moreover, the "genetic capital" determining IQ is eroding, in large part due to the greater reproduction of low IQ individuals and, to a smaller extent, due to immigration.

The Bell Curve's pessimistic argument about declining intelligence in society, isolated ruling castes, and over-reproduction of the least able is not new to Western social thought. These concerns were first voiced in an age of emerging mass democracy in the second half of the nineteenth century. Because Murray and Herrnstein are more interested as social scientists in testable hypotheses than in historical and cultural context, they present little of this history. But it may help our readers to understand how these arguments were articulated in the past and how they have fared over time.

The History of the Argument

The British Eugenics Movement and Statistics

The roots of Herrnstein and Murray's argument in *The Bell Curve* can be traced back to Francis Galton in *Hereditary Genius, its Laws and Consequences*.[8] Galton was the central figure in the founding of the eugenics movement and the study of the relationship of heredity to race and talent.[9] For Galton, reputation was an index of ability, independent of social background. From his analysis of biographical dictionaries and encyclopedias, he became convinced that talent in science, the professions, and the arts ran in families, so that it would be "quite practicable to produce a highly gifted race of men by judicious marriages during several consecutive generations."

Galton coined the term *eugenics* in 1883 in *Inquiries into Human Faculty and its Development*, defining it as "the study of the agencies under social control that may improve or impair the racial qualities of future generations, either physically or mentally."[10] Influenced in his thinking about evolution by his cousin Charles Darwin, whose *Origin of Species* appeared in 1859, Galton believed that man's evolution would be accelerated through eugenics: "... [W]hat Nature does blindly, slowly, and ruthlessly, man may do providently, quickly, and kindly." Through statistics, he tried to understand the laws of inheritance. Assuming a normal distribution of talent, he tried from limited data to establish the distribution of talent in the British population. To go further, he needed more data. To gather this information, he set up the Anthropometric Laboratory and collected detailed measurements on families.

Galton is also responsible for helping to move statistics in the biological and social sciences beyond descriptive data gathering to more formal methods of inference. Collecting meteorological, botanical, and human physical data, he looked for statistical relationships. To this end, in 1888 he formulated what we now know as the correlation coefficient to capture such relationships. Galton did not offer mathematical proofs, and he did not solve the problem of heritability of talent, but he recognized the major role that statistics plays in the study of the nature of intelligence.

Karl Pearson is recognized as one of the founders of modern statistical methods. In statistical matters, as well as in eugenics, he was strongly influenced by Galton. Pearson joined University College London as a professor of mathematics and then later became Galton Eugenics Professor and Head of the Department of Applied Statistics. The department Pearson built became the English center of statistics and biometrical methods, and in it he and his collaborators published some three hundred works dealing with the relationship between population traits and social behaviors, occupations, and diseases. Included in these publications was a series entitled *Studies in National Deterioration*, published as *Drapers' Company Research Memoirs* by Cambridge University Press from 1907 to 1923.

Pearson shared Galton's concern with under-reproduction of the cultured classes and high reproduction rates among the least "fit." The unfit were "the habitual criminal, the professional tramp, the tuberculous, the insane, the mentally defective, the alcoholic, the diseased from birth or from excess." The state, he argued, had not recognized the need for a public policy that paid attention to biological factors. In an early discussion of the nature versus nurture controversy he argued, "We have placed our money on Environment, when Heredity wins in a canter." For intelligence, "No training or education can create [it].... You must breed it" (quoted in Ref. 9, p. 33). In 1901, to focus on the relationship of heredity and evolution in animals and plants, and to promote the development of statistical theory applicable to biological problems, Pearson and the eminent biologist Walter Weldon (in consultation with Galton) founded the journal *Biometrika*. Half of the first wave of subscribers were

American, driven far more by their attraction to eugenics than by their interest in statistical methods.

The early volumes of *Biometrika* show Pearson, his colleagues, and their students at work on central issues in the eugenics movement. But Pearson also wrote at length about the problems associated with spurious correlation and the need to take great care in inferring causation from correlation, one of the principal tools in the studies of heredity in *Biometrika*. It is perhaps for this reason that the studies of heredity in *Biometrika* were soon displaced by more technical articles on statistical methods and the journal became a focal point for statistical rather than eugenic innovation.

It was just after the First World War that Ronald Aylmer Fisher emerged as a statistical theorist and a leader in the eugenics movement. Fisher's first major statistical paper, on the distribution of the correlation coefficient, appeared in *Biometrika* in 1915. In 1918 he introduced the analysis of variance as a tool in genetics studies, and then later showed that the same tool could be used in the context of randomized experiments. Fisher was also the founder of modern quantitative genetics. The statistical tools that underpin the studies of genetics and IQ originate from Fisher. At the same time as he was revolutionizing both the fields of statistics and genetics, Fisher continued to write about eugenics, and for decades was the editor of the *Annals of Eugenics*, a second journal founded with Karl Pearson that was devoted to careful quantitative genetics studies and statistical methods. (See Ref. 11 for further details.)

In recent years, much has been written about the eugenics movement and the evils it fostered.[9,12] A recent report from the Committee on Science, Engineering, and Public Policy at the National Academy of Sciences, in discussing values in science, states: "The history of science offers a number of episodes in which social or personal beliefs distorted the work of researchers. The field of eugenics used the techniques of science to show the inferiority of certain races."[13] While in part true, this critique of eugenics as a field ignores the commitment to public health and progressive causes of those who supported "positive" eugenics. It essentially equates British views to the much cruder American and German race-thinking and does an injustice to Galton, Pearson, and Fisher.

Modern readers need to recognize that they were social progressives, as were most of those involved in the early English eugenics movement. Thus to label their eugenics views as racist as we now use the term, with today's sensitivities, misses the mark. Galton did believe, however, that group differences, familial and racial, could be studied scientifically, and he set about doing so in a systematic fashion. Not only was he looking to improve the breed of mankind, but when he attempted to look at racial differences scientifically, he was reluctant to see differences when they did not appear to exist in clear and convincing form. The case of Karl Pearson is decidedly mixed, although we see limited ill effect of his eugenics views on his statistical methodological work.

Fisher was an exemplary scientist, both as a statistician and as a geneticist. But he had a fascination with eugenics that he pursued throughout his career. Even as an undergraduate at Cambridge in 1911, Fisher had speculated about breeding a superior race of Englishmen (see the reproduction of Fisher's essay in Ref. 14, pp. 53–54):

> Suppose we knew, for instance, 20 pairs of mental characters. These would combine in over a million pure [homozygous] mental types. In practice, each of these would naturally occur ... in a country like England, about once in 20,000 generations; it will give some idea of the excellence of the best of these types when we consider that the Englishmen from Shakespeare to Darwin (or choose whom you will) have occurred within ten generations; the thought of a race of men combining the illustrious qualities of these giants, and breeding true to them, is almost too overwhelming, but such a race will inevitably arise in whatever country first sees the inheritance of mental characters elucidated.

That Fisher's eugenics views motivated the genetics and statistical problems he chose to tackle is clear, but, with some minor exceptions, these views rarely encroached upon the scientific content of his work. The major exception is Fisher's pathbreaking 1930 book, *The Genetical Theory of Natural Selection*, the last half of which consists of essays on eugenics themes with limited empirical support.[15]

The modern history of statistics, even with these qualifications, remains intertwined with that of eugenics. The eugenics goals of Galton, Pearson, and Fisher clearly influenced them in the development of new statistical methods. Their statistical work and their scientific investigations are admirable, even if we do not share their passion for eugenics.

Statistics and Mental Testing

The history of statistics is also intertwined, if not entangled, with the the history of testing, especially mental testing.[16] The term *mental tests* was introduced by J. McKeen Cattell in a celebrated article in *Mind* in 1890.[17] It addressed and gave a generic name to the German, French, and English tasks devised to established differences in the mental capacities of individuals. Galton had tried at the Kensington Museum to create a set of tests on which those who were endowed with exceptional talents could show their abilities. But these were largely tests of speed, reflex time, quickness of perception, and strength.

Experimentation with intellectual tests, individually administered, was more developed in France and Germany than it was in England. In Germany, significant work on memory, involving strings of nonsense syllables, can be traced to Ebbinghaus in the 1880s. (See the related discussion in Ref. 18). In France, the effort to identify intelligence through performance on intellectual tasks such as pattern building with blocks, dates from the first decade of the nineteenth century, when Dr. Itard worked to understand the capacities of the "wild boy" of the Aveyron.[19] The search for ways to identify those who were retarded in their development and to develop appropriate educational programs for them was continued in the work of Esquirol, Seguin, and Alfred Binet.

The first intelligence tests, described by Alfred Binet and a colleague in 1896, involved counting objects in pictures, noting similarities in familiar objects, filling in missing words in a sentence, and describing how terms had different meanings. In 1905, and in revised form in 1908 and 1911, Binet and Simon published tests in sets appropriate for different age groups. These tests were to be individually administered, and to indicate whether individual children, largely between the ages of three and eleven, were performing at the appropriate level for their age.

American translators and revisers of the Binet tests between 1908 and 1925 took an artisanal, individually administered, and still experimental product, for which the French had very limited immediate use, and turned it into a group-administered pencil-and-paper measure whose administration in schools, the army, and industry became associated with smart and efficient use of human resources. To do that, they created a one-dimensional IQ score to replace the clumsy and difficult-to-interpret mental and chronological age levels reported by Binet; sampled more systematically the domains of verbal, mathematical, and spatial knowledge; standardized test administration; and made the development of testing and measurement integral to the growth and health of psychology as a profession. Henry Goddard, known later for his work on the generational transmission of criminality, drunkenness, and promiscuity in studies of *The Kallikak Family*,[20] made the first American version of the 1908 scale, administering it to 2,000 public school children in Vineland, New Jersey.

But Lewis Terman, who helped to prepare the individually administered Stanford Revision of the Binet in 1916, had the larger impact. He recognized that as long as the IQ test had to be administered one-on-one by a trained psychologist, it would be too expensive and recondite to enjoy widespread adoption and commercial success. Lewis Terman recognized that an accessible pencil-and-paper IQ test could be used both to reorganize schools with multiple tracks and homogeneous classrooms, and for the selection and placement of army recruits.

Terman encouraged his graduate student William Otis to develop a pilot pencil-and-paper measure. Both Terman and Otis then traveled to Vineland, New Jersey, to join in the creation of Army Alpha, the World War I group intelligence test administered to 1.7 million army recruits in 1917 through 1919 after trials with school, university, and institionalized populations.[21] That test is also the prototype for the Armed Services Vocational Aptitude Battery

(ASVAB) administered in World War II and still in use. The ASVAB cognitive tests of verbal and mathematical reasoning provide Herrnstein and Murray with the measure for IQ in the National Longitudinal Survey.

As author/editor of the National Intelligence Tests and the Terman Group Tests, the most widely sold intelligence tests in the 1920s, president of the American Psychological Association, and a strong advocate for the gifted, Terman long represented a voice for nature in the nature–nurture controversy.[22,23] If studies could show, he argued in 1924, that differences in intelligence were largely the result of heredity, the implications for education would be radical:

> If this, or even a half of it, should be found true, the practical consequences would be well nigh incalculable. Eugenics would deserve to become a religion. Educational effort, while it would deserve to continue, would largely have to be redirected. The first task of the school would be to establish the native quality of every pupil; second, to supply the kind of instruction suited to each grade of ability. Either group instruction would be abandoned, or, if it were retained, a three-track or five-track plan would operate from the first grade on, each with methods and a curriculum peculiar to itself. (Ref. 24, p. 336)

Eugenics in America

The racist strain in the eugenics movement was more pronounced in the United States than in Great Britain. England was a fairly homogeneous society, where the cleavage lines were of social class. And for England, as for France, population growth was essential to counter the size and power of the German empire. The United States, however, was made up of many different peoples. Voluntary European and Asian migration, and the forced migration of Africans had created a very heterogeneous society. Population growth within the society was heavily influenced by immigration, which, increasingly after 1880, brought in nationals from central, eastern, and southern Europe. In 1907, the peak year for pre–World War I U.S. immigration, 1.2 million new immigrants arrived. Eugenic arguments in the United States began to reflect native fears about the displacement of Nordic peoples by the "New Immigration." The post–World War I curtailment of immigration by quotas tied to national origin reflected those fears.[25]

Race at that time was associated with ethnic background as well as skin color. White Northern European Protestants, as Nordics, distinguished themselves in the popular race literature of the 1920s from lesser Alpines, Mediterraneans, Irish Catholics, blacks, and Jews. But how did they all compare on the dimension of intelligence? And how was intelligence to be measured? Galton had assumed that social position was a good proxy for intelligence, and Pearson, in one survey, had tried to use teachers' judgments of student quality as a stand-in. American psychologists in World War I were the first to offer scaled test scores as an indicator for intelligence.

Intelligence on Army Alpha was assessed by item batteries on language and mathematical and spatial reasoning, with questions that have become familiar on intelligence tests ever since. Questions about synonyms and antonyms, likes and unlikes, were joined with arithmetic problems and queries of practical judgment. The goal was to test native intelligence, but the language of the questions, the assumed knowledge base, and the fact that test-taker recruits required the ability to read and follow instructions privileged non-immigrants and those with more years of schooling. Almost a quarter of the recruits could not read a newspaper or write a letter, and the average white testee on the scale used at the time had the "mental age" of a 13–14 year old. The results were widely touted as demonstrating the superiority of Nordic stock over the rest. Nordics, lesser Alpines, Mediterraneans, and blacks were ranked in a hierarchy, with Nordics as the most intelligent. Carl Brigham's analysis of the results in *A Study of American Intelligence* was published with overlapping bell-shaped curves.[26]

Although analysis of test results also showed that scores followed the lines of social class and occupation, relatively little attention was given to those findings. Americans were not interested in class, one historian has concluded. They were interested in race.[27] Most of the leading psychologists associated with the army testing program assumed that intelligence was innate, unequally distributed, and concentrated in the racial stock of educated professional groups and families. Famous figures in the history of American psychology—such as Robert Yerkes, Carl Brigham, and Lewis Terman—were hereditarians. When the focus turned in the 1920s to the intelligence of school children from different social backgrounds, it became a common observation that parents from working-class and very poor homes were less likely to produce children who scored high on the tests. How could this be? Wasn't environment influencing the test results? Lewis Terman, still in the thrall of the hereditarian paradigm, argued that biology was in the saddle. He wrote in 1922, "The common opinion that the child from a cultured home does better in tests by reason of his superior home advantages is an entirely gratuitous assumption.... The children of successful and cultured parents test higher than children from wretched and ignorant homes for the simple reason that their heredity is better." (See also Refs. 22 and 28).

But in the course of the 1920s, true believers became skeptics, and it appeared that the mainstream of American psychology had made a major paradigm shift, from race to culture, and from nature to nurture. The explanation for this shift among psychologists can be attributed, in part, to the accumulating weight of scientific argument in the psychology journals (see Ref. 27, pp. 187–192). Thomas Garth, a psychologist at the University of Texas, surveying the literature, found that more than 250 studies on racial difference had been published between the end of the world war and 1930.

In this work, there was mounting evidence that environmental influences had not been equalized in comparisons of the measured intelligence of racial groups.

Some of this work was based on a reanalysis of the wartime army test data; other studies described fresh experiments. William Bagley, an educational psychologist, reviewing the army data, noted that blacks from Illinois had higher scores than whites from nine southern states, a finding difficult to reconcile with the hypothesis of racially based differences. In an experimental study comparing scores of native white, black, and immigrant Italian children on the same test, Bryn Mawr psychologist Ada Arlitt found "there is more likeness between children of the same social status but of different race than between children of the same race of different social status."[29] Commenting on the pattern of these findings in a book published in 1931, Garth wrote the following: "we have never with all our searching ... found indisputable evidence for belief in mental differences that are essentially racial" (quoted in Ref. 27, p. 178).

The work of Otto Klineberg also contributed to this fundamental revision of position. In a remarkable set of comparative studies that began at Columbia in 1925 when he was only twenty-six and extended over a decade, he set out to test hypotheses about the influence of race on intelligence. In a study of Southern-born black children in New York City, for example, he showed that the longer the children lived in New York, the higher their scores. By the end of the decade, the position of mainstream psychologists came into line with those of anthropologists such as Franz Boas, Irving Kroeber, and Ruth Benedict, for whom the paradigm of culture had replaced that of race.

Evidence of this striking reversal of opinion among psychologists can be found in the work of Carl Brigham, who had authored the major study of army test results in 1923. The author, who had argued for the tight linkage of intelligence and race, shifted his position markedly by the end of the decade. He had learned, he wrote, that tests "in the vernacular must be used only with individuals having equal opportunities to acquire the vernacular of the test." And, he continued, "... comparative studies of various national and racial groups may not be made with existing tests." His view of his own contribution on the subject of intelligence and race was extremely negative: "... one of the most pretentious of these comparative racial studies—the writer's own—was without foundation" (Ref. 30, p. 65).

The hereditarian paradigm, with its construction of nature and nurture as distinct and independent, was on the way out by 1930, or so it seemed. But it was not just scholarly research that made the difference. It was also the case that immigration restrictions in the 1920s had stemmed the flow of new migrants, making them a less visible target for scapegoating, and that poverty and unemployment were so widespread in the Great Depression that it was difficult to still associate them with any particular ethnicity.

Racist thories of superior and inferior intelligence also fell from favor because of backlash against the sterilization programs of the interwar period. Senseless, inhumane, and technocratic programs of sterilizaion were launched in the United States and Germany. By the end of the 1920s, in a program that had no British counterpart, twenty-four American states passed sterilization laws, and by the mid-1930s sterilized against their will and sometimes without their knowledge some twenty thousand Americans. In Nazi Germany in 1934 to 1939, 320,000 people suffered the same fate.[31]

Mobilization for the Second World War brought twelve million American men and women of different racial, religious, national, and class backgrounds into uniforms for their country. The "good war" revived the commitment to democracy and pluralism, glorified the common man in "G.I. Joe," and encouraged more open opportunities for employment and education.[32] For many, hereditarian theories seemed as archaic as they were un-American.

When hereditarian doctrines were driven from favor, they were not replaced by a naive environmentalism. Instead, there was increasing focus on the ways in which heredity and environment interacted. The nature–nurture conflict had been noticeably quiescent since the early 1940s, Ann Anastasi, the noted psychologist, argued at the American Psychological Association meetings in 1957. The reason, she said, was that geneticists, psychologists, and social scientists had become convinced that nature and nurture were interactive and interrelated, not independent and additive in their impact on human development. The focus had to be on "interaction."[33] And she called on researchers to focus on how nature and environment interacted in specific situations.[23,33]

The Modern Nature–Nuture Debate

The verdict, however, about the death of the nature–nurture argument was premature. Within the field of psychology, it had a continuing life. To no other specialization within the discipline was the construct of IQ so central as it was to measurement psychology. Most measurement psychologists continued to work within the paradigm established by Spearman for general intelligence, even when they believed that they were able to identify a variety of different kinds of intelligence. (See the chapters in the present volume by Carroll and Hunt.) Most mental testers had little doubt that IQ tests were measuring what they were intended to measure, that intelligence was relatively stable over time, and that IQ was probably a natural endowment. School systems organized with tracked programs produced students whose school results seemed to mirror the assumed distribution of natural ability. There was no necessary scientific reason for the testers to also identify with hereditarians, but the disproportionate representation in "gifted" programs of children of the wealthy and better educated, and among school dropouts and weak performers of the poor and racial minorities, made judgments of this kind frequent and common (See Ref. 28).

Geneticists, however, have also considered their research relevant to educational practice. The major papers of Burks on the nature–nurture relationship in the resemblances of parents/foster parents and children/foster children and of Tryon on the maze-learning behavior of different strains of rats were published first in the *Yearbook of the National Society for the Study of Education*.[34] Geneticists, who were not all in agreement on the particulars, came to share with testing psychologists a belief in the significance of heredity, but one that was more qualified and conditional, and based on interactionist models (see the chapters in this volume by Daniels, Devlin, and Roeder, and by Wahlsten). With psychologists who studied learning, geneticists shared an acceptance of the importance of environment, but rejected radical environmental determinism. The tensions and alignments between the two disciplines at the end of the 1960s are visible in the proceedings of a workshop that brought researchers together from both camps.[35]

The 1960s was a cataclysmic period for America, at home and abroad. In response to political opportunities created by a new public awareness of poverty in the nation[36] and the pressure of the civil rights movement, legislators intervened to reshape educational opportunity, employment, voting rights, and the distribution of income. Interventionist public policy did help to create a more level playing field in many areas, but it also generated resistance. Civil rights, compensatory education, affirmative action, job training, housing subsidies, legal aid, and income support each had a constituency, but each also added to the backlash. Some of the resistance was social and racial, but there was also taxpayer opposition to the cost of social programs.

In this period of backlash, which has not yet spent its force, two beliefs of the earlier eugenics movement have again gained currency. They are that the limits of native endowment, unequally distributed across races and social groups, explain failures in education and employment; and that state policies that try to alter this fact are themselves doomed to failure. Arthur Jensen's 1969 article in the *Harvard Educational Review*, "How Much Can We Boost IQ and Scholastic Achievement?" contributed significantly to the revival of research and polemic on the relationship of race to intelligence,[37] and to argument about the dysgenic consequences of public spending in the United States as well as in Britain (See Ref. 9, pp. 269–70).

Jensen argued, reversing a position he had taken earlier, that compensatory education programs weren't working both because of the way they were conceived and because of the limits of IQ in the targeted school population. In the debate that followed, Richard Herrnstein (1973) joined in, largely on Jensen's side, to argue the power of IQ to predict not only school success but success in life. A flood of counterargument (See Ref. 6) made clear that there was more than simple disagreement about the positions of both Jensen and Herrnstein.[38] The critics drew attention to serious errors in empirical analysis and in technical and scientific reasoning.

This debate, which raged from 1969 to 1975, raised questions about the meaning of IQ, the role of environment, the theory of intelligence, and the use of evidence. The debate was about theory and method in the nature versus nurture conflict. For the public and the popular press, however, it was about race, affirmative action, social programs, and whether the legislation on compensatory education and affirmative action could be maintained and defended. And the contributors often did not try to distance themselves from these issues.

Our Response to *The Bell Curve*

In *The Bell Curve*, Herrnstein and Murray revisit the arguments of the earlier period with what they see as a new body of compelling evidence drawn in part from the intervening literature and in part from their own analyses performed on data extracted from the National Longitudinal Study of Youth (NLSY), conducted by the National Opinion Research Center on behalf of researchers at Ohio State University and with funding from the Bureau of Labor Statistics.[39] The NLSY is a rich database drawn from a population that was born in the 1960s and entered peacetime employment in the late 1970s and 1980s. The approximately four million Americans who turned eighteen in 1980 are in this cohort, with information about their demographics, wealth, poverty, education, test scores (IQ), and occupational choices. Social scientists have tried to learn as much as they can about this generation from the data, testing hypotheses about the effects of social opportunity as well as native endowment on life outcomes.

Herrnstein and Murray also draw heavily on a body of research on the nature and heritability of intelligence. Some of that research comes from genetics and some from measurement psychology. In Part II of this volume, Daniels, Devlin, and Roeder lead off with a reassesment of the data on the heritability of intelligence, and Wahlsten follows with a discussion of the malleability of intelligence and how this relates to heritability. Singer and Ryff then present biologic arguments distinct from those rooted in heritability that might serve as an alternative framework for looking at the impact of intelligence and other explanations of social success. Then in Part III, Carroll and Hunt reassess the history of tests for intelligence and the argument for a single factor, known as *g*, in light of modern theories of cognition.

The *Bell Curve's* argument is, in large part, statistical. In Part IV, we reexamine Herrnstein and Murray's statistical analyses of the data from the NLSY. Cawley, Conneely, Heckman, and Vytlacil (Chapter 8) reexamine the relationship between cognitive ability and wages, and Cavallo, El-Abbadi, and Heeb (Chapter 9) look at the role of gender in the relationship between ability, race, and earnings. Winship and Korenman (Chapter 10) consider the relationship between years of schooling and IQ. Manolakes (Chapter 11) explores the assumed ties between low IQ and crime, a common assumption for early researchers in the eugenics movement. Next, Glymour (Chapter 12) concludes this part by offering a critical overview of causal thinking within quantitative social science. He pays particular attention to the assumptions implicit in statistical models that play an important role in regression, factor analysis, and *The Bell Curve*.

In Part V, Zigler and Styfco (Chapter 13) consider the policy implications of *The Bell Curve* for investment in children, challenging the conclusions drawn by Herrnstein and Murray. They are joined by Lemann (Chapter 14), who reassesses Herrnstein and Murray's argument about a growing cognitive elite. In the Afterword, Fienberg and Resnick review how well the quantitative scientific claims in *The Bell Curve* have withstood the critique offered in this volume, and they consider the larger question of the relationship among public policy, quantitative research, and personal conviction.

Herrnstein and Murray devote much space to calming the reader's fear of technical statistical arguments, and they segregate the technical material in footnotes and appendices. Indeed, their Appendix 1 is a primer on statistics, but it doesn't get much beyond means,variances, correlations, and some simple ideas on regression. The devil, however, is as always in the details. One not versed in the technical aspects of statistical methods and their application might suppose that these materials contain the answers that statisticians and quantitative social scientists might have about data sources, methods used, etc. Such semitechnical details rarely suffice, however, to answer searching questions, as several contributors to this volume help to point out.

Our volume of response, *Intelligence, Genes, and Success*, is particularly designed for quantitative readers who want to gauge for themselves the soundness of the statistical argument. They will want to question Herrnstein and Murray's treatment of the statistical data, look at how hypotheses are stated and tested, reflect on the limits of the data that have been used, consider the effects of exploring other hypotheses with the same data, weigh the effect of using different statistical procedures to establish the relationship between variables, and argue about whether the authors' understanding of heritability respects the findings of modern genetics.

But there is also a story here about scientific research, public policy, and America's future. Herrnstein and Murray tell the story as they see it, and very skillfully, moving from scientific findings to policy recommendations. In fairness to the authors, and to establish a narrative context, we need to provide an opportunity to review that argument as a whole before considering its separate parts. In the next chapter, Terry Belke presents a detailed and nonjudgmental summary of the *The Bell Curve*. Readers who are ready to begin the interrogation can move ahead at their own pace.

References

1. Herrnstein, R.J. and Murray, C. (1994), *The Bell Curve: Intelligence and Class Structure in American Life*, The Free Press, New York.
2. Jencks, C., (1972), *Inequality; A Reasseessment of the Effects of Family and Schooling in America*, Basic Books, New York.
3. Coleman, J., Hoffman, T., and Kilgore, S. (1982), *High School Achievment: Public, Catholic and Private Schools Compared*, Basic Books, New York.
4. Wilson, W.J. (1987), *The Truly Disadvantaged: The Inner City, The Underclass, and Public Policy*, University of Chicago Press, Chicago. Wilson, W.J. (1996), *When Work Disappears: The World of the New Urban Poor*, Alfred A. Knopf, New York.
5. Murray, C. (1984), *Losing Ground: American Social Policy, 1950–1980*, Basic Books, New York.
6. Herrnstein, R. (1973), *I.Q. in the Meritocracy*, Little, Brown and Co., Boston.
7. Taylor, R.M. (1995), "*The Bell Curve*," *Contemporary Sociology*, 24, 153–158.
8. Galton, F. (1869), *Hereditary Genius, its Laws and Consequences*, Macmillan, London.
9. Kevles, D.J. (1985), *In the Name of Eugenics: Genetics and the Uses of Human Heredity*, Alfred A. Knopf, New York.
10. Galton, F. (1883), *Inquiries into Human Faculty and its Development*, Macmillan, London.
11. Box, J.F. (1978), *R. A. Fisher, The Life of a Scientist*, Wiley, New York.
12. Mackenzie, D.A. (1981), *Statistics in Britain, 1865–1930. The Social Construction of Scientific Knowledge*, Edinburgh University Press, Edinburgh.
13. Committee on Science, Engineering, and Public Policy (1995), *On Being a Scientist. Responsible Conduct in Research*, 2nd ed., National Academy Press, Washington, DC.
14. Bennett, J.H. (1983), *Natural Selection, Heredity, and Eugenics, Including Selected Correspondence of R. A. Fisher with Leonard Darwin and Others*, Oxford University Press, New York.
15. Fisher, R.A. (1930), *The Genetical Theory of Natural Selection*, Oxford University Press, Clarendon; 2nd revised ed., (1958), Dover Publications, New York.
16. Resnick, D.P. (1982), "History of Educational Testing," in A.K. Wigdor and W.R. Garner, eds., *Ability Testing: Uses, Consequences and Controversies*. Part II, pp. 173–194, National Academy Press, Washington, DC.
17. Cattell, J.M. (1890), "Mental Tests and Measurements," *Mind*, 15, 373–381.
18. Fienberg, S.E. (1991), "The History of Statistics in Three and One-Half Chapters," *Historical Methods*, 24 (No. 3), 124–135.
19. Itard, J.M.G. (1962) *The Wild Boy of Aveyron*, translated by George and Muriel Humphrey, Appleton-Century-Crofts, New York. Lane, Harlan L. (1962), *The Wild Boy of Aveyron*, Harvard University Press, Cambridge.

20. Goddard, H.H. (1912), *The Kallikak Family*, Macmillan, New York.
21. Chapman, P.D. (1988), *Schools as Sorters: Lewis M. Terman, Applied Psychology, and the Intelligence Testing Movement, 1890–1930*. New York University Press, New York.
22. Terman, L.W. (1922) "Were We Born That Way?," *World's Work*, 44: 660.
23. Cravens, H. (1978), *The Triumph of Evolution: American Scientists and the Heredity-Environment Controversy 1900–1941*, University of Philadelphia Press, Philadelphia.
24. Terman, L.W. (1924). "The Possibilities and Limitations of Training," *Journal of Educational Research*, 10, 335–343.
25. Haller, M. (1963), *Eugenics: Hereditarian Attitudes in American Thought*, Rutgers University Press, New Brunswick, NJ. Taylor, P. (1971), *The Distant Magnet: European Emigration to the U.S.A.*, Harper & Row, New York.
26. Brigham, C. (1923), *A Study of American Intelligence*, Princeton University Press, Princeton, NJ.
27. Degler, C. (1991), *In Search of Human Nature: The Decline and Revival of Darwinism in American Social Thought*, Oxford University Press, New York.
28. Margolin, L. (1994), *Goodness Personified: The Emergence of Gifted Children*, Walter de Gruyter, New York.
29. Arlitt, A. (1921), "On the Need for Caution in Establishing Race Norms," *Journal of Applied Psychology*, 5, pp. 179–83.
30. Brigham, C. (1930), "Intelligence Tests of Immigrant Groups," *Psychological Review*, 37, 158–165.
31. Weindling, P. (1989), *Health, Race and German Politics Between National Unification and Nazism, 1870–1945*, Cambridge University Press, Cambridge.
32. Fussell, P. (1989), *Wartime: Understanding and Behavior in the Second World War*, Oxford University Press, New York.
33. Anastasi, A. (1958), "Heredity, Environment, and the Question 'How?'," *Psychological Review*, 65, 197–208.
34. Burks, B.S. (1928), "The Relative Influence of Nature and Nurture Upon Mental Development," *Yearbook for the National Society for the Study of Education*, 27, 219–316. Tryon, R.C. (1940), "Genetic differences in maze-learning ability in rats," *Yearbook for the National Society for the Study of Education*, 39, 111–119.
35. Ehrman, L., Omenn, G., and Caspari, E. (Eds.) (1972), *Genetics, Environment and Behavior: Implications for Educational Policy*, Academic Press, New York.
36. Harrington, M. (1963), *The Other America: Poverty in the United States*, Macmillan, New York.
37. Jensen, A.R. (1969), "How Much Can We Boost IQ and Schoastic Achievement?" *Harvard Educational Review*, 39, 1–123.
38. Block, N.J., and Dworkin, G. eds. (1976), *The IQ Controversy*, Pantheon Books, New York, pp. 410–540.
39. Bock, R.D., and Moore, E.G.J. (1986), *Advantage and Disadvantage: A Profile of American Youth*, Lawrence Erlbaum Associates, Hillsdale, NJ.

chapter two

A Synopsis of *The Bell Curve*

TERRY W. BELKE

The Bell Curve: Intelligence and Class Structure in American Life, by Richard J. Herrnstein and Charles Murray has engendered considerable debate.[1] Readers bring to their reading of this book an initial attitude toward the subject that colors their interpretation of the message that Herrnstein and Murray are trying to convey. The reader who is vehemently opposed to any claim that "intelligence" has a genetic basis dismisses the words as rantings of racists. At the other end of the spectrum, readers consumed with hatred and bigotry toward individuals of other races and ethnicity interpret the contents of this book as justification for their biases and actions. In either case, these readers are doing this work a disservice. This chapter offers a concise summary of arguments and materials in *The Bell Curve*. It is a factual reporting of what Herrnstein and Murray say and should be read before the commentary in the chapters that follow.

Introduction

The introduction begins with an overview of the history of the concept of intelligence, beginning with Galton and tracing the developments surrounding the IQ controversy to modern times. Galton first used tests of perceptual and motor skills to discriminate differences in intelligence. Later, Binet developed tests of reasoning, drawing, analogies, and pattern recognition that form the basis of modern intelligence tests. Spearman's contribution was the concept of a general intelligence factor (*g*) underlying correlations between tests of intelligence. Early advances in the study of intelligence were reversed by advocacy of testing for racial policies (e.g., sterilization laws). Finally, the 1960s heralded a fundamental shift away from causes within the individual as the source of social ills to causes outside the individual. Social factors that could be redressed by the government were considered the source of deficiencies. In this context of egalitarianism, recognition of biological bases of individual differences was and remains anathema.

The introduction closes with an outline of Herrnstein and Murray's perspective. First, the authors warn the readers against committing the ecological fallacy, that is, generalization from the aggregate to the individual, given that the analyses to be presented are for aggregate data. Second, the authors state that the importance of intelligence among human virtues has been inflated and that the assumption that a person's intelligence can be inferred from casual interactions is erroneous. Third, the authors note that the identification of IQ with attractive human qualities is wrong. Most importantly, readers should keep the following statement in mind as they read the book:

> Measures of intelligence have reliable statistical relationships with important social phenomena, but they are a limited tool for deciding what to make of any given individual. Repeat it we must, for one of the problems of writing about intelligence is how to remind readers often enough how little an IQ score tells you about whether the human being next to you is someone whom you will admire or cherish. (p. 21)

The Emergence of a Cognitive Elite

This part documents the transformation of society during the twentieth century from one in which social standing was largely determined by birth and cognitive ability was evenly spread throughout social strata to a society in which social mobility is a function of cognitive ability.

In the chapter entitled "Cognitive Class and Education, 1900–1990," Herrnstein and Murray document the opening of colleges to the general population, which led the way to cognitive partitioning of the American population. Three processes were initiated: growth of the college population, a more efficient recruitment of cognitive ability, and further sorting of cognitive ability by colleges. From 1900 to 1990, the percentage of 23 year olds with bachelor's degrees increased from 2% to 30%. However, this increase in the probability of going to college was not evenly spread across the range of cognitive ability. As the authors show, the probability of entering college increased dramatically for students in the upper half of the IQ distribution but decreased marginally for those in the lower half. Finally, this influx of cognitive ability into colleges was not evenly distributed among colleges. Colleges of greater prestige harvested a greater yield of top students than did colleges of lesser prestige. This process was aided by long-distance travel becoming commonplace and an increase in the number of families who could afford to send their offspring to elite colleges. They close this chapter with a discussion of the isolation of different segments of society created by cognitive partitions.

In the chapter entitled "Cognitive Partitioning by Occupation," Herrnstein and Murray document a second process of occupational sorting by cognitive ability. High-IQ professions have grown tremendously since 1940, and the proportion of individuals in the top decile of IQ in these professions has increased. Another line of evidence for occupational cognitive segregation offered by the authors is the decline in CEOs with only high school degrees and the concomitant increase in CEOs with graduate degrees. The point of this chapter is that

> in mid-century, America was still a society in which a large proportion of the top tenth of IQ, probably a majority, were scattered throughout the population, not working in a high-IQ profession and not in a managerial position. As the century draws to a close, some very high proportion of that same group is concentrated within those highly screened jobs. (p. 61)

In "Economic Pressures to Partition," the third chapter, the authors argue that worker productivity is directly linked to intelligence and that this relationship is sufficient to have economic consequences. This thesis directly contradicts the received wisdom that "the relation between IQ scores and job performance is weak, and, second, whatever weak relationship there is depends not on general intellectual capacity but on the particular mental capacities or skills required by a particular job" (p. 66). Herrnstein and Murray suggest that this received wisdom is repudiated by meta-analyses that show that job performance is well predicted by broadly based tests of intelligence, that the general intelligence factor is a better predictor of job performance than tests of specific skills, and that correlations between tested intelligence and job performance are higher than previously considered (p. 70). Tests of cognitive ability have been shown to be better predictors of job performance ratings than biographical data, reference checks, education, interview, college grades, interest, or age. In terms of economic consequences, Herrnstein and Murray illustrate through example the difference in the value of the productivity of an employee at the 50th and the 84th percentile of the IQ distribution. The magnitude of the difference in productivity varies with the complexity of the job. Estimates of the costs entailed in disallowing hiring based on intelligence range from a maximum loss of 80 billion to a minimum loss of 13 billion. The main point of the chapter is that intelligence is directly related to job performance and that "getting rid of intelligence tests in hiring—as policy is trying to do—will not get rid of the importance of intelligence" (p. 88).

In "Steeper Ladders, Narrower Gates," Herrnstein and Murray attempt to forecast the social consequences of continued cognitive partitioning in education and occupations. First of all, the value of intelligence in the marketplace has increased. Wages for individuals in high-IQ occupations have grown more rapidly than the salaries of low-IQ occupations. According to the authors, this increasing disparity in wages reflects an increasing economic demand for intelligence. "The more complex a society becomes, the more valuable are the people who are especially good at dealing with complexity" (p. 99). Along with this cognitive partitioning comes segregation of individuals of different cognitive abilities in both the workplace and community. Finally, Herrnstein and Murray discuss the implications of the conclusion that success in life is based to some extent on inherited differences in cognitive ability among people (p. 108). As the social environment becomes more uniform, the proportion of variance attributable to inherited differences in cognitive ability increases; therefore, success in a society stratified by cognitive ability depends increasingly on inherited differences and decreasingly on the social environment (pp. 109–110). In addition, Herrnstein and Murray suggest that there may be an increasing tendency for individuals of similar levels of cognitive ability to marry, aided by the feminist revolution. In summary, Herrnstein and Murray state that three phenomena are the result of these sorting processes:

1. The cognitive elite is getting richer, in an era when everybody else is having to struggle to stay even.
2. The cognitive elite is increasingly segregated physically from everyone else, in both the workplace and the neighborhood.
3. The cognitive elite is increasingly likely to intermarry. (p. 114)

Cognitive Classes and Social Behavior

Part II documents the relationship of cognitive ability to social behaviors. According to the authors, "high cognitive ability is generally associated with socially desirable behaviors, low cognitive ability with socially undesirable ones" (p. 117). In this part, Herrnstein and Murray argue that intelligence rather than socioeconomic status is responsible for group differences in social behavior. Furthermore, they maintain that the potential causal role of cognitive ability has been neglected and that even if cognitive ability is not the cause of differences in social behaviors, viewing the problem from the perspective of cognitive ability will contribute to our understanding of these social problems. The analyses in this part were done using the National Longitudinal Survey of Labor Market Experience of Youth (NLSY). All analyses were conducted with non-Latino white individuals to demonstrate that the relationship of cognitive ability to social behavior is independent of race and ethnicity (p. 125).

In Chapter 5, entitled "Poverty," Herrnstein and Murray show that low intelligence is a stronger precursor of poverty than coming from a low socioeconomic status background. The proportion of Americans living below the poverty line decreased from 50% in 1940 to 15% in 1970 and has remained around 15% since 1970. Using the NLSY, the percentage of white youths living in poverty is shown, broken down by parents' socioeconomic status (SES) and by cognitive class of the individual. In both analyses, the percentage of individuals living in poverty increases with decreasing parental SES and decreasing cognitive ability. When both parental SES and intelligence are placed in the same analysis, the relationship between IQ and the probability of being in poverty is stronger than the relationship between parental SES and the probability of being impoverished. Herrnstein and Murray conclude that "cognitive ability is more important than parental SES in determining poverty" (p. 135).

When the relationships are further broken down into educational categories, the relationship between IQ and poverty is stronger than the relationship between parental SES and poverty for individuals with a high school diploma. However, for individuals with a bachelor's degree, neither IQ nor parental SES shows a relationship with poverty. For children, the probability of living in poverty is higher the lower the IQ of the mother, independent of marital status. When marital status is considered, the relationship between mother's IQ and the probability that a child will live in poverty is stronger for mothers who are separated, divorced, or never married than for married women. The relationship between parental SES for mothers and the probability that a child will live in poverty is weaker. In conclusion, the authors state that "the high rates of poverty that afflict certain segments of the white population are determined more by intelligence than by socioeconomic background" (p. 141).

In the chapter entitled "Schooling," according to Herrnstein and Murray, dropping out of school is a recent phenomenon that developed with "the assumption that it is normal to remain in school through age 17" (p. 144). The proportion of people who obtain a high school diploma rose throughout the century from fewer than 10% in 1900 to around 80% in 1964, and then leveled off at around 75%. Early in the century the IQ gap between those who dropped out and those who completed high school was marginal; however, since the 1950s the gap between those who complete and those who do not increased to approximately 15 points. Analysis of the NLSY showed that students who dropped out were almost entirely from the bottom quartile of the IQ distribution. Furthermore, the relationship between the probability of dropping out and IQ score was stronger than the relationship to parental

SES. For temporary dropouts, the relationship between parental SES and the probability of getting a GED (General Educational Development) instead of a high school diploma was stronger than the relationship with IQ. Conversely, the probability of getting a bachelor's degree increased with IQ and the relationship with IQ was stronger than the relationship with parental SES. In sum, both socioeconomic background and cognitive ability play a role in performance in school.

In the chapter "Unemployment, Idleness, and Injury," analyses of the NLSY data by Herrnstein and Murray showed that the probability of being out of the labor force for a month or more decreased as IQ scores increased but increased as parental SES increased. Low cognitive ability also increased the risk of being off work due to disability. In conclusion, Herrnstein and Murray qualify these relations by noting that "most white males at every level of cognitive ability were in the labor force and working, even at the lowest cognitive levels" (p. 165).

"Family Matters," the next chapter, explores the relationship of cognitive ability to marriage, divorce, and illegitimacy. Analyses of the NLSY confirm the general view that the average age at first marriage tends to be high for individuals higher in cognitive ability than for those lower in cognitive ability. In terms of the probability of marriage by age 30, for individuals with a high school educational level, the probability increases with IQ and declines marginally as parental SES goes from low to high. For individuals with college, the probability does not vary with either parental SES or IQ. In terms of divorce, analysis of the NLSY showed that the probability of divorce within the first 5 years of marriage decreased as IQ went from low to high, but increased as parental SES went from low to high. In terms of educational levels, for individuals with a college education, the probability of divorce decreased as IQ scores increased. No relation was observed for high school graduates. The rate of illegitimate births has increased markedly in the past 30 years from 11% in the 1960s to 30% in the 1990s. Analyses of the NLSY showed that the percentage of white women who had ever given birth to an illegitimate child increased from 2% to 32% as cognitive class moved from very bright to very dull. For both IQ and parental SES, as values went from low to high, the probability of an illegitimate birth declined; however, the relationship of IQ to illegitimacy was stronger than the relationship of parental SES. Other analyses showed that the odds of a child being born out of wedlock were lowest if either the mother or both parents were absent by age 14 and highest if only the father was absent. In addition, the probability of the first child being born out of wedlock for white mothers living in poverty decreased markedly as IQ went from low to high and increased markedly as parental SES went from low to high. Herrnstein and Murray conclude that "low intelligence is an important independent cause of illegitimacy" (p. 189). Furthermore, comparing the top and bottom quartiles of IQ, the authors note that in the top quartile of cognitive ability, the percentage of households that consist of a married couple is higher, whereas the percentage of households that have experienced divorce and the percentage of children born out of wedlock are lower than for the bottom quartile of cognitive ability.

In "Welfare Dependency," Herrnstein and Murray showed that the probability of mothers going on Aid to Families with Dependent Children within a year of the first birth decreased as both IQ and parental SES went from low to high, with a stronger relationship found for IQ. When the analysis focuses only on chronic welfare recipients rather than all welfare recipients, the probability of a white woman becoming a chronic welfare recipient decreases as both IQ and parental SES go from low to high; however, in this case the relationship to parental SES is stronger. Education played a role in that women with a high school or less education were at risk for becoming chronic welfare recipients, not women with more than high school education. Herrnstein and Murray conclude that "having a baby without a husband is a dumb thing to do. Going on welfare is an even dumber thing to do, if you can possibly avoid it" (p. 200).

In "Parenting," Herrnstein and Murray ask "is the competence of parents at all affected by how intelligent they are?" (p. 203). The authors review literature claiming that parenting styles differ with social class. Middle-class parents reason with their children and appeal to abstract principles, whereas working-class parents use physical punishment. Middle-class parents encourage children to ask questions and give explanations; working-class parents expect compliance without question. In terms of abuse and neglect, both are concentrated in lower socioeconomic classes. Herrnstein and Murray suggest that these studies have ignored the possibility that parental IQ rather than status plays a role in parenting style. To support this thesis, analysis of the NLSY showed that the probability of a white mother having a low-birthweight baby decreased as the mother's IQ went from low to high. In contrast, variation in the mother's socioeconomic background did not affect the probability of a low-birthweight baby. Poverty did not play a statistically significant role. Another analysis showed that the probability that a child will live in poverty during the first 3 years of life varied with both mother's IQ and socioeconomic background. Preexisting poverty accounted for the relationship with mother's socioeconomic background, but not cognitive ability. Both mother's IQ and socioeconomic background were found to have a moderate relationship with developmental problems in children. Finally, Herrnstein and Murray showed that the probability of having a child in the bottom decile of IQ decreased markedly as the mother's IQ went from low to high, and moderately as mother's SES background went from low to high. The conclusion offered by the authors is that "people with low cognitive ability tend to be worse parents" (p. 232).

In "Crime," Herrnstein and Murray open the chapter with a discussion of the dichotomy of theories of criminal behavior: psychological and sociological. Sociological theories were prevalent during the 1950s through the 1970s. Psychological theories were prevalent early in the century. Herrnstein and Murray argue that crime is probably a function of both, yet public perceptions tend to be at the sociological pole. Previous research has suggested that criminals have a lower than average IQ. The authors used the NLSY to explore the role that cognitive ability may play in creating criminals. Consistent with previous research, the mean IQ of white males sentenced to a correctional facility was lower than the mean IQ of white males who either had no involvement with the law or were stopped by the police. Other analyses showed that variance in IQ was more strongly related to the probability of being in the top decile of self-reported crime and ever being interviewed in a correctional facility than was parental SES. A low IQ was a significant risk factor. In conclusion, Herrnstein and Murray offer the caution that "despite the relationship of low IQ to criminality, the great majority of people with low cognitive ability are law abiding" (p. 251) and the conclusion that "in trying to understand how to deal with the crime problem, much of the attention now given to problems of poverty and unemployment should be shifted to another question altogether: coping with cognitive disadvantage" (p. 251).

In "Civility and Citizenship," Herrnstein and Murray argue that cognitive ability makes a contribution to the capacity for civility and citizenship (p. 253). In particular, behavior such as voting indicates an involvement in the welfare of the community. Voting varies with socioeconomic class: "college graduates vote more than high school graduates; white-collar workers vote more than blue-collar workers; and the rich vote more than the poor" (p. 258). Indirect evidence suggests that there is an effect of cognitive ability. Analysis of the middle-class values index in the NLSY was conducted as a proxy for involvement in civility. The analysis showed that as both IQ and parental SES went from low to high, the probability of a score of "yes" on the middle-class values (MCV) index increased. In conclusion, Herrnstein and Murray suggest that "a smarter population is more likely to be, and more capable of being made into, a civil citizenry" (p. 266).

The National Context

In this section, Herrnstein and Murray describe "ethnic differences in cognitive ability and social behavior, the effects of fertility patterns on the distribution of intelligence, and the overall relationship of low cognitive ability to what has become known as the underclass" (p. 267). Given the controversial nature of these topics, the authors request that this section be read carefully.

In Chapter 13, "Ethnic Differences in Cognitive Ability," the authors convey the message that ethnic differences in cognitive ability are real and have consequences (p. 269). The purposes of the authors in this chapter are twofold:

> our primary purpose is to lay out a set of statements, as precise as the state of knowledge permits, about what is currently known about the size, nature, validity, and persistence of ethnic differences on measures of cognitive ability. A secondary purpose is to try to induce clarity in ways of thinking about ethnic differences, for discussions about such differences tend to run away with themselves, blending issues of fact, theory, ethics, and public policy that need to be separated. (p. 270)

Herrnstein and Murray begin by reviewing evidence about the size of ethnic differences in cognitive ability. In general they conclude that Asians and Asian Americans have a higher average IQ than white Americans, and white Americans have a higher mean IQ than black Americans. The difference in IQ between white and black Americans is given as approximately 1.08 standard deviations. In the NLSY, "a person with the black mean was at the 11th percentile of the white distribution, and a person with the white mean ... the 91st percentile of the black distribution" (p. 278).

The most commonly held view of this black/white difference is that it represents bias in the test. Herrnstein and Murray suggest that the difference is not a function of either predictive or cultural bias. First, "overwhelmingly, the evidence is that the major standardized tests used to help make school and job decisions do not underpredict job performance" (p. 281). Second, "the B/W difference is wider on items that appear to be culturally neutral than on items that appear to be culturally loaded" (p. 282). Nor does a lack of motivation to try appear to account for the difference. Herrnstein and Murray discuss the possibility of a uniform background bias, "in other words, the tests may be biased against disadvantaged groups, but the traces of bias are invisible because the bias permeates all areas of the group's performance" (p. 285). Finally, the authors consider the role of SES in generating the black/white difference in cognitive ability. First, when the two groups are matched on SES, the difference in IQ scores between the groups shrinks; however, Herrnstein and Murray point out that because SES is a result of cognitive ability, controlling for SES in this manner is guaranteed to reduce IQ differences. Second, although black IQ scores increase with SES, the magnitude of the black/white IQ difference does not decrease.

The answer to the question of whether the black/white difference is diminishing is "yes." A review of the white/black difference in standard deviations from longitudinal data of the National Assessment of Educational Progress showed a decrease in the difference on tests of science, math, and reading across 9, 13, and 17 year olds. Herrnstein and Murray speculate that the convergence may reflect rising investments in education and improvements in nutrition, shelter, and health care that disproportionately benefit cognitive development at the low end of SES (pp. 292–293). Despite the narrowing of the difference, whether convergence will continue is a source of speculation.

Next Herrnstein and Murray discuss the nature of the difference in cognitive ability between the races. The reasons given by the authors for tackling this controversial topic are to confront the assumption of genetic cognitive equality among the races, which has practical consequences for society, and to bring to the level of public discussion a topic that is taboo. "Taboos breed not only ignorance but misinformation" (p. 297). Finally, the authors acknowledge that

> the evidence about ethnic differences can be misused, as many people say to us. Some readers may feel that this danger places a moral prohibition against examining the evidence for genetic factors in public. We disagree, in part because we see even greater dangers in the current gulf between public pronouncements and private beliefs. (pp. 297–298)

The first point that Herrnstein and Murray emphasize is "that a trait is genetically transmitted in individuals does not mean that group differences in that trait are also genetic in origin" (p. 298). From this point, the authors estimate that if the group differences are a function of environmental differences, then "the mean environment of whites is 1.58 standard deviations better than the mean environment of blacks and .32 standard deviation worse than the mean environment for East Asians" (p. 298). In the opinion of the authors, environmental differences of this magnitude are implausible yet not impossible; however, for those who hold that the IQ difference is due to environmental differences, "explanations have to be formulated rather than simply assumed" (p. 299).

Herrnstein and Murray provide a case for thinking that genetics may be involved in the group differences. Profile differences between the groups are cited as evidence for a genetic factor. East Asians typically score at the same level or slightly lower than whites on verbal IQ subtests, but score much higher on subtests of visuospatial IQ. Blacks typically score higher than whites on subtests involving arithmetic and immediate memory, whereas whites typically score higher than blacks on subtests of spatial-perceptual ability. Furthermore, observations consistent with Spearman's hypothesis that the greater the degree to which a test measures *g*, the greater the black/white difference in test scores have been shown. The conceptual linkage to a genetic explanation comes from the statement that *g* has a high degree of heritability. Arguments against a genetic factor in group differences are reviewed. Briefly, these include uniform cultural bias, the Flynn effect (i.e., test scores tend to rise in the intervals between standardization), and transracial adoption studies. In a final section, Herrnstein and Murray summarize the preceding discussion with the following:

> If the reader is now convinced that either the genetic or environmental explanation has won out to the exclusion of the other, we have not done a sufficiently good job of presenting one side or the other. It seems highly likely to us that both genes and environment have something to do with racial differences. What might the mix be? We are resolutely agnostic on that issue; as far as we can determine, the evidence does not yet justify an estimate.
>
> We are not so naive to think that making such statements will do much good. People find it next to impossible to treat ethnic differences with detachment. That there are understandable reasons for this only increases the need for thinking clearly and with precision about what is and is not important. In particular, we have found that the genetic aspect of ethnic differences has assumed an overwhelming importance. One symptom of this is that while this book was in preparation and regardless of how we described it to anyone who asked, it was assumed that the book's real subject had to be not only ethnic differences in cognitive ability but the genetic source of those differences. It is as if people assumed that we are faced with two alternatives: either (1) the cognitive difference between blacks and whites is genetic, which entails unspoken but dreadful consequences, or (2) the cognitive difference between blacks and whites is environmental, fuzzily equated with some sort of cultural bias in IQ tests, and the difference is therefore temporary and unimportant. (pp. 311–312)

Further on, Herrnstein and Murray discuss the hypothetical consequences of learning that ethnic differences in intelligence are genetic in origin.

> If it were known that the B/W difference is genetic, would I treat individual blacks differently from the way I would treat them if the differences were environmental? Probably, human nature being what it is, some people would interpret the news as a license for treating all whites as intellectually superior to all blacks. But we hope that putting this possibility down in words makes it obvious how illogical—besides utterly unfounded—such reactions would be. (p. 312)

Second, the authors confront the common assumption that "if the differences are genetic, aren't they harder to change than if they are environmental?" (p. 313). The error here, according to the authors, is the underlying assumption that environmentally induced deficits are less hardwired and less real than genetically induced deficits (p. 313). In sum, Herrnstein and Murray state that knowledge that the differences are wholly genetic as opposed to wholly environmental does not justify treating another individual differently.

In Chapter 14, "Ethnic Inequalities in Relation to IQ," Herrnstein and Murray offer a view of ethnic differences in the social behaviors previously discussed in Part II when cognitive ability is taken into account. Specifically, the probabilities of various outcomes are compared for whites, blacks, and Latinos before and after controlling for IQ. After controlling for IQ, ethnic differences in higher education, occupations, and wages diminish. Differences in the probability of living in poverty and of being unemployed also shrink when controlling for IQ. Ethnic differences in marriage rates do not change. For illegitimacy, the difference between whites and Latinos decreases; however, the difference between blacks and whites remains large. Disparities between the different ethnic groups in the probability that a woman has been on welfare also diminish when controlling for IQ. The disparity between blacks and whites in the probability of giving birth to a low-birthweight baby also decreases. Ethnic differences in a child living in poverty for the first 3 years of life also diminish. Differences in the probability of incarceration between the different groups also decline. This chapter portrays the probability of various outcomes for members of different ethnic groups of similar cognitive ability. From this perspective, many ethnic differences in social and economic indicators diminish while others remain.

In Chapter 15, "The Demography of Intelligence," Herrnstein and Murray explore the downward pressure exerted on the national distribution of cognitive ability by differences in birth rates across the range of cognitive ability. In particular,

> if women with low scores are reproducing more rapidly than women with high scores, the distribution of scores will, other things equal, decline, no matter whether the women with the low scores came by them through nature or nurture. (p. 342)

Prior to modernization, higher status females had a reproductive advantage over lower status females. More children of higher status females survived than did children of lower status females. During this period large families were the norm. With modernization, the birth rates of privileged females declined disproportionately as these women put off marriage and reproduction to take advantage of educational and career opportunities (p. 344). Birth rates among less educated and unprivileged females remained high due to the intrinsically rewarding nature of motherhood. Consequently, "if reproductive rates are correlated with income and educational levels, which are themselves correlated with intelligence, people with lower intelligence would presumably be out reproducing people with higher intelligence and thereby producing a dysgenic effect" (p. 345).

"Foretelling the future about fertility is a hazardous business, and foretelling it in terms of IQ points per generation is more hazardous still" (p. 348). Although Herrnstein and Murray decline to forecast the future, they do comment on the current status of the number of children currently born to women at various IQ levels, the age at which they bear children, and the cognitive ability of immigrants. To illustrate the relation between IQ and fertility, Herrnstein and Murray show that in 1992, the mean number of children ever born to women ages 35 to 44 decreased as the level of educational attainment increased. Furthermore, 71% more births occurred among high school dropouts than among college graduates (p. 349). The dysgenic effect is also influenced by the low-IQ group having children at younger ages than the higher IQ group. Consequently, the low-IQ group will produce more generations per unit time than will the high-IQ group. In the NLSY, the mean age at first birth for mothers at the lowest level of cognitive ability was 7 years younger than for mothers at the highest level.

Second, the mean ages at which women from different ethnic groups have children also differ. From the NLSY, the mean ages for whites, Latinos, and blacks were 24.3, 23.2, and 22.3 years, respectively. Herrnstein and Murray predict that if these age differentials persist, they will increase the disparity in the cognitive ability of successive generations (p. 355). Finally, the authors suggest that immigration that constituted 29% of the population growth in the 1980s also contributes to the downward pressure on the national distribution of intelligence. Herrnstein and Murray estimate that the mean IQ of immigrants from the 1960s through the 1980s was less than the mean IQ of the native-born American population.

> Putting the pieces together—higher fertility and a faster generational cycle among the less intelligent and an immigrant population that is probably somewhat below the native-born average—the case is strong that something worth worrying about is happening to the cognitive capital of the country. (p. 364)

In Chapter 16, "Social Behavior and the Prevalence of Low Cognitive Ability," Herrnstein and Murray illustrate the prevalence of low cognitive ability among people who suffer most from the social problems outlined in Part II. Using the NLSY, the percentages of individuals in poverty, permanent high school dropouts, men who worked 52 weeks in 1989, able-bodied men who did not work, men ever interviewed in jail, women ever on welfare, women on welfare for 5 or more years, children born out of wedlock, mothers with low-birthweight babies, children living in poverty, children in the bottom decile of IQ, and percent of people scoring "yes" on the MCV index were plotted for each decile of IQ with the cumulative percentage function also drawn. Most of the cumulative percentage functions are negatively accelerated monotonic functions. This means that the largest percentages were in the lowest decile of IQ and that the percentages decreased with increasing deciles.

Living Together

In this final section of the book, Herrnstein and Murray attempt to outline the relevance of cognitive ability to understanding some major domestic issues in

America today. In particular, the implications of cognitive ability for education and affirmative action are discussed.

In Chapter 17, "Raising Cognitive Ability," Herrnstein and Murray review the record of attempts to raise cognitive ability through environmental improvements. "Taken together, the story of attempts to raise intelligence is one of high hopes, flamboyant claims, and disappointing results" (p. 389). Attempts to raise cognitive ability through nutrition have shown that vitamin and mineral supplements may produce increases in scores on nonverbal subtests. However, as Herrnstein and Murray point out, this finding should be regarded with caution as many variations of successful studies failed to replicate the effect on IQ. Most attempts to raise cognitive ability have focused on improvements in education. Of such attempts, Herrnstein and Murray conclude that "although they make some differences in IQ, the size of the effect is small" (p. 394). Studies of natural variation in quality of education support this conclusion. The Coleman report concluded that "variations in teacher credentials, per pupil expenditures, and other objective factors in public schools do not account for much of the variation in cognitive abilities of American school children" (p. 395). Compensatory education programs designed to improve the cognitive functioning of disadvantaged students generally produced no effect of narrowing the gap of cognitive ability. According to Herrnstein and Murray, evidence of interventions that produce measurable improvements in IQ comes from two sources. A controlled experiment in Venezuela in which additional lessons were given to students in an experimental group over a period of a year increased their IQ scores between 1.6 and 6.5 points compared with a control group. Second, studies of the effects of commercial coaching for the SAT show a relationship between hours of study and increases in the SAT math and verbal scores. In sum, Herrnstein and Murray offer that "as of now, the goal of raising intelligence among school-age children more than modestly, and doing so consistently and affordably, remains out of reach" (p. 402).

Preschool programs designed to raise cognitive ability such as Head Start produce short-term increases in cognitive ability that fade out over time. A naturalistic intervention that has shown promise for raising cognitive ability is "adoption out of a bad environment into a good one" (p. 410). Herrnstein and Murray review two studies that suggest that adoption from low SES to high SES homes produces a 12-point gain in IQ. In conclusion, the authors advocate good nutrition for all children, that Head Start programs be recognized for "rescuing small children from unsuitable, joyless, and dangerous environments" (p. 415), and that "the school is not a promising place to try to raise intelligence or to reduce intellectual differences" (p. 414). Advances in the endeavor to raise intelligence must await new insights about the development of cognitive ability.

In Chapter 18, "The Leveling of American Education," Herrnstein and Murray propose that the problem with American education is reflected in the declining SAT scores among the most gifted students. The educational system has been "dumbed down" to meet the needs of average and below-average students, but as a result the potential of the most gifted students remains undeveloped. Yet society depends on the skills of those in the top level of cognitive ability to "create its jobs, expand its technologies, cure its sick, teach in its universities, administer its cultural and political and legal institutions" (p. 418).

Herrnstein and Murray begin by showing that the common perception that the academic performance of the average student is much worse today is not substantiated by longitudinal measures of academic achievement. However, for the pool of youths in the top 10% to 20% of cognitive ability, the opposite is true. SAT scores declined markedly from 1963 to 1980. The explanation most commonly offered for this decline is an expansion of the pool of students taking the test to include students who previously did not consider going to college. According to Herrnstein and Murray, this commonly held explanation is mistaken. As evidence the authors point out that "throughout most of the white SAT score decline, the white SAT pool was shrinking, not expanding" (p. 426). Instead, the authors suggest that the decline reflects "a downward trend of the educational skills of America's academically most promising youngsters toward those of the average student" (p. 427). Further analysis of the SAT scores of the most gifted youths reveals that the number of youths scoring 700 or higher on the verbal SAT decreased by 41% from 1972 to 1993. Math scores rebounded from the decline with an increase in the proportion of 17 year olds scoring 700 or more during the 1980s.

Herrnstein and Murray suggest that the decline in the 1960s reflects a "dumbing down" of education.

> One of the chief effects of the educational reforms of the 1960s was to dumb down elementary and secondary education as a whole, making just about everything easier for the average student and easing the demands on the gifted student. (p. 430)

Traditional criteria of rigor and excellence gave way to "the need to minimize racial differences in performance measures, and enthusiasm for fostering self-esteem independent of performance" (p. 432). Other forces that eroded verbal skills were television replacing newsprint as a source of news and information, and the telephone replacing letter writing as the major form of communication. Math scores were less susceptible to these pressures than were verbal scores. Dilution of the curriculum benefitted the mediocre student, but depressed the development of the intellectual skills of the talented student.

In concluding this chapter, Herrnstein and Murray suggest that "critics of American education must come to terms with the reality that in a universal education system, many students will not reach the level of education that most people view as basic" (p. 436). One tendency is to blame students who do not work hard enough for the shortcomings of the educational system. However, the authors suggest two reasons why students are less to blame for not working harder. One is that "most American parents do not want drastic increases in the academic work load" (p. 437), and the second is that "the average American student has little incentive to work harder than he already does in high school" (p. 437). With respect to government interventions, Herrnstein and Murray recommend that "the federal government should actively support programs that enable all parents, not just affluent ones, to choose the school that their children attend" (p. 440), that the government should establish a federal scholarship program for students earning the top scores on standardized tests of academic achievement, and that the government should "reallocate some portion of existing elementary and secondary school federal aid away from programs for the disadvantaged to programs for the gifted" (pp. 441–442). Finally, Herrnstein and Murray recommend the resurrection of the classical idea of the educated person:

> To be an educated person must mean to have mastered a core of history, literature, arts, ethics, and the sciences and, in the process of learning those disciplines, to have been trained to weigh, analyze, and evaluate according to exacting standards. (p. 444)

In the next chapter, "Affirmative Action in Higher Education," Herrnstein and Murray provide an overview of affirmative action in practice and argue that the "current practice is out of keeping with the rationale for affirmative action" (p. 451). In general, most people agree that affirmative action in education is needed, but differ when asked to say what they mean by it. The authors offer the following definition: "perfectly practiced, affirmative action means assigning a premium, an edge, to group membership in addition to the individual measures before making a final assessment that chooses some people over others" (p. 450).

Herrnstein and Murray examine the magnitude of this edge in SAT scores for blacks, whites, and Asians at 16 of 20 top-rated universities. The median difference between the white and black means was 180 SAT points, whereas the median difference between the Asian and white means was 30 points. On the Law School Admission Test (LSAT), the difference between the white mean and the means for blacks, Latinos, and Asians were -1.49, -1.01, and $-.32$ standard deviations, respectively. Differences of similar magnitude were found for Medical College Admission Test (MCAT) and Graduate Record Examination (GRE) scores.

To answer the question of whether these differences that reflect affirmative action in practice are good or bad, Herrnstein and Murray examine the logic of college admissions. "College admission is not, has never been, nor is there reason to think that it should be, a competition based purely on academic merit" (p. 459). Rationales for affirmative action in the admissions process fall into the categories of institutional benefit, social utility, and just desserts. *Institutional benefit* refers to the rationale to admit students from racial and ethnic minorities to enrich the campus by adding to its diversity (p. 459). *Social utility* refers to the rationale to admit minority youths to increase minority representation at higher socioeconomic and professional levels, so that in the future it will be easier for minorities to pursue such professions. *Just desserts* refers to taking into account how well an applicant has done given the environment that it was accomplished in.

> The applicant who overcame poverty, cultural disadvantages, an unsettled home life, a prolonged illness, or a chronic disability to do as well as he did in high school will get a tip from most admissions committees, even if he is not doing as well academically as the applicants usually accepted. (p. 461)

To examine how these rationales would operate in theory in a selection process, Herrnstein and Murray set up a table for deciding between a white candidate and a minority candidate with SES for each candidate either low or high. For the decision between a high SES white and a low SES minority, the rationales favor assigning a large preference for the minority candidate. For the case of a high SES white and a high SES minority candidate, the just desserts and social utility rationales are in opposition, but a small preference is assigned to the minority. For the decision between a low SES minority and a low SES white candidate, the rationales favor both candidates; therefore, little to no premium is given to the minority candidate. For the case of the high SES minority over the low SES white candidate, social utility favors both, but just desserts favors the white candidate, so a modest premium is assigned to the white candidate. Using the NLSY, the authors show that for affirmative action in practice, the differences in cognitive ability in standard deviation units between a black candidate and a white candidate for the four cells described earlier were +1.25, +.91, +1.17, and +.58, respectively. In each case, the analysis indicates a large edge in cognitive ability is given to the minority candidate. Herrnstein and Murray ask the reader in a properly run system of affirmative action, "How big an edge is appropriate?" (p. 467).

Another consideration is the cost of affirmative action. "How much harm is done to minority self-esteem, to white perceptions of minorities, and ultimately to ethnic relations by a system that puts academically less able minority students side by side with students who are more able?" (p. 470). Students commonly observe the racial mix of the student population, students who stand out because they seem out of place in college, and students who stand out because they are especially smart (p. 471). Minority students may stand out due to poor academic performance. Dropout rates from college by black students are twice the rate for whites. "Getting discouraged about one's capacity to compete in an environment may be another cost of affirmative action" (p. 473). Growing racial animosity on campuses may be yet another cost. In conclusion, Herrnstein and Murray advocate a return to the original conception of affirmative action: "to cast a wider net, to give preference to members of disadvantaged groups, whatever their skin color, when qualifications are similar" (p. 448).

In Chapter 20, "Affirmative Action in the Workplace," Herrnstein and Murray evaluate the impact of affirmative action legislation on the workplace. The authors state that job discrimination legislation was developed based on

> the assumptions that (1) tests of general cognitive ability are not a good way of picking employees, (2) the best tests are ones that measure specific job skills, (3) tests are biased against blacks and other minorities, and (4) all groups have equal distributions of cognitive ability. (p. 483)

Herrnstein and Murray state that although these assumptions were defensible back in the 1960s, today it is well established that tests of cognitive ability are related to job productivity; that these tests have more predictive power than grades, education, or a job interview; that this predictive power arises from their measure of general cognitive ability rather than specific skills; that these tests are not biased against blacks; and that different ethnic groups have different distributions of cognitive ability (pp. 483–484). The authors recommend that "the government should scrap the invalid scientific assumptions that undergird policy and express policy in terms that are empirically defensible" (p. 484).

Next Herrnstein and Murray evaluate the effects of affirmative action in the workplace without and with controlling for cognitive ability. Without controlling for IQ, there has been an increasing trend in the percentage of blacks employed in clerical, professional, and technical jobs from 1960 to 1990. No trend shows an effect of the antidiscrimination law on hiring and promotions. When the same trend lines are adjusted for the known difference in IQ between blacks and whites, the trend lines show that both in clerical and in professional and technical positions, for individuals in the same IQ range, blacks were being hired at higher rates than whites since the 1960s, with both trends increasing into the 1980s. Herrnstein and Murray suggest that when adjusted for IQ, evidence for the impact of affirmative action becomes apparent. In the 1950s, blacks were systematically and unjustly excluded from these occupations. In the mid-1960s the underrepresentation of blacks in technical and professional occupations disappeared, and since the 1960s the representation of blacks in these professions has gone beyond parity (pp. 491–492).

In conclusion, Herrnstein and Murray advocate that the goal of an affirmative action policy should be equality of opportunity rather than equality of outcome. Furthermore, they state that

> if the quality of performance fairly differs among individuals, it may fairly differ among groups. If a disproportion is fair, then "correcting" it—making it proportional—may produce unfairness along with equal representation. We believe that is what happened in the case of current forms of affirmative action. (p. 501)

From this point, Herrnstein and Murray offer four policy alternatives that they think will produce fair treatment in hiring and promotions. One alternative is to create tests that meet the current requirements but still predict job performance. The problem with this alternative is that the relation of general ability to job performance will produce disparate impacts across groups. A second alternative is to allow employers to use educational credentials to narrow the pool of qualified applicants. The problem with this approach is that equivalent degrees do not mean equivalent cognitive ability due to affirmative action in universities. The third alternative is to produce norms for each group. Scores would be converted to percentiles based on the distribution for each group, and employers could hire top down from each distribution. According to the authors, race norming was used in the early 1980s but was outlawed in 1986. Finally, the fourth proposal, which is favored by the authors, is based on the proposition that "if tomorrow all job discrimination regulations based on group proportions were rescinded, the United States would have a job market that is ethically fairer, more conducive to racial harmony, and economically more productive, than the one we have now" (p. 505). As with education, Herrnstein and Murray advocate that the government "get rid of preferential affirmative action and return to the original conception of casting a wider net and leaning over backward to make sure that all minority applicants have a fair shot at the job or the promotion" (p. 505).

In the next chapter, entitled "The Way We Are Headed," Herrnstein and Murray predict a pessimistic future for American society based on the following trends: "an increasingly isolated cognitive elite, a merging of the cognitive elite with the affluent, a deteriorating quality of life for people at the bottom end of the cognitive ability distribution" (p. 509). Throughout this century, American society has been transformed into a society stratified by cognitive ability. The cognitive elite has grown to become a new class of individuals in high-IQ occupations that has displaced previous socioeconomic elites. The authors state that "the invisible migration of the twentieth century has done much more than let the most intellectually able succeed more easily. It has also segregated them and socialized them" (p. 513). Furthermore, the very bright have become more affluent and the affluent increasingly comprise the very bright, which has led to a blending of the interests of the affluent and the cognitive elite (p. 515). According to the authors, this will lead to what Robert Reich has referred to as a "secession of the successful" (p. 517). A scenario is presented in which 10% to 20% of the population are wealthy enough to bypass social institutions that they do not agree with and to constitute a political bloc with considerable clout. Herrnstein and Murray suggest that this new coalition of the affluent and the intelligent will fear the growing underclass in American society.

For the underclass, the future predicted by the authors is grim.

> People in the bottom quartile of intelligence are becoming not just increasingly expendable in economic terms; they will sometime in the not-too-distant future become a net drag. In economic terms and barring a profound change in direction for our society, many people will be unable to perform that function so basic to human dignity: putting more into the world than they take out. (p. 520)

As the illegitimacy rate among whites increases, a white underclass will emerge that will be subject to much the same fear and resentment from the cognitive elite that the black underclass experiences. Meanwhile, the outmigration of the ablest blacks from the black inner city will lead to an increasing concentration of blacks with limited cognitive ability and the attendant social problems (p. 522).

Finally, Herrnstein and Murray foresee that "the cognitive elite, with its commanding position, will implement an expanded welfare state for the underclass that also keeps it out from underfoot" (p. 523). Some of the features of that welfare state will be "child care in the inner city will become primarily the responsibility of the state" (p. 523), "the homeless will vanish" (p. 523), "strict policing and custodial responses to crime will become more acceptable and widespread" (p. 524), "the underclass will become even more concentrated spatially than it is today" (p. 524), "the underclass will grow" (p. 525), "social budgets and measures for social control will become still more centralized" (p. 525), and "racism will reemerge in a new and more virulent form" (p. 525). To avoid this future, Herrnstein and Murray recommend that the issues of a society increasingly dominated by a cognitive elite and a growing underclass need to be addressed now.

The final chapter of the book is entitled, "A Place for Everyone." Herrnstein and Murray state that "our central concern since we began writing this book is how people might live together harmoniously despite fundamental individual differences" (p. 528). To achieve this goal, the authors advocate a return to the original conception of human equality and the pursuit of happiness. The original conception of equality of rights that arose from the philosophies of Hobbes and Locke is not seen as synonymous with the assumption that "individuals are both equal and empty, a blank slate to be written upon by the environment" (p. 529). "They are equal in rights, Locke proclaimed, though they be unequal in everything else" (p. 530). Furthermore, the authors argue that this original conception of equal rights was what the Founders of America espoused. "The Founders saw that making a stable and just government was difficult precisely because men were unequal in every respect except their right to advance their own interests" (p. 531). In contrast, contemporary political theory works with a conception of equality that seeks to suppress social and economic inequalities that result when people are free to behave differently (p. 532).

Herrnstein and Murray suggest that for society to come to terms with the realities that people differ in intelligence and that intelligence bears on how well people do in life, it should operate in such a manner as to allow individuals throughout the range of cognitive ability to find valued places for themselves (p. 535). In the past,

> when the responsibilities of marriage and parenthood were clear and uncompromising and when the stuff of community life had to be carried out by the neighborhood or it wouldn't get done, society was full of accessible valued places for people of a broad range of abilities. (pp. 537–538)

In this traditional context, it was easier for an individual with low cognitive ability to find a valued place. In the contemporary world, the increased costs of the valued roles of spouse, parent, and neighbor, as well as the stripping of traditional functions from the neighborhood and the community, has made it increasingly more difficult for individuals of modest cognitive ability to find valued places in society (p. 538). From this observation, Herrnstein and Murray propose that "a wide range of social functions should be restored to the neighborhood when possible and otherwise to the municipality" (p. 540) in order to increase the valued places that people can fill.

The second set of policy prescriptions advocated by Herrnstein and Murray involve a simplification of rules generated by the cognitive elite that make life more difficult for everyone else (p. 541). "As the cognitive elite busily goes about making the world a better place, it is not so important to them that they are complicating ordinary lives" (p. 541). Such complexity serves as a barrier to people who are not cognitively equipped to struggle through the bureaucracy and, according to Herrnstein and Murray, should be removed. Another prescription concerns making the justice system simpler so that the rules about crime and the consequences for crime are simpler.

> The number of acts defined as crimes has multiplied, so that many things that are crimes are not nearly as obviously "wrong" as something like robbery or assault. The link between moral transgression and committing crime is made harder to understand. (p. 543)

Such simplification will make living a moral life simpler for persons of lower cognitive ability. For marriage, Herrnstein and Murray argue that the sexual revolution and the state have made it "much more difficult for a person of low cognitive ability to figure out why marriage is a good thing, and, once in a marriage, more difficult to figure out why one should stick with it through bad times" (p. 544). As a policy prescription, Herrnstein and Murray advocate returning marriage to its historic legal status to restore the rewards of marriage by validating the rewards that marriage naturally carries with it (p. 546).

Herrnstein and Murray conclude the book with the following words:

> Cognitive partitioning will continue. It cannot be stopped, because the forces driving it cannot be stopped. But America can choose to preserve a society in which every citizen has access to the central satisfactions of life. Its people can, through an interweaving of choice and responsibility, create valued places for themselves in their worlds. They can live in communities—urban or rural—where being a good parent, a good neighbor, and a good friend will give their lives purpose and meaning. They can weave the most crucial safety nets together, so that their mistakes and misfortunes are mitigated and withstood with a little help from their friends.
>
> All of these good things are available now to those who are smart enough or rich enough—if they can exploit the complex rules to their advantage, buy their way out of social institutions that no longer function, and have access to the rich human interconnections that are growing, not diminishing, for the cognitively fortunate. We are calling upon our readers, so heavily concentrated among those who fit that description, to recognize the ways in which public policy has come to deny those good things to those who are not smart enough and rich enough.
>
> At the heart of our thought is the quest for human dignity. The central measure of success for this government, as for any other, is to permit people to live lives of dignity—not to give them dignity, for that is not in any government's power, but to make it accessible to all. That is one way of thinking about what the Founders had in mind when they proclaimed, as a truth self-evident, that all men are created equal. That is what we have in mind when we talk about valued places for everyone. (p. 551)

References

1. Herrnstein, R. J. and Murray, C. (1994), *The Bell Curve: Intelligence and Class Structure in American Life*, The Free Press, New York.

PART II

THE GENETICS–INTELLIGENCE LINK

Parents across the country know that children in affluent neighborhoods are likely to have high test scores and those in poor neighborhoods low scores. This holds for IQ tests and whatever other ability tests schools administer. But why? Are the genes of the parents responsible for their affluence and for the success of their children? What do we know about the heritability of intelligence? How much of test score success can be attributed to genetic inheritance, and how much to environment? These are some of the questions that investigators in Part II try to answer.

There is general agreement among researchers that both genetics and environment explain familial similarity in IQ test scores. Is there any way we can make a case for one as more important than the other, and therefore as dominant? Herrnstein and Murray argue that genetics is the dominant factor—more important than environment. In their view, "Changing cognitive ability through environmental intervention has proved to be extraordinarily difficult." (p. 314) But how sound and current is their understanding of genetics, and how compelling is their argument?

Our best data on the genetic contribution to functioning intelligence comes from the family and twin studies recommended by Francis Galton in his 1869 work, *Hereditary Genius*, and pursued in the last half-century with the techniques of multivariate analysis and quantitative genetics. Although geneticists have made headway in identifying genes for discrete physical traits and some diseases, they have as yet made little progress identifying complex behaviors, like intelligence.

Twin, sibling, and parent-offspring studies, in which there is an effort to control for the role of both genetics and environment, remain our most important data source for the heritability of human intelligence. Daniels, Devlin, and Roeder analyze the more than 200 IQ studies that have been published to see the extent to which genes and environment account for variance in IQ scores between biological parents, adoptive parents, and their respective children. They argue that Herrnstein and Murray, who defend the dominance of genes, are employing a broad and inappropriate model of heritability that assigns to genes an influence that is unwarranted.

Their analyses and their understanding of genetics lead the three researchers to find no reason for pessimism about the positive effects of environmental interventions such as nutrition and health care, particularly at the pre-natal stage and in the first year of life. Similarly, based on their analyses, they find no reason to accept *The Bell Curve*'s claims of growing cognitive castes and dysgenesis. The scientific explanation offered in this chapter of the mechanisms for the transfer and development of intelligence across generations is as compelling as the statistical analysis.

Wahlsten argues even more strongly for the malleability of intelligence. Contrary claims that environmental changes have no important effect on biologically limited behavior contradict findings of a long-term upward trend in IQ scores, the effect of adoption on IQ in France, the IQ gains reported in the Carolina Abecedarian Project, the effects on IQ of health care and educational daycare for low birth-weight infants, and an Edmonton study of the effect of an additional year of schooling on children of the same age. Arguments against social intervention, made by Herrnstein and Murray, appear unsound when there are known positive effects for intervention, particularly in infancy and early childhood.

Singer and Ryff find, like the authors of *The Bell Curve*, that "ethnic differences in cognitive ability are neither surprising nor in doubt" (p. 269). Physical and mental health studies also report considerable variation across groups in life contexts and health-related life outcomes, from South Africa to the American Midwest. Singer and Ryff hold, however, that the methods employed by Herrnstein and Murray will not uncover the reasons why different racial and ethnic groups have differing and unequal life outcomes, and why there is intragroup variability. Attending to a limited number of variables for large aggregates will not reveal the factors in individual and group environments that explain these different outcomes.

They remain unconvinced by simple genetic explanations, even for diseases like essential hypertension which in the United States is associated with a heritable but complex gene component and has a different incidence in different racial groups. There is, they argue, too much variance that cannot be explained by the presence of continuously distributed genes. Instead, an interaction of genes and environments explains the phenotypic expression of the disease. We need to understand that interaction in order to intervene effectively. They make the case for a more context-rich methodology, able to integrate individual cases, and to explain outcomes that are both biological and psychosocial. From a health standpoint, it would be far more useful to classify human populations in terms of their susceptibility to particular diseases and the environmental triggers for their genes than to seek differences in race and ethnicity.

chapter three

Of Genes and IQ

MICHAEL DANIELS, BERNIE DEVLIN,
AND KATHRYN ROEDER

The fundamental premise of Herrnstein and Murray's opus, *The Bell Curve*, is that intelligence and its proxy IQ are determined primarily by genes.[1] Economic and social success, according to their analyses, are determined primarily by IQ. Thus logic dictates that economic and social success are due largely to genes, which therefore explains why government interventions to improve the lot of the poor have been on the whole unsuccessful. In one eloquent volume, the authors serve up justification for that oft-heard refrain "the poor will always be poor."

Herrnstein and Murray's (H&M's) logic is compelling on its surface. The current thinking holds that IQ is partially determined by genes, although the mechanism by which genes affect IQ and the degree to which they affect IQ is by no means clear. No one would dispute that intelligence must drive economic and social success to some degree. So perhaps genes do contribute to economic and social success. We have no argument with that ambiguous statement. But H&M invest a significance in genes that goes far beyond anything supported by the data, and in that we do find fault.

To understand H&M's arguments, we have to delve into a particular area of genetics, namely, quantitative genetics. We can't possibly cover the entire field, but we will not need more than a small portion of it to see through *The*

Bell Curve. H&M's interest in the genetics of a quantitative trait (IQ) follows in a long tradition. In fact, as evidenced by the early domestication of crop plants and animals, the physical nature of inheritance must have fascinated humans long before the dawn of modern civilization. This fascination carried over to the inheritance of our own species' characteristics. For at least thousands of years, and probably many more, humans undoubtedly wondered what makes individuals look different, what makes some more successful than others, and so on. The answers they concocted, at least in historical times, were sometimes marvelous and invariably inaccurate.

That is quickly changing. Some of the questions about the nature of human inheritance can now be answered, especially those regarding discrete physical traits and simple genetic diseases. In contrast, the nature of complex quantitative traits, such as intelligence or IQ, are still poorly understood. Not for lack of trying, however. To the contrary, progress is slow because the genetic problem itself is very difficult.

Even if the genetic basis of normal variation in human intelligence is not quite within our grasp, part of the IQ story is available in outline. The outline is sketchy, with only a few details. H&M attempt to infer larger portions of this story but, like many social scientists before them, they just don't get it quite right. As a consequence, many of the most highly charged predictions in *The Bell Curve* have, at best, a limited foundation.

H&M assert that IQ is highly heritable. We will develop what it means to say that a quantitative trait, such as IQ, is highly heritable. For now we take their assertion to imply a faithful transmission of IQs through generations. That is to say, if a mother and father both have high IQs, their offspring are quite likely to have high IQs also.

Therein lies the crux of one of the societal dilemmas that H&M envision. They argue that society is becoming increasingly stratified into cognitive groups, such as the cognitive elite and the cognitive peasantry (our name for the latter group). These groups interact only superficially, at best, and therefore are unlikely to intermarry. Moreover, because IQ is highly heritable, they argue that the groups have the potential to become castes with little or no mixing. In other words, the progeny of the cognitive elite will themselves be cognitively elite; those unfortunate progeny of the cognitive peasantry are doomed to be peasants. According to H&M, it's all in the genes, and there is little that can be done about it. Indeed, if IQ and intelligence are highly heritable, H&M's vision is plausible; if they are not highly heritable, their vision is only a phantasm.

H&M also observe that people of low IQ produce more children more quickly than people of high IQ. Their observations are not original. Such observations were made off and on since the origination of IQ tests early in this century. In fact, during the middle nineteenth century, it was noted with alarm that those in the "lower class" far outreproduced those in the "upper class." The worry was more than demographic: the concern was that society itself was endangered when people of superior quality did not produce a sufficient number of quality offspring. In this century, the worry has moved from quality to the erosion of intelligence per se, as measured by IQ.

This is the crux of another societal dilemma that H&M envision, the erosion of our cognitive capital. We can easily see the scenario: the cognitive peasants outreproduce the cognitive elite; and the heritability of IQ is such that elites generally produce elites and peasants generally produce peasants. It follows automatically that our society is becoming "dumb and dumber," to steal a current cinematic theme. Could this be true? Perhaps. Its plausibility again rests on the heritability of IQ, assuming IQ measures intelligence.

The most controversial part of *The Bell Curve* is the chapter dealing with IQ differences across races. Many sources indicate that African-American IQs are on average notably lower than Caucasian-American IQs. If IQ is heritable does this imply that the differences are genetic? In fact, the answer is no. Heritability has little to do with differences across populations, as H&M themselves note.

We use much of this chapter to explain heritability and to estimate the heritability of IQ. Once we have a handle on the heritability of IQ, we evaluate how plausible H&M's societal dilemmas are in light of the current evidence.

H&M argue that IQ measures intelligence and that intelligence is something real. The validity of these assertions, while questionable, is not essential to our presentation. (For further discussion see Chapters 6, 7, and 12.) We focus on whether IQ is heritable, which is itself a valid question. If IQ is synonymous with intelligence, or approximately so, then the meaning of IQ's heritability is simply richer.

Inheritance versus Heritability

Many of the H&M's predictions about future societal dilemas are based on their belief that intelligence, which they equate with IQ, is highly heritable. What does this mean? Is heritability synonymous with inheritance? Most laymen would say yes. Geneticists, for whom this is an important research question, say no. The many genes that *potentially* influence IQ are inherited, but IQ itself is not. IQ develops as the child does. One child with good genes and a poor environment may grow up with a stunted IQ, while another with mediocre genes and an excellent environment may thrive. Under the prevailing environmental conditions for the Caucasian population, for which there is the most data, there is a correlation of IQ between parent and offspring, partially induced by the inheritance of genes. The correlation is not large. Thus we can speak of the predictable inheritance of genes, but the unpredictable heritability of IQ. Inheritance of genes is inevitable and unaffected by environmental conditions, while heritability is labile, depending critically on environmental conditions.

Historical Background

Many genes exhibit variation within a population. Different forms of a particular gene are called *alleles*. Therefore, for a complementary pair of genes, a mother might contribute a different allele to her child than did the child's father. The variation of complementary alleles in the population is called *genetic variation*. Without the tools of molecular genetics, we cannot know what alleles were inherited by a child, and we can only indirectly infer that there is genetic variation in the population.

Nevertheless, before the advent of molecular biology, scientists and laymen believed that there was genetic variability. Why? By studying individuals they observed variability in the phenotypes, the "visible" features of the organism. Moreover, they noticed that baldness tended to run in families, that tall parents tended to have tall children, that it was good livestock management to allow only the best bulls to breed, and so on. In other words, they were studying the phenotypic variation in the population and how it tended to run in families. The study of inheritance began in earnest with the experiments of Gregor Mendel during the mid-1800s with pea plants. Only after the rediscovery of his results during the first decade of this century, however, did the field begin to bloom.

Mendel was an excellent experimentalist. He studied simple observable characters in pea plants, such as wrinkled versus smooth seed coats. The beauty of his experiments was that he chose characters, undoubtedly deliberately, that exhibited simple inheritance: each character was determined by a single gene. There was therefore direct correspondence between genotype and phenotype, at least in his garden. The smooth versus wrinkled pea seed character provides a good example. If we signify the "smooth" allele by S and "wrinkled" allele by s, a progeny inheriting an S from both parents, or an S from one and an s from another, will produce only smooth seeds. Alternatively, progeny that inherit an s from both parents will produce only wrinkled seeds. That is to say, when two plants producing wrinkled seeds (ss) were mated, only offspring producing wrinkled seeds were generated; whereas when two smooth seeded plants were crossed, the outcome depended on the plants crossed. $SS \times SS$ and $SS \times Ss$ yields all smooth progeny; $Ss \times Ss$ yields three quarters smooth, and one quarter wrinkled, on average.

Naturally Mendel could do no more than infer the genetic system from his experiments because molecular genetics, even in rudimentary form, would not develop for almost another century. Mendel examined the association of characters or phenotypes in parents and their offspring, and inferred from the obvious association its cause, the inheritance of genes.

Mendel's observations and mode of inference are still in use today. In fact, similar types of associations or correlations are sought to infer that IQ is also genetically determined. For instance, identical twins are genetically identical, so one might naively expect a perfect correlation. Yet the observed correlation is less than 1, only about 0.8. Likewise, the IQs of parents and offspring are correlated, with a value of approximately 0.46. A parent and her child have 50% of their alleles in common, so the naive expected correlation is 0.5, in good agreement with 0.46. However, non-identical twins also share half their alleles, on average, and the correlation in their IQs is typically much larger, approximately 0.6.[2]

These correlations cannot be explained by a simple genetic model. IQ is obviously not analogous to the simple characters studied by Mendel. Nevertheless, the correlations of IQ among various types of relatives do not rule out a more complicated genetic model.

Things become even murkier for a simple genetic explanation of IQ when we consider the correlation of parents and their *adopted* children. That value is about 0.19. The expected value is zero because parents and adopteds share no alleles in common. For nontwin siblings (henceforth, siblings) raised together, the correlation in IQs is about 0.47; when they are raised apart, the correlation drops to 0.24. Clearly the genetics have not changed, only some portion of the environment has. Consequently, any model attempting to explain the correlation of IQs among related individuals must incorporate the effects of the environment. While there is an obvious impact of environment, the correlations, being greater than zero even for children raised apart, also suggest a role for genes. But what kind of role?

Actually the problem we have just described for IQ is one that has been studied since 1918, when the statistician and geneticist, R.A. Fisher, published his seminal paper on the subject. In that paper, Fisher developed a model for character variation that was determined by environmental variation and by many genes, each gene having a small effect. The model also explained the correlation of a quantitative character for various relatives. Since that time, an elegant body of theory, dubbed *quantitative genetics*, has developed through the combined efforts of statisticians and geneticists. The theory has been immensely useful to animal and crop breeders, as well as evolutionary biologists.[3]

The *sine qua non* of quantitative genetics is heritability. This is the quantity H&M use to describe the association of IQ for various relatives. IQ is a quantitative character and the IQs of relatives can be obtained, so it is amenable to quantitative genetic analysis. Because heritability is a complex subject, we cannot cover even a portion of its relevant portions in this chapter. Nevertheless, our summary of heritability should be sufficient for the reader to understand and challenge the arguments presented by H&M in *The Bell Curve*. On the other hand, the reader should be mindful that our summary cannot be fully rigorous, and we readily admit to glossing some of heritability's fine details.

An Example

The fundamental difference between inheritance and heritability can be illustrated by a hypothetical example. Most biochemical processes require a chain of reactions, in which each stage is driven by an enzyme. Below we illustrate such a process, in which compound A is broken down to B by enzyme E_{AB}, which in turn is broken down to C by enzyme E_{BC}, and so on. A single gene codes for each enzyme.

$$\begin{array}{ccccccc} A & \longrightarrow & B & \longrightarrow & C & \longrightarrow & D \\ & E_{AB} & & E_{BC} & & E_{CD} & \\ & \uparrow & & & & & \\ & Zn & & & & & \end{array}$$

Suppose there are two alleles for each enzyme in the population, a form that breaks down the compound quickly (coded for by the + allele) and a form that breaks down the compound slowly (coded for by the − allele). Every child receives two copies of each gene, and this combination is called the *genotype*. For instance, a child could inherit either the ++, +−, or −− genotypes. We say a population has *genetic variation* when both forms of one or more genes occurs therein. Under certain environmental conditions, a child having a genotype with the ++ forms for each gene will convert A to D more rapidly than a child with the −− forms for each gene. Other genotypes with some pluses and some minuses will convert A to D at intermediate speeds. The total population variation in metabolic rate (from A to D) is called *phenotypic variation*.

If genetic variation of this sort explains some or all of the phenotypic variation in the population, we say the metabolic rate of A to D is heritable. How could the metabolic rate *not* be heritable if genes determining metabolic rate are inherited? Let's add a wrinkle to our hypothetical scenario. Suppose zinc is a cofactor required by E_{AB}. Then, if zinc is scarce and therefore the rate-limiting element of the chain of reactions, its supply controls the overall metabolic rate by its impact on the breakdown of A to B. All phenotypic variation will be due to the variable intake of zinc; none of it will be explained by the genotypic variation.

Both the environmental and genetic basis for IQ is much more complex than our example of metabolic rate. Nevertheless, there is a direct analogy. For a specific population and environment, IQ could indeed be heritable; for the same population and another environment, heritability could be near zero.

To better understand the details of heritability, let us further consider our enzyme example in a zinc-rich environment. For each enzyme an individual possesses zero, one, or two +'s for E_{AB}. Suppose that the rate of breakdown from A to B is $\mu - r$, μ and $\mu + r$ for individuals with zero, one, or two + alleles, respectively. This is what is known as an *additive genetic model* because the genotype $+-$ metabolizes at a rate halfway between the other genotypes. Let σ_G^2 describe the variance in the population for the rate of breakdown of A to B. By extension, assuming no environmental effects, the total variation in the rate of breaking A down to D is computed by adding up the variance in rates due to each of the three enzymes in the pathway. In the case of entirely additive effects, the variance is referred to as σ_A^2.

Another type of genetic model is the dominance model. An example in Mendel's garden is the rough/smooth gene. An *Ss* pea is smooth—it is not half-smooth and half-wrinkled. Similarly, in our enzyme example an individual with a $+-$ genotype might process A to B just as fast as an individual with a $++$ genotype. When the model is nonadditive, due to dominance, then the total variability in the rate of breakdown of A to B, due to genes, is described by the sum of two variance terms, the additive and the dominance genetic variance: $\sigma_G^2 = \sigma_A^2 + \sigma_D^2$. To partition the total genetic variability into two components, simply fit a linear model with rates of processing as the dependent variable and number of $+$ forms of the gene as the independent variable. The additive effect, σ_A^2, is the mean square due to regression, and the dominance effect, σ_D^2, is the mean square error. To compute the genetic variance in the rate of breakdown from A to D, assuming no interaction between enzymes, one merely adds up the additive and dominance effects across enzymes. The dominance effect is often called the *nonadditive* or *interaction effect.*

Of course, one typically cannot observe the genotypes of the individuals, nor do we even know what genes are contributing to a complex trait such as IQ. Furthermore, for most complex traits it is well known that the environment does affect the expression of the trait. Consequently, the phenotypic variance equals the sum of the variance due to genetic effects, both additive and dominance, plus the variance due to environmental effects. Much research effort in quantitative genetics has been devoted to the study of how to partition phenotypic variance into genetic and environmental components.

Heritability in the Narrow and Broad Sense

Even among scientists the term *heritability* has different meanings. For instance, quantitative geneticists referring to the heritability of IQ would mean something quite different than do psychometricians, in general, or H&M in particular. This difference is critical to H&M's arguments about the implications of the "high heritability" of IQ. To quantitative geneticists heritability means something far narrower—and more important to H&M's arguments—than it does to most psychometricians. H&M argue that the "cognitive elite" will propagate itself by marriage among elites, as will the complementary group, the "cognitive peasantry." We might find the authors' argument compelling if the high heritability they describe was in fact the quantitative geneticists' heritability. It is not, however. Instead, the authors refer to a "heritability" that has far less predictive value.

In general, heritability represents the proportion of variability in a trait explained by genetic factors (usually specific genetic factors) and is therefore a ratio of variances: the numerator is the variability attributed to a genetic component, and the denominator is the total variability of the trait in the population. If we write σ_G^2 for the total variance in the phenotype due to genetics and σ_T^2 for the total variance in the trait in the population, then a certain type of heritability, called ***broad-sense heritability***, h_B^2, can be defined as $h_B^2 = \sigma_G^2/\sigma_T^2$. Total variability is due to genetic and environmental factors, as well as measurement error. The latter, usually small, is commonly lumped with environmental components. Consequently, we can picture the partitioning of phenotypic variance as a circle in which total phenotypic variance, the area of the circle, is divided into genetic and nongenetic components.

At a simple level, quantitative geneticists recognize two distinct components of genetic variance for quantitative traits: being additive genetic variance and nonadditive genetic variance. Of these, only additive genetic variance has predictive value. Nonadditive genetic variance is due to the interaction of genes. Because we do not know the genes involved, let alone the magnitude nor direction of their impact on a trait, knowledge of the amount of nonadditive genetic variance is of no predictive value. Returning to the enzyme example, suppose a slow-converting mother and a fast-converting father produce an offspring. Under the additive model, we would predict that the child would convert A to B at a rate equal to the mean of his parents. If, however, the + allele is dominant, then the child might be a fast converter like his father, or a slow converter, like his mother. The child will always be a fast converter if the father is ++. If, however, the father is +−, the mating $+ - \times - -$ yields +− (fast converter) and −− (slow converter) with equal probability. Thus, in general, although it is true that genes are inherited, it is far more difficult to predict the outcome of a given mating based on the knowledge of the phenotypes of the parents in the presence of large nonadditive effects.

Similarly, one might predict the IQ of an offspring of parents with IQs of 100 and 120 to be 110, assuming the trio experienced the same environment. If, however, there were substantial interactions between genes, the expected IQ of the child might be far higher or far lower than either parent. Conversely, a mother and father could both have quite high IQs, even if the additive effect of their genes is only average, due primarily to favorable interaction effects. In the jargon of plant and animal breeders, we would say that the parents have good *phenotypic values* (the observable IQs), but average *genotypic values* (the genes determining IQ). Of course, with humans one rarely if ever knows the genotypic values, especially for a trait as complex as IQ.

From the two sources of genetic variation, geneticists define two types of heritability. Genetic heritability in the "narrow sense" has only the additive genetic variance in the numerator, whereas heritability in the "broad sense" has variance due to both additive and nonadditive effects in the numerator. Defining σ_A^2 for the additive genetic variance and σ_D^2 for the nonadditive genetic variance, the broad-sense heritability is $h_B^2 = (\sigma_D^2 + \sigma_A^2)/\sigma_T^2$ and the narrow-sense heritability is $h^2 = \sigma_A^2/\sigma_T^2$. Thus broad-sense heritability is always larger than narrow-sense heritability. When quantitative geneticists refer to heritability, they mean narrow-sense heritability; when psychometricians refer to heritability, they are thinking of broad-sense heritability. The reasons are historical. Geneticists are interested in predicting the outcomes of mating, while psychometricians are usually interested in questions inherent in the nature versus nurture framework, that is, how much variability is "genetic."

Thus while broad-sense heritability is of theoretical interest, only narrow-sense heritability informs us about the predictability of the outcomes of matings in a population. It is narrow-sense heritability that is the critical quantity determining the likelihood of both of H&M's nightmarish genetic visions: cognitive castes and dysgenics. H&M miss this subtlety. In fact, as we discuss subsequently, cognitive castes become almost impossible and dysgenics unlikely when they are evaluated in light of realistic values for narrow-sense heritability. By relying on broad-sense heritabilities, H&M inflate the apparent reliability of transmission of IQ between parents and offspring. In fact, there are additional sources of inflation in H&M's estimates, which we discuss later. In brief, studies of IQ, and our reanalyses of them, suggest a narrow-sense heritability of 0.34 and a broad-sense heritability of 0.48. This is a far cry from H&M's maximum value of 0.8 or their middling value of 0.6. Consequently, H&M give the impression that IQ is highly "heritable," but it is not.

Another feature of heritability that is noteworthy is its dependence on implicit factors. Fisher, who invented the concept of heritability, recognized the weakness of this measure of heredity, scorning its "hotch-potch of a denominator."[4] Fisher was referring to the fact that heritability of a character is population and environment dependent, as dictated by its denominator. That is to say, for a given population, the heritability of a character could change substantially by fundamentally changing the environment, by making an environment more or less homogeneous, and so on. Moreover, the heritability of a character will not be the same for two different populations, even if they experience the same environment. Thus, as we shall discuss at length later, it is almost impossible to extrapolate the heritability of a character for one population from information about another—even if the populations enjoy the same environment—and especially if they do not.

Estimating the Heritability of IQ

A quantitative trait is determined by the genes passed on by the parents and the effect of the environment. In traditional quantitative genetic studies, the environmental component is divided into two constituents: the maternal environment and the external environment. For humans, the maternal environment refers to prenatal development, which is obviously affected by the nutrition received by the mother before and during pregnancy, as well as other factors, such as alcohol consumption and smoking habits. This definition presents a complication for some psychometric study designs, particularly those involving adopted individuals. For instance, adopted-out siblings share both the maternal environment and any early external environment they experienced prior to their separation by adoption. To adapt the traditional quantitative genetics environmental dichotomy, we lump any shared early external environment with the maternal environment, calling it the *preseparation environment*. For most adoption studies, the duration of postbirth overlap is small.

Although the maternal effect is known to be large in mammals,[5] this effect is essentially ignored in all the twin-adoption IQ studies. To see its possible impact, consider the total variance in human birth weight; 20% or more of the variance is explained by the maternal effect (Falconer, 1981). Moreover, for the development of IQ, it is clear that the environment plays an important role and that the early environment is particularly important. Because some of the twin-adoption studies assume the preseparation effect is zero, they may grossly overestimate heritability of IQ. Yet these are the studies that H&M rely on most heavily.

We review later our analysis of published IQ studies, which evaluated the effect of the earliest environmental conditions on IQ.[6] The results of our analysis reveal that the effect of the preseparation may be substantial—about 20% of the twin IQ variability appears to be explained by the preseparation effect. As we would expect, the estimated preseparation effect for siblings, who have not shared the womb simultaneously, is smaller (about 5%). We draw two conclusions from these results: First, early environment, even that in the womb, appears to have a large impact on IQ. Hence it is plausible that interventions aimed at improving the preseparation environment could lead to a significant increase in IQ. Secondly, ignoring the preseparation environmental effect (and the distinction between narrow- and broad-sense heritability), as H&M do, causes heritability to be overestimated.

The Data and a Model

To determine the effect of genetics and environment on IQ, we simultaneously analyzed the data from 212 IQ correlations. The analysis included 204 studies examined by Bouchard and McGue, with the subsequent corrections and additions by Bouchard.[7] We supplemented this set with the twin studies that have appeared since 1981. These new IQ results derive from a study of monozygotic twins reared apart,[8] the Swedish Adoption/Twin Study of Aging (SATSA): monozygotic twins reared together and apart and dizygotic twins reared together,[9] and two studies of monozygotic and dizygotic adult twins reared together.[10] The data gleaned from each study consist of the correlation among relatives and the sample size. The studies are classified into categories based on familial relationship. These categories include, for example, identical twins reared apart and siblings raised together.

To analyze the data from these studies we used a Bayesian meta-analysis, a standard technique for combining information across studies.[11] See the appendix to this chapter and Ref. 6 for details of the analysis. To partition the effect of environment and genetics on IQ, we used a standard quantitative genetic model for the components of variance (e.g., Ref. 5). For example, consider identical twins who were adopted. Each twin pair shares exactly the same alleles at all of their genes, as well as the same preseparation environment: hence, the covariance between twins is σ_G^2 plus the effect due to a shared maternal twin environment. Identical twins who were raised together also have a component due to a common shared environment occurring after the perinatal stage. The covariance between parent and child is considerably less because, in this pairing, only half of the genes are shared and there is no shared maternal effect. If parents do not select their mates according to IQ, the expected covariance between a parent and child is equal to half of the additive genetic variance plus a component due to a common shared environment occurring after the perinatal stage. In fact, people do tend to select mates with IQs similar to their own, a phenomenon known as *assortative mating*. In the appendix we give a table of expected covariances for each study type; these covariances are adjusted for the effect of assortative mating. We also describe four competing models that were fit to the data.

For each type of study, our model allowed for variability across studies in the observed correlations among relatives of the same type. This variability is due to measurement error and variability in the studies themselves. For example, the studies varied by type of IQ test given, age of the participants in the study, the range of environments experienced by the participants, as well as other factors.

Results

To achieve a better understanding of IQ heritability, we focus on two issues: (1) the magnitude of the additive and nonadditive genetic components, commonly hypothesized to explain in total at least 60% of the variation in IQ; and (2), the magnitude of maternal effects (preseparation environment), usually assumed to be negligible. To examine these issues, we fit various statistical models to the IQ data, with the aim of predicting the average correlation for different kinds of relative pairs. Our best-fitting model includes maternal effects—for twins and siblings—and assumes a common shared nonmaternal environmental component. This model differs from one commonly used in psychometric studies that allows for different external environmental components for twins, siblings, and parent-offspring.

The weighted average correlation for each of the nine study designs obtained from the 212 observed correlations is given in Table 3.1. To examine the fit of our model we compared the observed average correlations with the posterior mean correlations predicted by the model: it is clear that this model provides a good fit to the data. To formally compare the competing models, we used Bayes factors. The Bayes factor results strongly favor the model presented here.

Our model estimates the additive genetic variance at 0.34 and the nonadditive at 0.15 (Table 3.2), so the narrow-sense and broad-sense heritabilities are estimated to be 34% and 48%, respectively. Other meta-analyses of these data

Table 3.1. Posterior means for IQ correlations by study type

Relationship	Raised	Observed	Expected
Monozygotic twins	Together	0.85	0.85
Monozygotic twins	Apart	0.74	0.68
Dizygotic twins	Together	0.59	0.59
Siblings	Together	0.46	0.44
Siblings	Apart	0.24	0.27
Midparent/child	Together	0.50	0.50
Single-parent/child	Together	0.41	0.40
Single-parent/child	Apart	0.24	0.22
Adopting parent/child	Together	0.20	0.17

The column labeled "Observed" contains the weighted average of the observed correlations for studies of the same type, and the column labeled "Expected" contains the predicted values of these correlations from our model.

Table 3.2. Posterior means and 95% credible intervals for standardized covariances derived from our model

Effect	Estimate
Additive genetic	0.34 (0.27, 0.40)
Nonadditive genetic	0.15 (0.09, 0.20)
Total genetic	0.48 (0.43, 0.54)
Maternal (twins)	0.20 (0.15, 0.24)
Maternal (siblings)	0.05 (0.01, 0.08)
Environment	0.17 (0.13, 0.21)

A posterior mean is an estimate of the parameter of interest. A credible interval is roughly analogous to a confidence interval: the probability the parameter of interest lies in a 95% credible interval is 0.95, provided the model from which the interval is derived is correct.

have yielded similar results. Using a different method of analysis, Chipuer et al. estimated narrow-sense heritability at 32%.[12] Similarly, using path analysis, Rao et al. estimate narrow-sense heritability at about 34%.[13] Although these analyses agree with ours in their estimates of narrow- and broad-sense heritability, their models differ in that they exclude, *a priori*, maternal effects.

The notable difference between our model and those fit by others is how the shared environment is partitioned. The magnitude of the maternal twin effect is strikingly large. Our model estimates a large maternal effect for twins and a smaller effect for sibs: 20% of the variability is attributed to the former and 5% for the latter. The difference between the two estimates is intuitively appealing. Twins share the womb concurrently, whereas siblings share the same womb serially. Mothers may have similar personal habits from one pregnancy to another, but the temporal separation between progeny presumably ensures a diminished correlation of sibling IQ.

The apparent good fit of our model, which forces a common shared external environment term, is potentially surprising to many researchers in the field. A competing model often used by psychometricians, which does not include maternal effects, predicts the strongest environmental term for twins, then sibs, and finally parent-offspring. Once the environmental component is partitioned into a maternal and external shared environment in our model, a common shared environment appears to be sufficient to explain the data. Furthermore, the sum of maternal and external environments for twins, sibs, and parent-child has the expected pattern 0.37, 0.22, and 0.17.

Plomin and Loehlin present an overview of some of the studies we analyzed here.[14] They estimate heritability using "direct" and "indirect" methods.

Direct estimates are obtained by studying the correlation in IQs of adopted children with either their parents or their biological siblings. An example is the Minnesota-twin adoption study from which they estimate broad-sense heritability of $h_B^2 = 0.78$ (See Ref. 8). The direct estimates are currently receiving a great deal of attention due to a belief that twins separated at birth are the perfect study specimens. Falconer, on the other hand, argues that the most accurate estimate of heritability is obtained as the difference between the identical and nonidentical twin correlations—the indirect estimation method. Direct estimation methods assume all environmental correlations are zero because the children were reared apart. Of course, as we saw earlier, the twins do share a preseparation environment, but this has previously been assumed to have no effect on IQ heritability. The studies that estimate heritability indirectly do not make this assumption. By appropriate contrasts, these studies cancel out the effects of such terms. Plomin and Loehlin note that the direct studies usually yield a much greater estimate of heritability than the indirect studies, an anomaly they could not resolve. By accounting for maternal effects and nonrandom mating, it appears that this puzzle can be resolved and disparate direct and indirect heritability estimates can largely be rectified (see Ref. 6 for details). Furthermore, maternal effects, which were previously assumed to be negligible, play a particularly important role in this resolution.

It is not difficult to garner biological evidence to support the importance of the maternal environment. Notably, brain growth occurs during both the prenatal and perinatal period, with substantial growth occurring *in utero* and the majority of brain development completed by age one. Moreover, diverse kinds of data indicate that IQ can be diminished or enhanced by prenatal environmental effects, such as nutrition and exposure to drugs and alcohol (see references in Ref. 6).

In summary, we combined all the information available for all IQ studies to determine the effects of genetics and environment. Our analysis supports a substantial maternal effect on the variability in twins' IQ (20%) and a much lower estimate of broad-sense heritability than obtained by direct estimates (48% vs. 74%). Narrow-sense heritability is estimated at 34%. Our model indicates that narrow-sense heritability is likely to be less than 40%. Broad-sense heritability is likely to be less than 54%. Thus heritability is a far cry from the 40 to 80% claimed by H&M, and 60% is surely not a "middling value." Clearly 43 to 54% is a better bound for broad-sense heritability and 27 to 40% for narrow-sense heritability.

Social Implications

In their volume, H&M go well beyond the analysis of genetic, economic, and social science data. Using such analyses, they forecast the future structure of

American society, such as predicting the development of cognitive castes and the erosion of cognitive capital, and they touch on a number of socially charged issues, such as the underlying basis for African-American versus Caucasian-American IQ differences. Their forecasts for American society depend on both quantitative genetics and evolutionary theory. Because they ignore the latter and overvalue the former (i.e., heritability), their forecasts cannot be taken too seriously. We also find their treatment of socially charged issues wanting.

Cognitive Castes

H&M's argument for cognitive castes goes as follows. There is little doubt that economic and educational forces are stratifying the society by intelligence (or IQ). This stratification reinforces the natural tendency for assortative mating by IQ. Combining these factors with the high heritability of IQ, it is apparent that society is in the process of being divided into cognitive castes—one elite, another dull, the remainder middling—with little or no mixing among castes.

Their argument for the development of cognitive castes pivots on ever-increasing assortative mating, for which there is no quantitative evidence to date, and the high heritability of IQ. But the heritability they refer to in their book is based on direct estimates and twin-adoption studies. Even if the preseparation effect is ignored, which our results suggest would be a substantial oversight, such studies yield estimates of broad-sense heritability only. The critical parameter for the evolutionary process they invoke is narrow-sense heritability because it is the genetically based between-generation correlation of IQs that is critical for their argument.

All current meta-analyses place narrow-sense heritability around 0.34. This value is small enough that cognitively elite parents are by no means guaranteed to produce elite offspring. Instead, because of the relatively small value for narrow-sense heritability, the statistical effect of regression to the mean, and the relative scarcity of elites, the greatest portion of the next generation of cognitive elites should spring from parents of middling IQ rather than from the cognitive elite of today. From the genetic perspective, in lieu of cognitive castes, we would expect cognitive mixing.

There is a caveat, however. We have assumed, as Herrnstein and Murray also do, that the environment is relatively homogeneous across cognitive groups, not highly positively correlated. Otherwise, cognitive castes of a sort might be possible because those with limited environment, no matter how great their genetic potential, are not apt to achieve elite status.

Dysgenics

H&M also claim, "Mounting evidence indicates that demographic trends are exerting a downward pressure on the distribution of cognitive ability, that the pressures are strong enough to have social consequences (p. 341)," and that "something worth worrying about is happening to the cognitive capital of the country (p. 364)." This is occurring, they argue, because birth rates are higher for uneducated women than for the educated. Because education is closely linked with cognitive ability, this tends to produce a dysgenic effect, or a downward shift in the ability distribution. Furthermore, they claim that "blacks and Latinos are experiencing even more dysgenic pressures than whites, which could lead to further divergence between whites and other groups in future generations (p. 341)." As with their notion of castes, the authors have again ventured into the realm of evolution, describing the temporal change in distributions of genes (cognitive capital) in a population or set of populations as a response to differential fertility. What is missing from their arguments is a clear statement of how such evolution proceeds.

At a simple level, the "response to selection" to which H&M refer is calculated as the product of narrow-sense heritability times the reproductive differential, which is also known as "Fisher's fundamental theorem of natural selection."[15] This product yields the change in the population's *genetically based* IQ distribution. Using their estimate that, at present, the average IQ of women who gave birth is 98, the response will be a decline of one third of an IQ point per generation. [Because the average IQ of all women is approximately 99, the reproductive differential is approximately one.[16] For Caucasians, we estimate the narrow-sense heritability to be 0.34.] Extrapolating from this estimate, the mean of the *genetically based* IQ distribution would drop approximately one standard deviation per millennium. H&M incorrectly estimate the drop to be almost three times that magnitude, at least.

But these calculations are misleading in several ways. For instance, in deriving this simple relationship between reproductive differential and response to selection, geneticists make assumptions about environmental variation. This simple approach assumes that both genetically high- and low-IQ individuals have equal probability of experiencing the range of possible environments. The response to selection can be unpredictable when this assumption fails because the lack of independence may decouple the relationship between phenotypic and genotypic values. This assumption, it seems to us, is highly unrealistic for current populations, particularly non-Caucasian populations.

One can best understand the assumptions and implications of this evolutionary model by imagining a simple experiment with wheat. Suppose a plant breeder plants two varieties of wheat (*A* and *B*) in each of two different environments, one poor and the other excellent. Suppose that variety *A* is larger than variety *B* in both environments. Even so, we find that plants of variety *B*, grown in the excellent environment, tend to be larger than plants of variety *A*, grown in the poor environment. If the plant breeder selects only the largest half of the plants in each environment to reproduce, then, over time, he will eventually have nothing but plants containing the "growth genes" from variety *A* in either environment. His fields would have experienced a strong response to selection. On the other hand, suppose he had merely selected all relatively large plants to reproduce, ignoring the differential environments. Under this scenario, he might have collected seeds from all of the plants in the excellent environment and none from the poor environment. With this system, the plant breeder would realize no response to selection.

The analogy with IQ should be apparent. Suppose persons in low-quality IQ environments have similar genetic IQ potential to those in high-quality IQ environments. Regardless of their genetic potential, the observed IQ of a person raised in a poor environment will tend to be lower than the IQ of a person raised in the high-quality environment. Now suppose there is a reproductive differential whereby individuals of low IQ enjoy the greatest reproductive success. Under this scenario, it may appear as if the population as a whole is undergoing dysgenics when in fact it may not be. The actual dysgenic response will depend on the relative frequency of the subpopulations in the low- and high-quality environments. That is not to say that the IQ distribution will necessarily be stable over the generations, even if there is no dysgenic response; it is important to remember that the phenomenon of dysgenics depends on the inherent *genetic* values, not the phenotypic values, of the individuals in the population.

Is there any empirical evidence for IQ dysgenics? The analyses of the National Longitudinal Survey of Labor Market Experience of Youth (NLSY), cited as support for a dysgenic effect in Caucasians (pp. 348–356),[17] are flawed because they use samples of women who have not completed childbearing. Women who bear children later in life typically have higher IQs and are wealthier, the latter presumably enabling those families to provide an environment that fosters IQ. The impact of deleting observations from the upper tail of the normal distribution is usually tremendous. The women in the NLSY ranged in ages from 25 to 33 for H&M's analyses, and they were younger for the Ree and Earles study. Hence, insofar as we are aware, there is nothing but anecdotal evidence for dysgenics.

Is there any empirical evidence against IQ dysgenics? Instead of a dysgenic effect, the opposite seems to have occurred during this century. IQ scores have risen 15 to 20 points throughout the world since 1950. This pattern, known as the *Flynn effect*,[18] is widely acknowledged. Of course, the Flynn effect cannot be due to genetics alone. Genetic changes of this magnitude are virtually impossible in this short a time span. Specifically, for the mean of a quantitative trait to change substantially in one or two generations, due to genetic changes, the directional selection on IQ would be draconian at best. Applying Fisher's fundamental theorem, an increase of 11 to 12 IQ points per generation would only occur if the average IQ of parents was 134. Thus, for this to be a genetic phenomena, only individuals with unusually high IQs could reproduce, and no others. Such limits to reproductive freedom would be noticeable by the least observant citizen. Thus, we can rule out a dominant role for genetics in the recent increase in IQ. The Flynn effect might appear to be antithetical to IQ dysgenics. However, because it is likely to be largely an environmental effect, it cannot be used to refute unequivocally the possibility that dysgenics is occurring or has occurred. Hence, insofar as we are aware, there is nothing but anecdotal evidence against dysgenics either (see Epilogue in this chapter for further discussion).

Race and IQ

H&M say they are firmly agnostic on the issue of whether the origin of the IQ differences among races and ethnic groups is genetic, environmental, or both. They rightly point out "That a trait is genetically transmitted in individuals does not mean that group differences are also genetic in origin (p. 298)." We have some difficulty reconciling these statements with the rhetoric in the book. From our reading, it appears the authors favor a genetic explanation: H&M critically evaluate studies that support the environmental hypotheses, while failing to note some of the glaring deficiencies in the studies purported to support the genetic hypothesis. Regardless of their intent, it is important to evaluate this politically potent issue.

It is very unlikely that the IQ differential between races can be explained by genetics only. As discussed shortly, scores on IQ tests do not rise or fall on the basis of the racial admixture of the population. And IQ scores have risen across the world in the post–World II period, suggesting that IQ differentials are predominantly due to environmental effects.

While the majority of human genes show only minor differences among the races, some genes affecting visible differences among the races show much larger differences. The most obvious are the genes for skin color. Investigation of such loci invariably reveal that the traits they determine have been under directional natural selection, defined as the differential multiplication of variant types.[19] It is not clear to us why IQ would be positively selected in Caucasians but not in Africans. It seems more plausible that there was little selection in either population, or positive selection in both.

African Americans vary in the relative proportions of African and Caucasian ancestry. This variation was exploited in research by Scarr et al.[20] Scarr's group reasoned that if the IQ differential was due to genetics, then African Americans with higher degrees of Caucasian admixture would have higher IQs. To quantify the degree of admixture, the Scarr group used a battery of genetic markers and standard statistical methods. Contrary to the genetic supposition, they found no relationship between the level of Caucasian admixture and IQ. The correlation was extremely low, about 0.01. Similar results, again indicating no impact of admixture, were obtained by Loehlin et al. and Tizard.[21] This type of association was indeed found for another quantitative trait, blood pressure. Using methods quite similar to the Scarr group, MacLean et al. found a significant negative correlation between blood pressure and degree of Caucasian admixture.[22]

From the admixture studies one should conclude that the IQ differential is not due to genetics or that the impact of genetic differences is small. H&M discredit these studies because the Caucasian genes in the African-American population are not from a random sample. However, this logic contradicts their own argument regarding the historical distribution of IQs. H&M argue that prior to the middle of this century, IQ was more or less equitably distributed among the classes of society. Because the sample only needs to be random with respect to IQ, it is difficult to construct plausible scenarios in which the sampling of "IQ genes" was biased.

What explains the difference in mean IQ between Caucasians and African Americans? Although IQ is a partially heritable trait, claims that the difference is primarily due to genetics are merely speculations with miniscule likelihood that they are correct. Claims that the difference is due primarily to environmental effects are more plausible, but are not supported by an irrefutable mass of empirical evidence. The evidence for the environmental hypothesis includes the fact that IQ increased by more than one standard deviation in the past 30 years. Moreover, results of studies examining the effects of adoption, increased nutrition, and enhanced education all indicate that IQ is malleable to varying degrees. It would be a mistake, however, to claim that the races exhibit no differences in any genes critical to IQ, as this is almost impossible. The most likely scenario is that the IQ differential is predominantly due to environmental effects. Notably, under this scenario, it is unreasonable to posit the direction of any IQ differential if environments were homogenized for all groups. That is to say, on a level playing field and with the data at hand, we cannot predict whether African Americans or Caucasians would have higher average IQs.

How Does the Environment Affect Heritability?

Heritability is environment dependent, even though inheritance of genes is inevitable and unaffected by environment conditions. The extent to which genes influence IQs can change from generation to generation, varying with the environment. In essence, the environment enables the expression of genes. Therefore, the degree to which nature determines traits is dictated by the environment in which the organism develops. Even when two populations have exactly the same genetic variability, they will not have the same heritability if the environment is not the same or environmental variability is not equal.

H&M believe, as do many people in their field,[23] that heritability necessarily increases as the environmental variation decreases. H&M say, "The irony is that as America equalizes the circumstances of people's lives, the remaining differences in intelligence are increasingly determined by differences in genes (p.91)." This argument is somewhat misleading because, in general, no such prediction can be made. In an almost homogeneous environment for IQ, heritability could be almost zero, almost one, or anywhere in between. The realized heritability is determined by the complex interplay of environment and genes, which will vary depending on the "environment" that is being homogenized.

Epilogue

We began this chapter by stating that heritability was a complicated technical concept and that we could do no more than give the reader a taste of the subject. We hope we did not boil it down so much that the taste was disagreeable or that the subject appeared trivial. Quantitative genetics models are a rich and complex blend of genetics and statistics.

Quantitative genetics practitioners also recognize that their models are a bit of a sham. They are elegant simplifications of nature, and when their genetic underpinnings are examined closely, they are invariably found to be inaccurate. So why do plant and animal breeders, as well as evolutionary biologists, find the models so interesting? It is because, in experienced hands, the models are useful for producing higher agricultural yields and yielding deeper evolutionary insights. In inexperienced hands, however, the models are no more useful than a blunt chisel.

IQ studies don't fit into an agricultural framework and aren't very interesting evolutionarily either. Therefore, the value of quantitative genetic studies of IQ could be questioned, and some have claimed they have no value whatsoever. We do not share this narrow viewpoint. The impact of genes on cognitive function, be it normal or abnormal function, is of fundamental interest to humankind. IQ studies, although imperfect, contribute knowledge regarding the genetics of cognitive function. The problem lies not in the studies, but in how their results have been interpreted in popular writings, some of which have been authored by scientists themselves.

In particular, the subtle interplay of environment and genes rarely comes across in these writings, either because the authors judge the subject too complex for their readership or because they don't grasp it themselves. The nature–nurture paradigm strikes us as a perfect example. As witnessed by *The Bell Curve*, this paradigm still seems to arouse interest in certain arenas despite a half-century of genetics research demonstrating that gene expression is environment dependent. In our view, such research makes the nature–nurture paradigm an illogical construct.

Field and laboratory studies yield valuable coarse-scale information on gene-environment interaction (see Chapter 4). As the field of molecular genetics matures, the interplay of genes and environment is increasingly understood at a fine scale. While some of this knowledge derives from human studies, most of it has been generated by research on organisms such as nematodes, *Drosophila* flies, and mice, to name a few model organisms. Regardless of the study subject, the results of these studies are pertinent to humans, revealing that gene expression can be exquisitely sensitive to the environment and to genetic background, that is, the other genes the organism carries. How important these results are to a trait such as IQ, which is presumbly affected by many genes, is anybody's guess. What is clear, however, is that the nature–nurture paradigm is antiquated.

How these subtleties can be communicated to the public is an important question that goes well beyond genes and IQ. Misleading headlines and articles appear almost weekly regarding studies of human genetic disorders. One reads, for instance, about the gene for breast cancer. Cancer cannot develop because of one mutated "breast cancer" gene; only a suite of communicating genes, working within the framework of the organism, can produce a cancer. And not all individuals who have the "breast cancer" gene develop breast cancer. Why? Because the impact of the mutation depends on its environmental and genetic background.

To some extent, we are also guilty of oversimplification, especially regarding quantitative genetics issues. So we are sympathetic to the difficulty of presenting complex concepts in limited space. Nevertheless, we hope that our presentation is sufficient for the reader to understand the principal arguments of *The Bell Curve* and to conclude, as we do, that the arguments have no basis in science. H&M's cognitive castes are unrealistic in light of small narrow-sense heritability and the scarcity of elites. Their argument about the erosion of cognitive capital (IQ dysgenics) is presented without acknowledging its evolutionary framework or its environmental assumptions. When viewed from a more rigorous perspective, sizable IQ dysgenic effects are implausible.

Many critics of *The Bell Curve* have been no more rigorous than H&M and often are much less so. For instance, numerous critics have cited the Flynn effect as proof that H&M's worry about IQ dysgenics is unfounded. It is a tempting and facile argument. However, as we described previously, the Flynn effect must be wholly or largely the result of environmental influence: it tells us nothing about the genetics of populations, as they pertain to IQ, and how that might be changing over time. Hence the presence of the Flynn effect cannot obviate the possibility of IQ dysgenics. Conversely, in view of the Flynn effect, how does one estimate the selection differential required for this evolutionary argument? It makes little sense to compare IQs across generations if, in fact, the IQ scale is entirely different for those generations. In that case, what appears to be a drop in IQ by simple subtraction could really be an increase in IQ after adjusting for the scale change inherent in the Flynn effect.

Some critics even accuse H&M of claiming racial IQ differences are genetic, suggesting the critics had not even read the book. H&M were far more subtle. What they suggested was that there appeared to be greater weight for genetically driven as opposed to environmentally driven differences. Our evaluation of the literature, which is not particularly abundant, suggests just the opposite.

Based on their genetic and social analyses, H&M argue that the limited efficacy of interventions to raise IQ, together with their substantial costs, prohibits society from improving the cognitive function of its young people by more than a minor degree. We are not convinced by their arguments, and our analysis of IQ heritability may provide a direct contradiction. Specifically, because our results indicate that the maternal environment has a notable impact on cognitive function, it is plausible that interventions aimed at improving that environment could lead to a significant increase in IQ. Moreover, it seems unlikely that enhanced maternal care, in the form of education and/or dietary supplements, would be be prohibitively expensive (see Chapter 13). While the data in hand only hint at the efficacy of such an intervention, the hypothesis itself can and should be tested by experimental and comparative methods.

Appendix: Model for Bayesian Meta-Analysis of IQ Studies

Our statistical model for heritability is built upon two levels of distributional assumptions. In the first level, we assume a likelihood model for the observed correlations among relatives in each type of study. The likelihood depends on a number of parameters that describe the proportion of the variability in IQ that can be ascribed to each genetic and environmental component in the model. For each type of study, the likelihood model allows for variability across studies in the observed correlations among relatives of the same type.

In the second level of the model, we specify a prior distribution for the parameters of the model. Ideally, the prior distribution is a vehicle for inputting all that the investigator knows about the parameters of the model before the study was performed. The likelihood updates the prior, yielding the posterior distribution. Consequently, it contains all that is known *a posteriori* about the parameters in the model.

In a meta-analysis such as this one, it is usually desirable to choose a noninformative prior distribution because we are analyzing nearly all of the data available concerning IQ correlations between relatives. Thus there is little information available from which one could formulate a prior. For this reason, we assume any component of variance is *a priori* equally likely to lie between zero and one.

The notation for the various factors in our model contributing to IQ are summarized in Table 3.3. To partition the effect of environment and genetics on IQ, we use a standard quantitative genetic model for the components of variance (e.g., Ref. 5). We fit four models to the data. The expected covariance among relatives for the richest of our four models (Model IV) is summarized in Table 3.4. The remaining models differ from Model IV in that Models I and II constrain the maternal effects to be zero, and Models I and III constrain the external environmental effects to be equal. The model discussed in the chapter is Model III.

To appreciate the logic behind the expected covariances for sibs, consider a single gene. When sibs share only one allele from a parent, say, from their mother, then the phenotypic covariance between them can only be due to the additive effect of that allele; when sibs share two alleles, one allele from each parent, the covariance between their phenotypes can be due to additive effects and the interactions of those alleles. Of course, interactions of allele at different genes are possible and can be quite important. These effects are another source of nonadditive genetic variation. Because sibs share a single allele of a gene with probability 1/2 and two alleles with probability 1/4, the genetic covariance term is $1/2\sigma_A^2 + 1/4\sigma_D^2$, ignoring assortative mating. The $1/2\sigma_A^2$ follows by averaging phenotypic covariances due to additive effects over sibs who share 0, 1, or 2 alleles at a gene. In practice, even if raised apart, siblings tend to be more similar than predicted by this genetic model because people tend to chose a mate with an IQ similar to their own. The term r is used to account for this effect, which is known as *assortative mating*. Using Bouchard and McGue's estimate, r is set to 0.33. The effect of assortative mating is to inflate the apparent additive genetic component (see Table 3.4).

Table 3.3. Notation for the quantitative genetic models

Effect	**Notation**
Additive genetic	σ_A^2
Nonadditive genetic	σ_D^2
Total genetic	σ_G^2
Maternal (twins)	$\sigma_{M_T}^2$
Maternal (siblings)	$\sigma_{M_S}^2$
Environment (twins)	$\sigma_{ES_T}^2$
Environment (siblings)	$\sigma_{ES_S}^2$
Environment (parent/child)	$\sigma_{ES_P}^2$
Assortative mating	r

The σ^2's denote covariances and r is a parameter for assortative mating.

To demonstrate that our results are robust to the choice of r, we refit the models using $r \pm 2s$, where s is the standard deviation of r obtained from the set of studies compiled by Bouchard et al. The results were essentially identical.

From each study used in our analysis, we have two relevant pieces of information: the estimated correlation between relatives of the jth type in the ith study ($\hat{\rho}_{ij}$) and the number of relative pairs (n_{ij}). The expected correlation ρ_j is equal to the expected covariance, divided by the total variance in IQ for this type of study, σ_T^2. (For an identical twin study, the total variance is $\sigma_T^2 = \sigma_A^2 + \sigma_D^2 + \sigma_{M_T}^2 + \sigma_{ES_T}^2 + \sigma_E^2$, where σ_E^2 is the nonshared environmental effect.) Because the studies reported correlations, rather than covariances, we based our models on standardized covariances: $\sigma_k^\dagger = \sigma_k^2/\sigma_T^2$, $k \in \{A, D, M_T, M_S, ES_T, ES_S, ES_P\}$, that is, the proportion of variability in IQ explained by the k'th effect. Each of these standardized covariance components, $\sigma_k^\dagger$, is naturally constrained to the interval (0,1); hence these terms are assumed, *a priori*, to follow a uniform distribution, as mentioned earlier.

We use the Fisher transformation of the correlation between relative pairs to obtain approximate normality.[24] Define the transformed observation as $z_{ij} = 1/2 \log\{(1 + \hat{\rho}_{ij})/(1 - \hat{\rho}_{ij})\}$. The expected value of z_{ij} equals $1/2 \log\{(1 + \rho_j)/(1 - \rho_j)\}$. For each type of study, if the observed correlations differed only due to measurement error, then one would expect z_{ij} to follow a normal distribution with expectation given in the previous footnote and variance $1/(n_{ij} - 3)$. However, several of the studies appear to be outliers under a normal likelihood, suggesting that the studies are more heterogeneous than predicted by measurement error alone. Such heterogeneity is not surprising because the studies differ by a number of factors. To account for this heterogeneity, we

Table 3.4. Expected covariances between relatives

Relationship	Raised	Expected covariance
Monozygotic twins	Together	$\sigma_A^2 + \sigma_D^2 + \sigma_{M_T}^2 + \sigma_{ES_T}^2$
Monozygotic twins	Apart	$\sigma_A^2 + \sigma_D^2 + \sigma_{M_T}^2$
Dizygotic twins	Together	$\frac{1}{2}(1 + r\sigma_A^\dagger)\sigma_A^2 + \frac{1}{4}\sigma_D^2 + \sigma_{M_T}^2 + \sigma_{ES_T}^2$
Siblings	Together	$\frac{1}{2}(1 + r\sigma_A^\dagger)\sigma_A^2 + \frac{1}{4}\sigma_D^2 + \sigma_{M_S}^2 + \sigma_{ES_S}^2$
Siblings	Apart	$\frac{1}{2}(1 + r\sigma_A^\dagger)\sigma_A^2 + \frac{1}{4}\sigma_D^2 + \sigma_{M_S}^2$
Midparent/child	Together	$\sigma_A^2 + \sigma_{ES_P}^2$
Single-parent/child	Together	$\frac{1}{2}(1 + r)\sigma_A^2 + \sigma_{ES_P}^2$
Single-parent/child	Apart	$\frac{1}{2}(1 + r)\sigma_A^2$
Adopting parent/child	Together	$\sigma_{ES_P}^2$

See Table 3.3 for definitions of notation. "Relationship" refers to the kind of relative pairs examined by the IQ study. "Midparent" refers to the average of the two parents' IQs.

assume that the transformed value follows a t-distribution with 2 degrees of freedom. Such a distribution allows for substantial variability among studies. However, our results varied little with as many as 1,000 degrees of freedom, which is essentially a normal distribution.

To obtain the posterior distribution of the standardized variances we used the Gibbs Sampler with a Metropolis step.[25] To compute Bayes factors for the competing models, we implemented the method for nested models described by Verdinelli and Wasserman.[26]

References

1. Herrnstein, R.J. and Murray, C. (1994), *The Bell Curve: Intelligence and Class Structure in American Life*, The Free Press, New York.
2. Bouchard, T.J., Jr., and McGue, M. (1981), "Familial Studies of Intelligence: A Review," *Science*, 212, 1055–1059.
3. Fisher, R.A. (1918), "The Correlation Among Relatives on the Supposition of Mendelian Inheritance," *Transactions of the Royal Society, Edinburgh*, 52, 399–433.
4. Fisher, R.A. (1951), "Limits to Intensive Production in Animals," *British Agricultural Bulletin*, 4, 217–218.
5. Falconer, D.S. (1981), *Introduction to Quantitative Genetics*, Longman, New York.
6. Devlin, B., Roeder, K., and Daniels, M. (1997), "On the Heritability of IQ," *Nature*, in press.
7. Bouchard, T.J., Jr., personal communication.
8. Bouchard, T.J., Jr., Lykken, D.T., McGue, M., Segal, N.L., and Tellegen, A. (1990), "Sources of Human Psychological Differences: The Minnesota Study of Twins Reared Apart," *Science*, 250, 223–228.
9. Pedersen, N.L., Plomin, R., Nesselroade, J.R., and McClearn, G.E., (1992), "A Quantitative Genetic Analysis of Cognitive Abilities During the Second Half of the Life Span," *Psychological Science* 3, 346–353.
10. McGue, M., Bouchard, T.J., Jr., Iacono, W.G., and Lykken, D.T. (1993), "Behavioral Genetics of Cognitive Ability: A Life-Span Perspective," in *Nature, Nurture, and Psychology* (eds. Plomin, R., and McClean, G.E.), American Psychological Association, Washington, D.C., pp. 59–76.
11. Rosenthal, R. (1984), *Meta-Analytic Procedures for Social Research*, Sage Publications, Beverly Hills, CA. National Research Council (NRC) (1992), *Combining Information. Statistical Issues and Opportunities for Research*. National Academy Press, Washington, D.C.
12. Chipuer, H.M., Rovine, M.J., and Plomin, R. (1990), "LISREL Modeling: Genetic and Environmental Influences on IQ Revisited," *Intelligence*, 14, 11–29.
13. Rao, D.C., Morton, N.E., Lalouel, J.M., and Lew, R. (1982), "Path Analysis Under Generalized Assortative Mating. II. American IQ," *Genetical Research* (Camb.), 39, 187–198.
14. Plomin, R., and Loehlin, J.C. (1989), "Direct and Indirect IQ Heritability Studies: A Puzzle," *Behavior Genetics* 19, 331–342.
15. Fisher, R.A. (1958), *The Genetical Theory of Natural Selection*, second revised edition, Dover Publications, Inc. New York.
16. Sattler, J. (1988), *Assessment of Children's Intelligence and Other Special Abilities*, Allyn and Bacon, Boston.

17. Ree, M.J. and Earles, J.A. (1991), *Aptitude of Future Manpower: Consequences of Demographic Change*, Brooks Airforce Base: Manpower and Personnel Division, Air Force Systems Command.
18. Flynn, J.R. (1987), "Massive IQ Gains in 14 Nations: What Do IQ Tests Really Measure?," *Psychological Bulletin*, 101, 171–191.
19. Muller H.J. (1962), *Studies in Genetics*. Indiana University Press, Bloomington, IN.
20. Scarr, S. Pakstis, A.J., Katz, S.H., and Barker, W.B. (1977), "Absence of a Relationship Between Degree of White Ancestry and Intellectual Skills Within a Black Population," *Human Genetics*, 39, 69–86.
21. Loehlin, J.C., Vanderburg, S.G., and Osborne, R.T. (1973), "Blood Group Genes and Negro-White Ability Differences," *Behavior Genetics*, 3, 263–270. Tizard, B. (1974), "I.Q. and Race," *Nature*, 247, 316.
22. MacLean, C., Adams, M.S., Leyshon, W.C., Workman, P.L., Reed, T.E., Gershowitz, H., and Weitkamp, L.R. (1974), "Genetic Studies on Hybrid Populations. III. Blood Pressure in the American Black Community," *American Journal of Human Genetics*, 26, 614–626.
23. *Wall Street Journal*, December 13, 1994, p. A18.
24. Fisher, R.A. (1921), "On the 'Probable Error' on a Coefficient of Correlation Deduced from a Small Sample," *Metron*, 1; 3–32.
25. Geyer, G. (1989), "Practical Markov Chain Monte Carlo," *Statistical Science* 4, 473–482. Smith, A.F.M., and Roberts, G.O. (1993), "Bayesian Computation via the Gibbs Sampler and Related Markov Chain Monte Carlo Methods," *Journal of the Royal Statistical Society B*, 55, 3–24.
26. Verdinelli, I., and Wasserman, L. (1995), "Computing Bayes Factors By Using a Generalization of the Savage-Dickey Density Ratio," *Journal of the American Statistical Association*, 90, 614–618.

chapter four

The Malleability of Intelligence Is Not Constrained by Heritability

DOUGLAS WAHLSTEN

In *The Bell Curve*, Herrnstein and Murray claim that a high value for heritability of intelligence limits or constrains the extent to which intelligence can be increased by changing the environment.[1] In this chapter it is argued that the calculated numerical value of "heritability" has no valid implications for government policies and that evidence of a nonspecific genetic influence on human mental ability places no constraint on the consequences of an improved environment. On the contrary, a very small change in environment, such as a dietary supplement, can lead to a major change in mental development, provided the change is appropriate to the specific kind of deficit that in the past has impaired development. The results of adoption studies, the intergenerational cohort effect, and effects of schooling also reveal that intelligence can be increased substantially without the need for heroic intervention.

The Bell Curve continues a long tradition, exemplified by Malthus, of basing recommendations for government social policies on claims about limitations imposed by human psychology and biology. Although Herrnstein and Murray (H&M) do not advocate a complete reversion to the laissez faire ideal of Malthus and Herbert Spencer, some of their policy recommendations point in that direction. While advocating government-imposed income supplements for fulltime workers earning low wages who are trying hard, they also urge a wholesale decentralizing of government powers and a return to a greatly simplified common law and tort law, which would undoubtedly involve the dismantling of a vast array of existing programs. They argue that success in life in America depends strongly on intelligence and that intelligence is mainly hereditary; hence, further success of the nation requires improvement of the gene pool. Specifically, they propose (a) restoring the rewards of marriage (p. 546) by ending all child support payments for unmarried women, whom they note are often poor and black; (b) ending "the extensive network of cash and services for low-income women who have babies" (p. 548); (c) providing "birth control mechanisms that are increasingly flexible, foolproof, inexpensive, and safe" (p. 549) to reduce births among low income women; and (d) procedures to "shift the flow of immigrants . . . towards those admitted under competency rules . . . " (p. 549). H&M argue that everyone, including political leaders, should "try living with inequality, as life is lived" (p. 551), inequality that they believe is mainly biological.

Biological arguments concerning intelligence have often been invoked in the United States to influence current political policies. Just as Putnam warned of terrible things that would follow from any success of the Civil Rights movement, especially school integration,[2] Jensen, in his hotly debated *Harvard Educational Review* article, opposed the Head Start antipoverty program and warned rhetorically: "Is there a danger that current welfare policies, unaided by eugenic foresight, could lead to the genetic enslavement of a substantial segment of our population?"[3] He expressed special concern about biological reproduction by "Negro Americans." *The Bell Curve* strongly supports the arguments of Jensen, but its political conclusions go much further in the direction of Malthus, Spencer, and Galton.

Heritability

Although biology forms the foundation of their policy recommendations, the biology presented by H&M is itself rather impoverished. Only one biological concept is invoked, the population genetic notion of "heritability." They claim that correlational statistics can reveal the percentage of variance in intelligence that is caused by genetic differences among people. While acknowledging there is disagreement among experts about the true value of this percentage, they assert that the "most unambiguous direct estimate" indicates it is about 60% to 70%. However, the validity of the heritability coefficient itself, not just its true value, is still subject to debate, and the concept has been strongly criticized as biologically unrealistic.[4–8] Eminent geneticists and statisticians have questioned the methodology commonly used to compute the heritability coefficient.[9–12] R.A. Fisher, inventor of the analysis of variance and statistical genetics, lamented the common misuse of "the so-called coefficient of heritability, which I regard as one of those unfortunate short-cuts which have emerged from biometry for lack of a more thorough analysis of the data."[13] The quantitative geneticist Kempthorne advised a broad readership "... that most of the literature on heritability in species that cannot be experimentally manipulated, for example, in mating, should be ignored."[14] Thus, the fundamental biological assumptions and the very conceptual foundations of *The Bell Curve* remain insecure and controversial.

Heritability and Plasticity

H&M claim high "heritability" of IQ means that improving the environment of a poor child a modest amount will be ineffective because "such changes are limited in their potential consequences when heritability so constrains the limits of environmental effects" (p. 109). They are simply wrong on this point. They commit what Lewontin has termed the "vulgar error that confuses heritability and fixity."[15] A heritability estimate does not in any way "constrain" the effects of a changed environment. Bad genes and poor upbringing most certainly can impair mental development, but there is nothing about either genetic or environmental effects on the mind that render them impervious to advances in scientific knowledge.

Small treatments tend to have small effects unless the treatments directly and precisely ameliorate a specific difficulty that impairs development. If such a specific difficulty can be identified, a very small change in environment can lead to a dramatic improvement. This dictum applies equally to specific deficits arising from the environment or a gene.

For example, during the first few decades of the twentieth century, pellagra was quite common among the working poor of the southern United States. The eugenicist Davenport claimed the slow learning and health problems of pellagrins resulted from an infection combined with bad genes, while the experimental proof by the physician Goldberger that it was a vitamin deficiency disease caused by low wages leading to a poor diet was ignored in official government corridors.[16] Now we know that a small daily dose of the vitamin niacin can effectively prevent pellagra, just as vitamin C prevents scurvy. Niacin addresses the precise shortcoming of the diet that otherwise leads to pellagra. Although Goldberger was correct to say that a rise in wages would prevent the nutritional deficiency disease, increased earnings is a broad- spectrum treatment that improves many aspects of the environment, one of which is the culprit causing pellagra. Scientific knowledge of a specific cause may enable cheaper and sometimes more effective treatments to be devised, but lack of this specific knowledge does not imply that no other treatment will help.

Concerning genetic defects, it is firmly established that the often devastating effects are not in principle inevitable. An immense volume of data from physiological and developmental genetics has established beyond reasonable doubt that the action of a gene in the nucleus of a cell of an individual is regulated by the surroundings of the gene, including other genes in the cell nucleus, substances in the cytoplasm of the cell, and myriad influences from the environment outside the cell. The activities of many kinds of genes can be modulated or abruptly switched on or off by signals from the environment external to the organism. Likewise, the results of rearing an individual in a novel environment often depend strongly on the organism's heredity, which illustrates the concept of the norm of reaction.[4,17–18]

Consider obesity in laboratory mice. The recessive gene named diabetes (*db*), when inherited from both parents to yield the *db/db* genotype, results in an extremely plump animal with all the clinical signs of diabetes (Fig. 4.1). However, diabetes is the *phenotype*, the measured consequences of possessing that genotype in the typical mouse colony where animals have free access to rich food around the clock and little else to do but eat. The *db/db* genotype does not code for diabetes and the clinical disease is not inevitable. Simply preventing the *db/db* mouse from overeating by giving it only the amount of food consumed the previous day by its lean sibling (the "pair-feeding" control) prevents the onset of diabetes.[19] Similarly, rearing mice on a high-fat diet induces diabetes in the inbred mouse strain C57BL/6J but not the A/J strain, although neither strain is diabetic on a normal lab-chow diet.[20] Pair-feeding will not prevent all kinds of diabetes in mice, and a high fat diet will not cause diabetes in any mouse. The response to environmental change depends on the genotype.

It frequently happens that a genetic mutation renders the organism more sensitive to variations in the environment. Thus, the scientific truth is directly opposite the claim by H&M that genes limit or constrain modifiability. Consider the well-known example of the human genetic disease phenylketonuria (PKU). The originating cause is a defect in the nucleotide base sequence of a gene coding for the structure of the enzyme phenylalanine hydroxylase that acts in the liver to transform the amino acid phenylalanine derived from dietary protein into a useful substance, tyrosine.[21] Lacking an active form of this enzyme, the child suffers from high levels of phenylalanine in the blood and severe brain damage. Lacking knowledge of the cause, physicians and psychologists were unable to help these pitiful victims of a bad gene. Once the cause was better understood, a special diet low in phenylalanine was formulated to prevent the devastation, and PKU was rapidly eliminated from institutions for the mentally incapable in countries with adequate newborn screening programs.

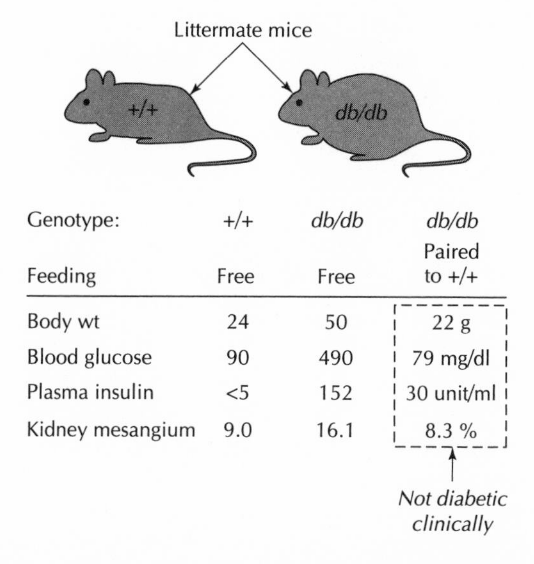

Genotype:	+/+	*db/db*	*db/db*
Feeding	Free	Free	Paired to +/+
Body wt	24	50	22 g
Blood glucose	90	490	79 mg/dl
Plasma insulin	<5	152	30 unit/ml
Kidney mesangium	9.0	16.1	8.3 %

Figure 4.1. Data for normal (+/+) mice and their genetically defective siblings with the *db/db* genotype. When the *db/db* mice are allowed to eat only the same amount of food as their lean siblings by pair-feeding them, the obesity and physiological symptoms of diabetes do not appear. Data from Ref. 19.

Sweeping statements about the ineffectiveness of environmental change denote helplessness and pessimism occasioned by ignorance rather than any inherent resistance of intelligence or any other phenotype to modification. Each of the 50,000 or more genes in the human chromosomes functions in a highly specific way as part of the biochemical system of a cell, and genetic knowledge can help to devise effective treatments only when a specific gene that impairs development is known. Vague statements about the supposed limits imposed by nonspecific heredity in general are groundless. They may impose limitations on the human imagination and quench optimism about the future. They may be politically expedient for those in positions of power, just as they were in the time of Malthus. Nevertheless, they do not have support from biological science.

Plasticity of Intelligence

The Bell Curve asserts confidently that "Changing cognitive ability through environmental intervention has proved to be extraordinarily difficult" (p. 314). On the contrary, available data indicate that a modest, short-term improvement can have substantial effects on mental ability test scores, whereas a large and lasting improvement can exert quite a large effect. Evidence comes from a variety of sources, including adoption studies, long-term (cohort) effects on IQ, and diverse influences of schooling on mental abilities. Evidence also indicates that intelligence requires persistent nurturing and is not fixed early in childhood.[22] It is much like muscle mass, where good diet and exercise are needed continually and flaccidity sets in quickly if essential nutrients or practice are denied.

Long-Term Changes in Environment

Well-controlled adoption studies done in France have found that transferring an infant from a family having low socioeconomic status (SES) to a home where parents have high SES improves childhood IQ scores by 12 to 16 points or about one standard deviation (see Fig. 4.2),[23,24] which is considered a large effect size in psychological research. Adoption can entail a major improvement in a child's environment, but the adoptive home is usually not off the scale of decent environments and therefore is not expected to yield a rich harvest of superior intellects, especially when the child's earlier experiences were very substandard. Nonetheless, the French adoption studies have observed changes in average test scores similar to the difference between means of nonadopted children living in high versus low SES homes. These results suggest that a major portion of the typical social class difference in intelligence originates in the environment.

Changing mental ability test scores a substantial amount is not so difficult. In fact, routine IQ testing reveals this commonly happens without deliberate

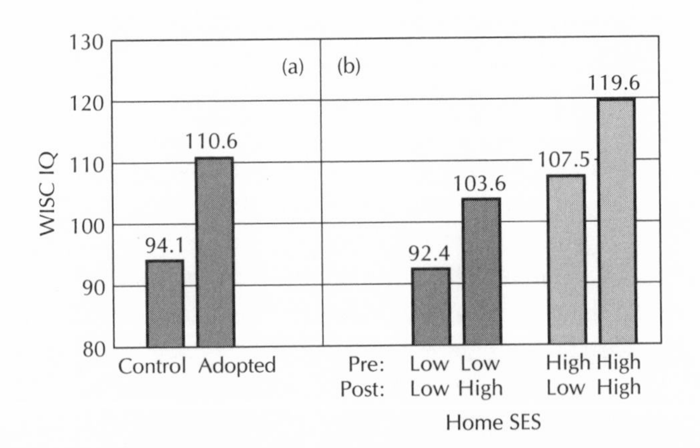

Figure 4.2. Mean Wechsler Intelligence Scale for Children (WISC) IQ score for children in two adoption studies done in France. (a) Schiff et al. (Ref. 23) compared two groups of children having the same biological mother and similar biological fathers. One group had been adopted into homes of well-educated professionals, whereas the control children had remained with the mother living in poverty. (b) Capron and Duyme (Ref. 24) tested children from four conditions categorized according to parental socioeconomic status (SES) prior to and after adoption.

intervention to enhance intelligence per se. The extent of this phenomenon tends to be obscured by the method of scoring the tests. The average IQ in a population should be about 100 and the standard deviation should be about 15, such that about 95% of people at the same age will score between 70 and 130 IQ points. One mechanism for scaling an IQ test is basically quite simple. The test is given in a particular year to a representative random sample of the population, and the numbers of items correct at any particular age have mean and standard deviation of M and S, respectively. Then each raw score X can be converted to a standard score Z that represents the number of standard deviations from the mean for the individual, using $Z = (X - M)/S$. The Z scores have a mean of 0 and standard deviation of 1. Finally, the Z scores can be transformed to IQ scores with the formula $IQ = 100 + Z(15)$. Thus, IQ indicates relative performance on a test rather than absolute degree of intelligence. Over a period of several years, it becomes necessary to restandardize the IQ test using more appropriate test items and a new sample of the population. This periodic restandardization of a test tends to keep the mean IQ close to 100, even if the underlying trait called *intelligence* is changing substantially in the population.

A large body of evidence reveals that raw, unstandardized intelligence has been gradually increasing for several decades in many industrialized countries.[25] Two kinds of data show this trend clearly. Perhaps the most persuasive comes from The Netherlands, an ethnically homogenous country, where almost all 18-year-old males are given the Ravens Progressive Matrices test as part of military induction. The test itself has not been modified for several decades. As shown in Figure 4.3, there is a very large cohort effect, amounting to a 21 IQ point increase in the population over three decades. A similar, albeit less dramatic improvement in the whole population was observed from 1956 to 1977 in Alberta, Canada, in 10-year-old children who took the identical test, the Raven Coloured Progressive Matrices.[26] Not only did the generational mean rise by about 0.45 standard deviation (see Fig. 4.4), equivalent to about 7 IQ points, but the significant superiority in 1956 of boys over girls vanished by 1977.

Another kind of evidence derives from the IQ restandardization procedure when people given the new version of a test then take the old version so that validity can be assessed. For example, when the Wechsler Intelligence Scale for Children (WISC) was revised in 1972, the sample of children scored 7 points higher on the previous version of the WISC that had been standardized in 1947. Combining these kinds of data for several IQ tests, a one standard deviation increase in mean intelligence in the United States is apparent over several decades (Fig. 4.5). The cohort effect is gradual and almost linear since the Second World War, but in terms of a population-wide change in intelligence manifested in one generation of Americans, it is a large effect indeed. It is especially thought provoking that the size of the cohort effect is

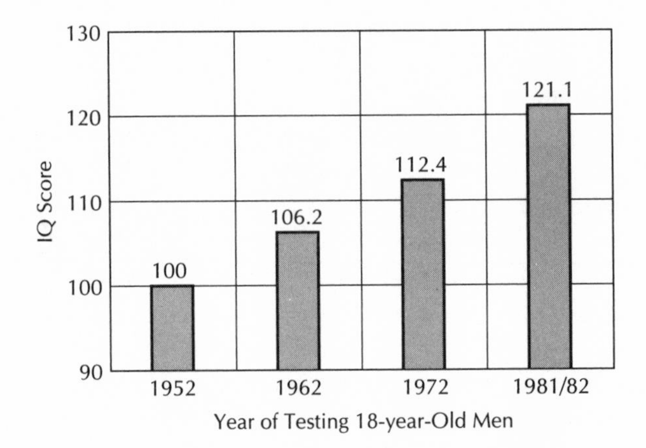

Figure 4.3. Mean Ravens Progressive Matrices scores converted to IQ scores for 18-year-old men in the Netherlands who were tested at the time of military induction in different years. Based on data in Table 1 of Ref. 25.

not much different than the widely publicized black/white IQ difference in the United States. That is, more recently born children exceed the raw intelligence of their own parents at a comparable age by almost the same average amount as Americans of European ancestry exceed Americans of African ancestry.

H&M mention the cohort effect in their discussion of group differences, but they lightly dismiss it as a mere improvement in "test-taking" skills or betterment of the living conditions of the disadvantaged. They muse that "one does not get the impression that the top of the IQ distribution is filled with more subtle, insightful, or powerful intellects than it was in our grandparents' day" (p. 308). Thus, faced with weighty evidence against their thesis, they are willing to dismiss the cohort effect with counterarguments that negate some of their own claims. If the cohort effect represents improvement by only the bottom half of the "bell curve" rather than the entire population, then their earlier claim (p. 314) that increasing IQ is extraordinarily difficult loses credibility. After all, if the population mean increases by 15 points but the top part of the distribution does not increase, the lower scores must have increased by much more than 15 points. In any event, achieving extraordinarily high levels of performance requires exceptional effort under the tutelage of expert instructors.[27,28] Outstanding achievement and brilliant creativity do not come "naturally" to anyone merely because of their genes.

H&M maintain assiduously in the first half of their book that IQ tests as we know them are very good measures of general intelligence. Yet the cohort effect causes them to revert to subjective impressions about their grandparents, who must have been children at a time when the IQ test was still embryonic in the mind of Alfred Binet. Actual experience in the United States, with the earliest

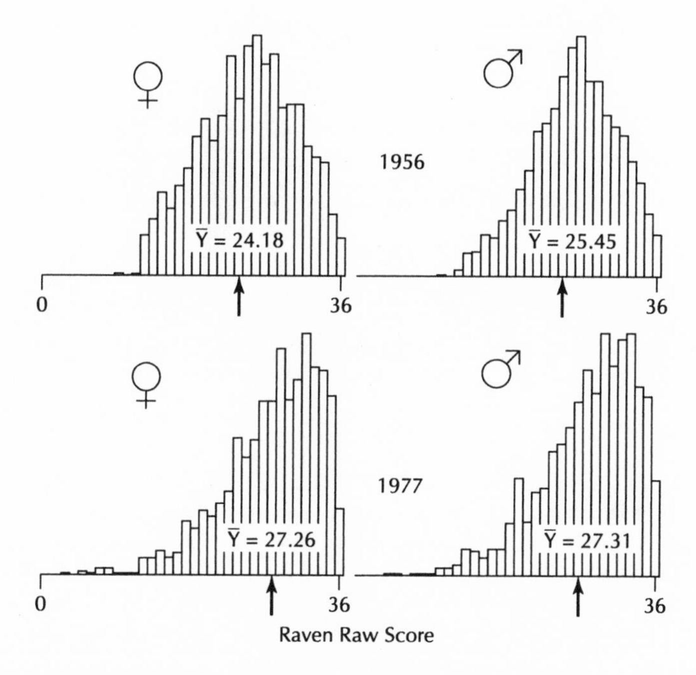

Figure 4.4. Distributions of Ravens Coloured Progressive Matrices raw scores for Edmonton, Alberta, children in grade 3 in 1956 and 1977. Data from Ref. 26.

administrations of IQ tests revealed that many men at the apex of American society were none too heavy under the helmet. The December 29, 1915 issue of the *Chicago Herald* trumpeted to its public: "Hear how Binet-Simon method classed mayor and other officials as morons" (reprinted in Ref. 16, p. 241). As for outstanding intellects, without a doubt they are products of their times and countries, but their achievements do not provide a valid measure of the intelligence of their lesser countrymen, who all too often failed to recognize genius in their midst, partly because of the prevailing political and social definition of genius.[29] Formal IQ tests were intended to supplant subjective impression and common prejudice with carefully constructed instruments administered in controlled conditions. For H&M to tiptoe around the cohort effect by suggesting that IQ tests do not really measure genuine intelligence but something more superficial and transitory is a negation of a fundamental thesis of *The Bell Curve*.

The cohort effect poses an even greater challenge to the raison d'être of *The Bell Curve*. H&M raise alarm about several worrisome social trends in the United States and argue that inadequate intelligence is the root cause of most social problems. They present striking graphs of social statistics over several decades that reveal a dramatic deterioration in American society, especially

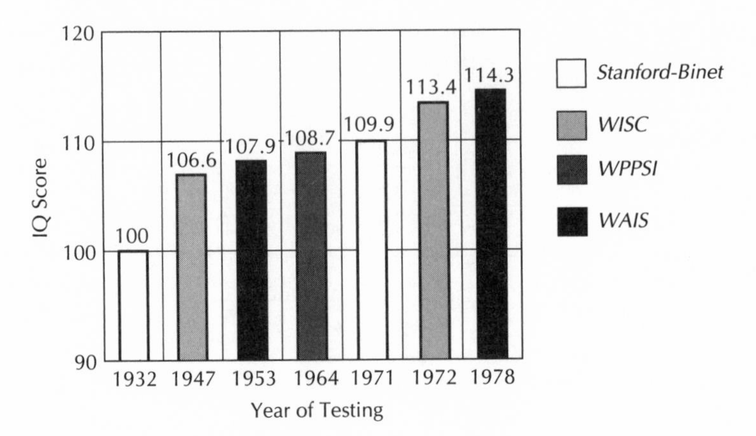

Figure 4.5. Mean IQ scores of standardization samples for four IQ tests given to Americans of European ancestry in various years. In each case two or more of the tests were taken by the same people in the same year, and averages were expressed in terms of the base score of 100 on the Stanford-Binet in 1932. Abbreviations: WISC, Wechsler Intelligence Scale for Children; WAIS, Wechsler Adult Intelligence Scale; WPPSI, Wechsler Preschool and Primary Scale of Intelligence. Based on Ref. 25.

from 1960 to 1990. Over this period, we are told, the marriage rate has declined while the divorce rate has increased from 7% to 20% and "illegitimate" births have increased from 5% to 30%, welfare caseloads have risen from 1.5% to 7%, and the rate of violent crimes is now five times higher than three decades ago. Nevertheless, the raw intelligence of American youth has apparently increased a substantial amount over this same interval. A national decline in intelligence could not possibly be the basis for these negative social trends.

Shorter Term Environmental Effects—Schooling

If a lifelong change in environment occasioned by adoption exerts a major influence on mental development, it would not be at all surprising to learn that lesser changes or those of shorter duration would have correspondingly less impact. A large research literature on the effects of formal schooling supports this expectation. Intelligence is not simply the wellspring of success in school; it is an important product of education. The two directions of causation are usually obscured in the typical classroom, and there is no convenient way to assess how much of the correlation (r^2 = 0.3) between IQ and school grades results from the two kinds of processes. Fortunately, there are situations where effects of the schooling experience can be seen more clearly.

Sometimes children begin school later than their peers of the same age in a nearby district. For example, this occurred in Virginia when schools were closed in certain districts from 1959 to 1964 to avoid the desegregation mandated by the U.S. Supreme Court. Reviewing this and related evidence, Ceci concluded that a child's IQ declines on average by about 5 points for every year of delay in starting formal schooling.[30] This does not mean that absolute intelligence declines. Rather, the relative performance of children lags further and further behind their peers in school.

In many districts there is a fixed birth date that decides when a child may begin formal education in state-supported schools. For example, in Alberta this date is March 1. Those who will not reach the age of 5 years by March 1 must wait until the following September to enter kindergarten. Consequently, there are children with birthdays on either side of the cutoff date that differ minimally in chronological age but differ by a full year of schooling. This effect is known to be important for the growth of intelligence (see Ref. 30). From the school starting data cited earlier, one would expect a "cutoff" effect of about 5 IQ points. In a recent series of four studies done by developmental psychologists in the Edmonton, Alberta school system, this gain from schooling was confirmed (Table 4.1). Although the difference in mean IQ between children of similar age in kindergarten and grade 1 was not statistically significant in any one study taken by itself, meta-analysis of the four studies pointed to a significant advantage of 4 IQ points for those in grade 1. (The Alberta government recently abolished mandatory kindergarten for 1 year to save money, which presents a further opportunity to investigate this effect.)

A somewhat smaller but well-established influence of schooling can be seen in the gradual decline in average IQ that occurs during the summer vacation period among children that spend this time exercising their muscles and social skills but not their academic minds.[31] Of course, most children recover from this decline when they return to the classroom. Because the vacation is so short,

Table 4.1. Mean IQ scores for Edmonton children with similar ages but differing by 1 year of classroom schooling

	Less schooled		1 year more school		Effect
Study	***N***	**Mean**	***N***	**Mean**	**size (*d*)**
Ferreira and Morrison (Ref. 32)	24	109.1	24	114.0	0.38
Varnhagen et al. (Ref. 33)	40	115.9	39	118.4	0.14
Morrison et al. (Ref. 34)	10	111.0	10	118.0	0.54
Bisanz et al. (Ref. 35)	19	100	19	105	0.42

Meta-analysis[a]: Weighted $d_+ = 0.30$.
95% confidence interval for $\delta = 0.01$ to 0.56.
[a] Meta-analysis by D.W. of the four separate data sets conducted from the published descriptions using the procedures of Hedges and Olkin (Ref. 36) for studies with two independent groups.

there is little gain in the intelligence of a nation to be achieved by keeping children in school all year. The importance of the effect is to demonstrate how a rather modest change in environment can exert an almost immediate influence on mental ability, just as strength of muscle begins to wane as soon as a broken limb is immobilized in a plaster cast.

Substantially larger and potentially very important gains in the intelligence of a nation may be achieved by starting children in formal education earlier and keeping them in school longer. Young people who drop out of high school suffer a gradual decline in IQ relative to their better schooled peers of the same age, and there is a very high correlation ($r^2 = .9$) between IQ and years of schooling completed, a major portion of which is a schooling effect on IQ (see Ref. 30).

Amelioration of Early Disadvantages

Of special relevance to the argument of *The Bell Curve* are the results of several recent studies that attempted to improve the lives of infants born with obvious disadvantages, including low birthweight and poverty. These conditions occur at much higher frequencies for Americans of African ancestry, a group singled out for attention by H&M. The evidence from these studies is especially convincing because all employed random assignment of children and families to treatment and control conditions.

The Carolina Abecedarian Project[37,38] and Project CARE[39] examined North Carolina children born healthy and with normal gestation but whose family environments posed a high risk for poor mental development because of low parental IQ and education as well as minimal financial resources. A large majority of their subjects was of African ancestry. Children assigned randomly to the experimental condition received enriched, educational day care outside the home every weekday from about 3 months after birth until they started public schooling. Controls received nutritional supplements and pediatric medical care or crisis intervention but no educational day care. As shown in Figure 4.6, gains in IQ at 4 to 4.5 years were almost as large as those found in the French adoption studies, even though the children returned home every day and spent holidays and weekends with their family (mostly unemployed, single mothers) in poverty-stricken neighborhoods. Furthermore, the mean IQ scores of the enriched groups appeared to be quite typical of healthy American children. Children benefitting from the Abecedarian project continue to show higher IQ scores than controls at age 12, after 7 years in the public schools (see Ref. 37).

Low birthweight infants typically show a deficit in mental ability when they later begin schooling. Two projects, the Infant Health and Development Program (IHDP) multicenter study at eight sites in the United States[40] and the Mother–Infant Transaction Program (MITP) in Vermont,[41] observed sub-

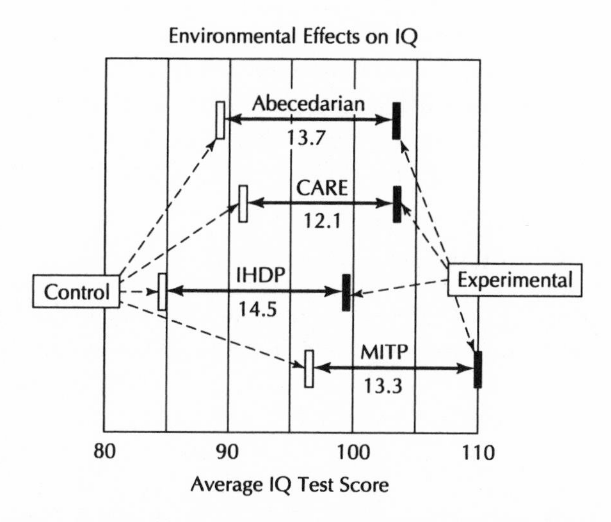

Figure 4.6. Mean IQ test scores for children randomly assigned to the control (open rectangle) and experimental (filled rectangle) conditions in four studies designed to enhance the development of disadvantaged children in the United States. The Abecedarian project represents Stanford-Binet IQ scores at 48 months (Ref. 38). Project CARE represents McCarthy Mental Ability scores at 54 months for children in the Control with no day care and Child Development Center day care groups (Ref. 39). IHDP represents Stanford-Binet IQ scores at 36 months for controls versus the intervention group that received an average of 500 home visits plus days in day care (Refs. 40 and 42). MITP represents the Kaufman Mental Processing Composite scores for children at age 9 years that were in the low birthweight control and experimental groups (Ref. 41).

stantial advantages on IQ tests for children assigned randomly to the more intensive intervention conditions. The IHDP study involved pediatric follow-up examinations, home visits, parental group support, and educational day care each weekday starting at the age of 12 months; controls received only the pediatric surveillance. For those children in the birthweight range 2001 to 2500 grams whose families participated most fully in the various programs, especially the day care, the average IQ gain over controls was about 14.5 points at the age of 3 years (see Fig. 4.6).[42] The Vermont MITP study emphasized training of mothers to optimize mother–infant interaction, presuming that better educated and responsive mothers would provide better home environments month after month and year after year. This was supported by mental tests of school children at the age of 9 years, in which the experimental group was 13 points higher than controls (see Fig. 4.6).

These four studies of disadvantaged children in the United States demonstrate conclusively that enriched educational experiences early in life can substantially improve performance on IQ tests. They indicate that a substantial elevation in the intelligence of children in an entire country could be achieved by a suitable program of universal day care, and they prove that the substandard school performance of children from certain minority groups could be enhanced considerably by better experiences.

Heritability as Impediment

Among the many cited studies of improved childhood mental ability through enhanced education, not one benefitted in even the slightest way from the large corpus of research on "heritability" of IQ in twins and adoptees. It does not matter whether the field of human behavior genetics finally decides that the heritability of IQ in the United States is 25%, 40%, 50%, or 70%. Any such estimate will be utterly useless to anyone seeking better ways to improve the intelligence of the nation through health care and education. Only knowledge about specific genetic effects can possibly be of any benefit to those who possess the genes in question or those who seek to assist them.

An estimate of heritability would be useful to those contemplating a program of eugenic selective breeding, as is commonly done in agriculture. The proportion of additive genetic variance in a population ("narrow-sense heritability") can be used to predict the initial response of a population to selection for an extreme phenotypic score. One might, for example, compute the expected gain in intelligence to be achieved by compulsory sterilization of everyone with an IQ below 100. Suppose the narrow-sense heritability is about 50%. The rise in average IQ score in the next generation from such a drastic policy would be at most a 5- or 6-point gain, which is equivalent to the gain expected from starting schooling 1 year earlier. Of course, this calculation assumes all else would remain the same. There would likely be some devastating net effects on the intelligence of a nation accruing from the kind of rapid dismantling of democracy and descent into fascism needed to allow wholesale selective breeding.

Perhaps a more immediate danger posed by claims about the heritability of IQ in *The Bell Curve* and elsewhere is that psychologists, educators, and policy makers will be swayed by arguments that high heritability means improving the environment is futile. Campbell and Ramey introduced their report on long-term benefits of the Abecedarian project with this statement: "Because no known genetic or physiological cause has yet been identified to explain most cases of mild retardation, the psychosocial environment has been implicated in its etiology." However, this either/or approach to understanding development gains no support from controlled studies of bona fide hereditary defects that are known to be responsive to environmental treatments and even psychosocial manipulations.

Many ways to improve the cognitive development of disadvantaged children are already well documented. Whether this scientific knowledge should be translated into effective social policy is a matter of political will, the social calculus of the cost/benefit ratio, and the question of costs to whom and benefits for whom. Claims that spending on social welfare will be ineffective because of the inertia of human heredity make about as much sense as arguments that airplanes cannot fly efficiently or a rocket could never reach Mars because the acceleration of Earth's gravity is too strong. Advances in aeronautical engineering dispelled doubts about flight. Gradually and against great resistance from behavior genetics, developmental psychology is discovering ways to lift humanity from the fetters of poverty.

NOTE: Portions of this chapter, including Figures 2, 3, and 5, appeared previously in Wahlsten, D. (1995), "Increasing the Raw Intelligence of a Nation Is Constrained by Ignorance, Not Its Citizens' Genes," *The Alberta Journal of Educational Research*, 41, pp. 257–264, and are reprinted with the generous permission of the editor and the Faculty of Education at the University of Alberta. This work was supported in part by grant OGP45825 from the Natural Sciences and Engineering Research Council of Canada.

References

1. Herrnstein, R.J., and Murray, C. (1994), *The Bell Curve: Intelligence and Class Structure in American Life*, The Free Press, New York.
2. Putnam, C. (1961), *Race and Reason. A Yankee View*, Public Affairs Press, Washington, D.C.
3. Jensen, A. (1969), "How Much Can We Boost IQ and Scholastic Achievement?," in *Environment, Heredity, and Intelligence*, Reprint series No. 2, *Harvard Educational Review*, 1–123.
4. Gottlieb, G. (1992), *Individual Development and Evolution. The Genesis of Novel Behavior*, Oxford University Press, New York.
5. Lewontin, R. (1974), "The Analysis of Variance and the Analysis of Causes," *American Journal of Human Genetics*, 26, 400–411.
6. McGuire, T. R., and Hirsch, J. (1977), "General Intelligence (g) and Heritability (H^2, h^2)," in I. C. Uzgiris and F. Weizmann (Eds.), *The Structuring of Experience*, Plenum, New York, pp. 25–72.
7. Wahlsten, D. (1990), "Insensitivity of the Analysis of Variance to Heredity-Environment Interaction," *Behavioral and Brain Sciences*, 13, 109–161.
8. Wahlsten, D., and Gottlieb, G. (1997), "The Invalid Separation of Effects of Nature and Nurture: Lessons from Animal Experimentation," in R. Sternberg and E. Grigorenko (Eds.), *Intelligence, Heredity, and Environment*,Cambridge University Press, New York, 163–192..
9. Goldberger, A.S. (1978), "The Nonresolution of IQ Inheritance by Path Analysis," *American Journal of Human Genetics*, 30, 442–445.
10. Kempthorne, O. (1978), "Logical, Epistemological and Statistical Aspects of Nature-Nurture Data Interpretation," *Biometrics*, 34, 1–23.
11. Roubertoux, P.L., and Capron, C. (1990), "Are Intelligence Differences Hereditarily Transmitted?," *Cahiers de Psychologie Cognitive*, 10, 555–594.

12. Taylor, H.F. (1980) *The IQ Game. A Methodological Inquiry into the Heredity-Environment Controversy*, Rutgers University Press, New Brunswick, NJ.
13. Fisher, R.A. (1951), "Limits to Intensive Production in Animals," *British Agricultural Bulletin*, 4, 217–218.
14. Kempthorne, O. (1990), "How Does One Apply Statistical Analysis to our Understanding of the Development of Human Relationships?," *Behavioral and Brain Sciences*, 13, 138–139.
15. Lewontin, R. C. (1991), *Biology as Ideology. The Doctrine of DNA*, Anansi, Toronto.
16. Chase, A. (1977), *The Legacy of Malthus. The Social Costs of the New Scientific Racism*, Knopf, New York.
17. Gottlieb, G., Wahlsten, D., and Lickliter, R. (in press), "The Significance of Biology for Human Development: A Developmental Psychobiological Systems View," in R.M. Lerner (Ed.), *Theoretical Models of Human Development, Vol. 1, Handbook of Child Psychology*, 5th ed., Wiley, New York.
18. Platt, S.A., and Sanislow, C.A., III (1988), "Norm-of-Reaction: Definition and Misinterpretation of Animal Research," *Journal of Comparative Psychology*, 102, 254–261.
19. Lee S.M., and Bressler, R. (1981), "Prevention of Diabetic Nephropathy by Diet Control in the *db/db* Mouse," *Diabetes*, 30, 106–111.
20. Surwit, R.S., Kuhn, C.M., Cochrane, C., McCubbin, J.A., and Feinglos, M.N. (1988), "Diet-Induced Type II Diabetes in C57BL/6J Mice," *Diabetes*, 37, 1163–1167.
21. Woo, S.L.C. (1991), "Molecular Genetic Analysis of Phenylketonuria and Mental Retardation," in P.R. McHugh and V.A. McKusick (Eds.), *Genes, Brain, and Behavior*, Raven Press, New York, pp. 193–203.
22. Clarke, A.M., and Clarke, A.D.B. (Eds.) (1979), *Early Experience: Myth and Evidence*, Free Press, New York.
23. Capron, C., and Duyme, M. (1991), "Children's IQs and SES of Biological and Adoptive Parents in a Balanced Cross-Fostering Study," *Cahiers de Psychologie Cognitive*, 11, 323–348.
24. Schiff, M., Duyme, M., Dumaret, A., and Tomkiewicz, S. (1982), "How Much Could We Boost Scholastic Achievement and IQ Scores? A Direct Answer from a French Adoption Study," *Cognition*, 12, 165–196.
25. Flynn, J.R. (1987), "Massive IQ Gains in 14 Nations: What IQ Tests Really Measure," *Psychological Bulletin*, 101, 171–191.
26. Clarke, S.C.T., Nyberg, V., and Worth, W.H. (1978), *Technical Report on Edmonton Grade III Achievement 1956–1977 Comparisons*. Alberta Education, Edmonton.
27. Ericsson, K.A., Krampe, R.T., and Tesch-Romer, C. (1993), "The Role of Deliberate Practice in the Acquisition of Expert Performance," *Psychological Review*, 100, 363–406.
28. Wagner, R.K., and Oliver, W.L. (1995), "How to Get to Carnegie Hall: Implications of Exceptional Performance for Understanding Environmental Influences on Intelligence," in D.K. Detterman (Ed.), *Current Topics in Human Intelligence, Vol. 5, The Environment*,Ablex, Norwood, NJ, pp. 87–102.
29. Weisberg, R.W. (1986), *Creativity, Genius and Other Myths*, Freeman, New York.
30. Ceci, S.J. (1991), "How Much Does Schooling Influence General Intelligence and its Cognitive Components? A Reassessment of the Evidence," *Developmental Psychology*, 27, 702–722.

31. Heyns, B.L. (1978), *Summer Learning and the Effects of Schooling*, Academic Press, New York.
32. Ferreira, F., and Morrison, F.J. (1994), "Children's Metalinguistic Knowledge of Syntactic Constituents: Effects of Age and Schooling," *Developmental Psychology*, 30, 663–678.
33. Varnhagen, C.K., Morrison, F.J., and Everall, R. (1994), "Age and Schooling Effects in Story Recall and Story Production," *Developmental Psychology*, 30, 969–979.
34. Morrison, F.J., Smith, L., and Dow-Ehrensberger, M. (1995), "Education and Cognitive Development: A Natural Experiment," *Developmental Psychology*, 31, 789–799.
35. Bisanz, J., Dunn, M., and Morrison, F.J. (1995), "Effects of Age and Schooling on the Acquisition of Elementary Quantitative Skills," *Developmental Psychology*, 31, 221–236.
36. Hedges, L.V., and Olkin, I. (1985), *Statistical Methods for Meta-Analysis*,Academic Press, Orlando.
37. Campbell, F.A., and Ramey, C.T. (1994), "Effects of Early Intervention on Intellectual and Academic Achievement: A Follow-Up Study of Children from Low-Income Families," *Child Development*, 65, 684–698.
38. Ramey, C.T., Yeates, K.O., and Short, E J. (1984), "The Plasticity of Intellectual Development: Insights from Preventive Intervention," *Child Development*, 55, 1913–1925.
39. Wasik, B.H., Ramey, C.T., Bryant, D.M., and Sparling, J.J. (1990), "A Longitudinal Study of Two Early Intervention Strategies: Project CARE," *Child Development*, 61, 1682–1696.
40. Ramey, C.T., Bryant, D.M., Wasik, B.H., Sparling, J.J., Fendt, K.H., and LaVange, L.M. (1992), "Infant Health and Development Program for Low Birthweight, Premature Infants: Program Elements, Family Participation, and Child Intelligence," *Pediatrics*, 89, 454–465.
41. Achenbach, T.M., Howell, C.T., Aoki, M.F., and Rauh, V.A. (1993), "Nine-Year Outcome of the Vermont Intervention Program for Low Birth Weight Infants," *Pediatrics*, 91, 45–55.
42. Brooks-Gunn, J., Klebanov, P.K., Liaw, F., and Spiker, D. (1993), "Enhancing the Development of Low-Birthweight, Premature Infants: Changes in Cognition and Behavior Over the First Three Years," *Child Development*, 64, 736–753.

chapter five

Racial and Ethnic Inequalities in Health: Environmental, Psychosocial, and Physiological Pathways

BURTON SINGER AND CAROL RYFF

In Chapter 13 of *the Bell Curve*, Herrnstein and Murray (H&M) note that "ethnic differences in cognitive ability are neither surprising nor in doubt."[1] This is largely true. What is not so obvious is (1) the causes of such differences, (2) their immutability, and (3) the consequences of ethnic and racial differences in cognitive ability. H&M argue that the causes are largely genetic, that cognitive ability is more or less fixed over an individual's lifetime, and that it effects a wide array of social consequences, ranging from unemployment to family income, etc. There are, of course, alternative explanations for ethnic and racial differences in cognitive ability and for their consequences. H&M's favorite statistical tool of multiple regression is not the only way to explore such values. In this chapter we offer an alternative approach focused on health outcomes, and we demonstrate how *nongenetic* biological, environmental, and social pathways can explain much of the racial and ethnic differences.

An extensive literature documents a strong association between position in a racial and/or ethnic hierarchy and age-specific chronic disease prevalence and mortality.[2–4] Explanations for these associational representations remain, however, elusive. The difficulty stems, in part, from the fact that human populations are heterogeneous within each level of a hierarchy. There are always people at the top who violate the general trend and have poor health outcomes. Analogously, there are always people at the bottom who have good health into late life. In addition, racial/ethnic differences in morbidity and mortality mask critical environmental and psychosocial factors implicated in the physiological and neurochemical changes that comprise pathways to disease.

We present a conceptual and empirical approach for explaining the associations between race/ethnicity and health. Our framework includes both stress-induced and infectious diseases from a longitudinal perspective. We argue that complex representation of life histories is the ultimate route to understanding these inequalities in health. Thus, the overall objective is the specification of diverse pathways through environmental, social, and psychological factors to disease incidence and death. Equally important, we argue, is the characterization of life histories of positive health and well-being, including physiological substrates that facilitate protection from illness/adversity and recovery from it. A similar approach might well offer a radically different explanation for racial and ethnic differences in social outcomes than that suggested by Herrnstein and Murray in the *Bell Curve*.

In the sections that follow, we first discuss U.S. black/white differences in morbidity and mortality in a community sample. This serves as a generic illustration of the heterogeneity in health status, both within and between, racial strata. To provide an international comparison and to expand the discussion to racial and ethnic categories, we then present findings on tuberculosis transmission in South Africa. In contrast to the prevailing emphasis on chronic disease, this example brings to the fore infectious disease outcomes, a neglected terrain in discussions of racial, ethnic and socioeconomic differences in health. This section also presents recent U.S. data on the distribution of tuberculosis among racially segregated minorities.

Prior attempts at explanation for these differences in disease have focussed largely on single-variable models (e.g., working conditions, environmental exposure, medical care) operating over short time segments. The life-history approach that we advocate emphasizes mechanisms that operate in multiple life domains through time. Critical to the perspective are cumulation processes, which focus on the piling up of adverse life experiences. Such cumulation is a fundamental feature of extant stress models of health, although they are rarely examined within life-history models. Also important are *exposure processes*, the realm of infectious disease. We argue that the environmental and social factors that govern exposure are stratified by the sociodemographic variables of race, ethnicity, and class.

To illustrate the life-history approach to health outcomes, we present findings from the Wisconsin Longitudinal Study (WLS), described in further detail later. The organizing principles and methodological steps for using such complex, multidomain, temporally expansive data are briefly summarized.

Mental health is the targeted outcome in these analyses, and our illustrative focus is on resilience, defined as those with high profiles of psychological well-being who have previously experienced major depression. We describe multiple life-history pathways of resilience among WLS women.

Our life-history approach is equally applicable to individual, case-study analyses. Accordingly, we use the framework to interpret the autobiographical account of a young black man growing up under apartheid in South Africa. The example also underscores our emphasis on resilience and positive health outcomes, that is, how to explain the well-being of individuals subjected to enormous adversity.

We contend that the elaboration of environmental and psychosocial life histories must ultimately be connected with explicit physiological pathways leading to health outcomes. To address these issues, we summarize recent advances in efforts to map these biological processes. Our objective therein is to build bridges between those involved in elaborating the internal mechanisms of disease (and positive health) and those studying the experiential processes (i.e., the life-history details) that catalyze and sustain the physiological processes.

A central conclusion is the essential role of *prevention*, which requires a focus on work, family, and neighborhood contexts, in contrast to illness and disease treatment orientations, which are essentially removed from the proximal origins of health problems. A preventive agenda also underscores the importance of construing health not just as illness or the absence thereof, but also as states of wellness.

Black/White Differences in Morbidity and Mortality

New Haven EPESE Study

The New Haven Established Populations for Epidemiological Studies of the Elderly is a probability sample of 2,806 noninstitutionalized men and women, aged 65 years and older, living in the city of New Haven, Connecticut in 1982.[5] The overall EPESE program consists of epidemiologic cohort studies in four locations: New Haven, CT; East Boston, MA; Washington and Iowa Counties, IA; and Durham, NC.[6] The aim of these studies is assessment of the general level of physical and mental health in a community population of older individuals, specifically with regard to the prevalence and incidence of certain chronic conditions, functional abilities, level of depressive symptomatology, and cognitive impairment. Another goal is the determination of behaviors, socioenvironmental conditions, and biologic variables that are predictive of future declines in health in these elderly populations, including incidence of morbid conditions, hospitalization, institutionalization, and mortality.

Although many investigators have acknowledged the multidimensional nature of constructs such as physical health,[7] most have analyzed health conditions by developing unidimensional scales of conditions along a continuum from optimal to poor health without adequate consideration of (1) the dimensionality of the conditions under consideration, (2) the heterogeneity of the populations being described by such indices, or (3) the diversity of purposes for which the measures are intended. The scientific thrust to arrange items in health status indices in a hierarchical order, rather than to conceptualize them along several vectors or dimensions, has led to extensive debate about whether items should be ranked according to their life-threatening nature or their degree of functional impairment, often resulting in some compromise between the two.

Berkman et al. present a full multidimensional characterization of the elderly black and white populations in New Haven. Each racial group is described by four profiles of conditions and the assignment of degree-of-similarity scores to individuals, representing the proximity of their conditions to these profiles.

> Profile I, *The Healthy Elderly*, shows no indication of any disability having a prevalence substantially above the marginal frequencies. Thus, "healthy" applies to populations in which hypertension, a limited set of functional disabilities (walking, pushing, lifting, and bending), overweight, and some kind of vision problem may be viewed as common rather than exceptional.
>
> Profile II, *The Cognitively Impaired Elderly*, shows strong indication of cognitive impairment conjoined with an inability to do heavy work and with hearing impairments. For blacks, but not whites, the preceding conditions are conjoined with hip fractures, arthritis, Parkinson's disease, amputation, and vision problems. In contrast, for whites, but not blacks, they are conjoined with functional disabilities in walking blocks and pushing. On the surface, these profiles for blacks and whites may seem intrinsically different. With the exception of Parkinson's disease, however, the conditions on which the profiles disagree are very prevalent in *both* populations.
>
> Profile III, *The Elderly with Minor Impairments in Physical Functioning and Selected Chronic Conditions*, comprises functional disabilities involving gross mobility functions (e.g., walk blocks) and physical performance (e.g., inability to push heavy objects, bend, or lift objects). Several chronic conditions are also associated with these disabilities, although only arthritis is common to both whites and blacks. For blacks, the other accompanying limitations include hypertension and stroke, cigarette smoking, alcohol consumption, and obesity. Whites are more likely to have cancer, Parkinson's disease, cataracts, glaucoma, and urinary incontinence associated with this cluster of conditions.
>
> Finally, Profile IV, *Major Limitations in Activities of Daily Living (ADL) and Functioning and Selected Chronic Conditions*, refers primarily to the inability to maintain independence in seven activities of daily living (walking across a small room, bathing, grooming, dressing, eating, transferring from a bed to a chair, and toileting) and eight more minor limitations. Although limitations in ADL and functioning are common to both races, the chronic conditions

accompanying dysfunction are not entirely overlapping. For both races, this profile includes the prevalence of myocardial infarction, stroke, diabetes, and hip fracture. For blacks, other chronic conditions are cancer, cirrhosis, arthritis, glaucoma, and urinary incontinence.

With these profile comparisons in hand, four key features are shown to differentiate blacks from whites. (1) There is a much higher concentration of healthy whites than healthy blacks. About 25% of whites are in Profile I, whereas only 12.6% of blacks are in this profile with a similar absence of conditions. (2) There are very few people with major limitations of ADL and comorbidity (Profile IV) in each population (0.7% of the whites and 1.2% of the blacks). (3) For persons who share the characteristics of only two profiles, the most frequently occurring combination is Profiles I and III. Profile III for whites, however, is *not* comparable with Profile III for blacks. Thus, blacks and whites who have combinations of disabilities but are also substantially like the healthy profile have distinct disabilities (hypertension and stroke for blacks; myocardial infarction, cancer, and/or vision problems, possibly accompanied by functional disabilities, for whites). (4) The highly heterogeneous nature of both populations is exemplified by the fact that 24.5% of whites and 32.5% of blacks exhibit some of the characteristics of three profiles.

To examine black/white differences in survival, the focus is only on Profile I because that is the profile in which blacks and whites are strictly comparable (e.g., no one in either group has reported limitations or conditions). Survival plots show no large differences, although there is a trend for whites to have better survival experience than blacks. When one stratifies the Profile I group by housing characteristics, however, some suggestive differences appear. In particular, blacks in public housing having lower death rates than whites with comparable housing and health conditions. The situation is reversed in other types of housing in the community. Thus, the indistinguishable survivor plots for all healthy persons result from a mixture of two populations defined by housing characteristics, in one of which blacks tend to live longer than whites and in the other of which the reverse is true.

This brief summary of findings from the New Haven EPESE study underscores three important points. First, it documents descriptive differences in health status between elderly blacks and whites. Second, it intimates, but does not fully explicate, the preceding life-history experiences that account for these differences. The findings thus call for greater understanding of the interrelationships among work, family, neighborhood/community experiences as they unfold over the life course. Such a focus on experiential histories also requires linkage to the physiological substrates of cumulating adversity in these domains. Third, looking to the future, the findings call for the need to characterize the most prevalent routes to death following from the baseline racial differences in health profiles and the population distributions relative to them. This is the need to examine the ongoing dynamics of persons relative to profiles, recognizing that an individual's proximity to a profile does not remain constant over time.

Racial/Ethnic Differences in Infectious Disease: The Case of Tuberculosis

In this section, we review the literature on racial and ethnic differences in the distribution of tuberculosis in South Africa. Our analysis includes a brief summary of the history of how such differences came about. Two key themes emerge from this historical overview. The first is that *racial discrimination* is a central social structural feature of the processes involved in the transmission of tuberculosis. The second is that such discrimination processes drive the *conjunction of factors* implicated in the etiology of this disease. It is the convergence of high crowding, dilapidated housing, airborne particulates, poor nutrition, and compromised immunity that are the requisite conditions for the spread of this disease. Additional intervening factors, as we elaborate, pertain to differences in health care and an extensive array of psychosocial factors, all of which ultimately ensue from racial discrimination. Finally, while the bulk of our discussion pertains to South Africa, we also provide illustration of similar processes in the United States via summary findings on residential segregation and related distributions of tuberculosis.

History of Tuberculosis in South Africa

Signs of spinal tuberculosis in remains of neolithic humans provide the earliest evidence of a long association between the tubercle bacillus and humans. Hindu physicians had a broad clinical knowledge of tuberculosis 3,000 years ago, and Hippocrates provided the first recorded clinical description of the disease in 400 B.C.[8,9] It is probable that members of the dynasty of Tutankhamun, in particular, Amenhotep IV and his famous Queen Nefertiti, were victims of tuberculosis. There is strong evidence that the tubercle bacillus was very active in the days of early Egyptian civilization.

Despite the long history of tuberculosis north of the Sahara, the present epidemic of the disease affecting the black people of Southern Africa is the first in their long history. Failure to develop a culture based on large-scale urban development and the lack of major outside intrusion into the interior of Africa are the key to the *absence* of any major tuberculosis epidemics in Southern Africa until the colonial era. There were no overcrowded cities or industrial complexes. There was only limited contact with traders in coastal areas and conditions favorable for the spread of tuberculosis did not exist. The main body of the population now occupying Southern Africa derives from a migration process occurring over the past 1,200 years.

The colonial mining enterprise was the source of major European migration starting in the early part of the nineteenth century and continuing thereafter. Many of the miners from Europe were tuberculin positive but at the time of migration did not express symptoms of the disease. These miners were the source of disease transmission; that is, the original transmission occurred *across* racial lines. Subsequent transmission occurred *within* race, both in local settings and via circular migration (i.e., movement back and forth between mining centers and home territories).[10] Living conditions of the miners were ideally suited to disease exposure and transmission: the mines themselves were abundant facilitators of the particulates carrying the tubercle bacillus, while the living quarters were crowded, poorly constructed, and unsanitary. In addition, the poor nutrition of the native miners, combined with their lack of natural disease antibodies, rendered them particularly vulnerable to acquiring and suffering from tuberculosis. These conditions were similar to the quality of life in industrial England, where tuberculosis was also epidemic (see Ref. 9), although in the latter the mechanisms of discrimination were operative not along racial but rather along class lines.

Apartheid and Racial/Ethnic Differences in Health

Historical medical records from Capetown and Johannesburg document persistent black/white differences in tuberculosis mortality from 1896 to 1955 (see Fig. 5.1).

These sharp racial disparities were, after the enactment in the late 1940s of apartheid laws, subsequently elaborated with a fourfold classification of distinct social groups: white, Asian, coloured, African. Each individual in the Republic of South Africa was required to carry an identity card designating him or her as "European" (Caucasian), "Bantu" (Negroid), "Coloured" (mixed descent), or "Asian" (usually Indian). Such classifications dominated every facet of life:

> And how he is classified determines where he lives; what education he receives and what work he is able to do; how much money he earns; whom he may marry, where his wife and children live; whether he has any political rights and where he may exercise them. The color of his skin and racial classification may even determine which ambulance picks him up when he is ill, to which hospital he will be taken, and where he will be buried when he dies (Ref. 2, p. 64).

South Africa thus officially became a society stratified by racial differences that were marked, in turn, by gradations in power and prosperity. The hierarchical system placed Indians lower in the social structure than whites, but coloureds were lower than Indians, and Africans were the most oppressed

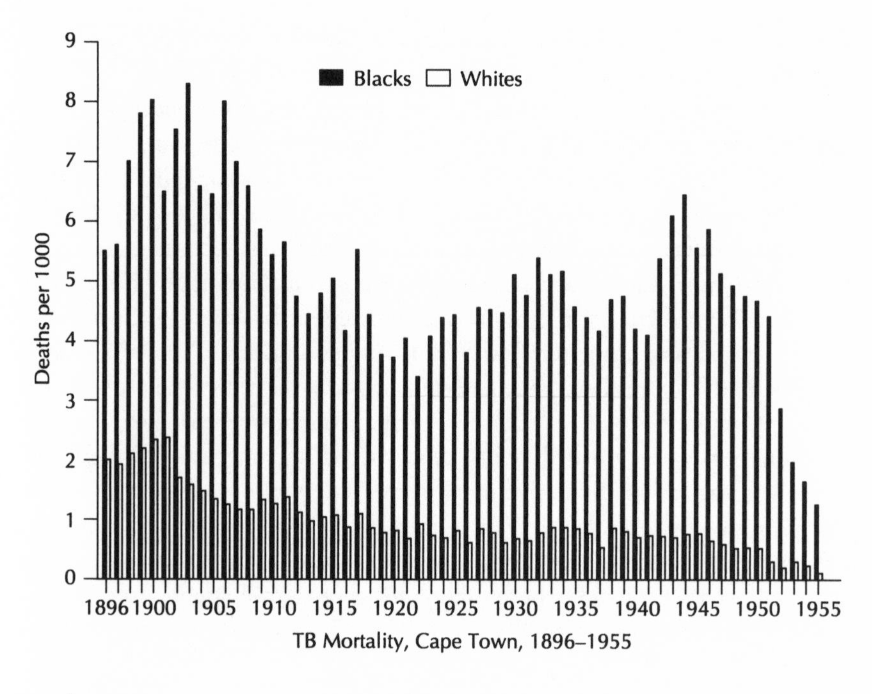

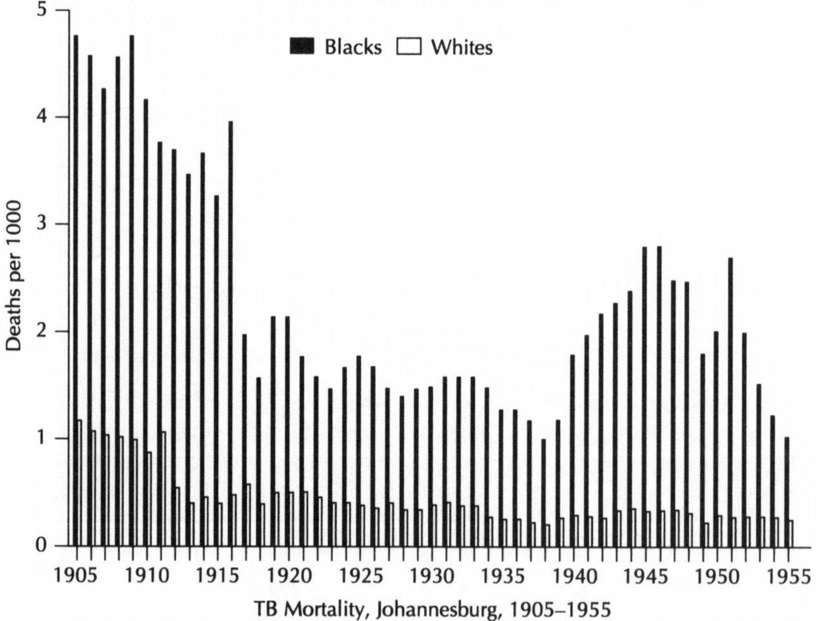

Figure 5.1. Black/white differences in tuberculosis mortality from 1896 to 1955.

of all. These differences in the social structure of South Africa translated to differential health profiles, of which tuberculosis is a dramatic example. Table 5.1 reports official South African statistics on notification rates of pulmonary tuberculosis.

The data point to clear differences in disease incidence across the four social groups. According to the annual report of the Medical Officer for Health of Cape Town, where notification machinery was more evenly applied than in the country as a whole, the incidence of *all forms* of tuberculosis in 1979 (per 100,000 people) was among whites 18, Asians 58, coloureds 215, and Africans 1465.[11] Mortality from tuberculosis demonstrates the same picture (Table 5.2), with the highest rates evident for coloureds and Africans.

At the national level there were between 12 and 17 times more deaths due to infectious diseases among coloureds than among whites. Among coloureds 70% to 85% of this mortality was attributable to tuberculosis (Ref. 2). Among whites, tuberculosis is not a major killer, accounting for less than 1% of mortality.

These differences do not, however, explicate the intervening mechanisms that produce such effects. We noted earlier a conjunction of factors implicated in the transmission of tuberculosis (i.e., crowding, dilapidated housing, airborne particulates), but others must be considered as part of the larger etiological story. *Medical treatments*, for example, are dramatically influenced by

Table 5.1. Pulmonary tuberculosis notification rates (number of notifications per 100,000 population)

	Year	White	Asian	Coloured	African	Risk ratio
Cape Town						
City	1979	16	58	199	365	72.3
Langa	1979	—	—	—	1,951	
Pietermaritzburg	1979	32	72	66	814	25.4
Port Elizabeth	1979	17	224	475	741	43.6
Durban	1977	16	146	73	333	20.8
Kimberley	1977	14	—	387	316	22.6
South Africa						
as a whole	1968–77	18	142	326	280	15.56
	1979–80	17	145	330	1,000	58.83

Source: Ref. 2.

Table 5.2. Tuberculosis mortality (number of deaths per million population)

	Year	White	Asian	Coloured	African
Cape Town	1979	15	80	120	750
Guguletu and Langa	1979	—	—	—	1,723
Port Elizabeth	1979	22	—	436	1,003
Johannesburg	1975–76	30	60	410	250
Kingwilliamstown	1978	0	—	900	470
Pretoria	1977	0	70	0	30
South Africa as a whole	1978	18	27	226	93

Source: Ref. 2.

position in the social order. At the diagnostic level, procedures for early detection (e.g., mass miniature radiography) showed limited, and even decreasing, use among blacks in the 1970s. Neonatal bacillus Calmette-Guérin (BCG) inoculation was also not evenly distributed among racial groups—in Kimberley in 1976–77, 90% of white babies were inoculated, compared with 61% of coloured babies and 47% of African babies. In addition, there was inconsistent availability of antituberculosis drugs, such as rifampicin.

Malnutrition is also implicated as an intervening factor in the social distribution of tuberculosis (and numerous other diseases). It is estimated that between 30% and 40% of all young black children suffered from inadequate food. With regard to heterogeneity of disease prevalence within racial groups, we note that among black workers, miners were generally the nutritional elite and had access to more diagnostic facilities, with both factors affording them greater protection from tuberculosis.

Among the most neglected of factors in the larger causal chain, however, is the wide array of *psychosocial influences* fueled by apartheid:

> Apartheid is a major source of physical and mental ill health because of the stress and tension it generates in the daily lives of millions of people; the destruction of social support systems such as the family and the community through mass uprooting; and the pervasive insecurity, harassment, and violence which characterize the psychosocial environment....
>
> The system of enforced migrant labour, police control and repression aims to dehumanize every aspect of life for the black majority, and to perpetuate its sense of powerlessness. The effects of this can be seen in the high rates of violent crime, alcoholism and drug abuse and suicide (Ref. 2, p.165).

At any one time, it is estimated that between 60% and 80% of the economically active adult male population of Bantustans lived away from home.[12,13] The absence of males not only reduced the labor needed to sustain much-needed food-producing agriculture among the Africans, it also created problems in the socialization of children, high rates of marital breakdown, desertion, and widowhood. Without husbands, women suffered constant anxiety and insecurity about having to manage households that lack basic resources. Life in the mines for men, in turn, was degrading and humiliating—having left hungry children behind, they were treated like "harried and trapped rats underground."[14] Such existence lacks any form of dignity and led many migrants to drugs, alcohol, prostitutes, or homosexuality for solace, with related antisocial activities and violence.

It is in the lives of African children that psychosocial pathways to ill health were tragically played out. Many were born to 12 and 13 year olds, who themselves born out of wedlock "seek an outlet for their own lack of affection, deprivation of maternal care, educational impoverishment, and boredom" (Ref. 2, p. 170). Such children grew up grossly deprived of food, shelter, instruction, nurturance, and attention. Many were exploited labourers from age 6 or 7 on. Given this cumulation of early life adversity, many grew into rootless, alienated adolescents, prone to crime and violence.

Thus, in multiple domains of life (work, family, neighborhood) and across multiple age groups, the strains of racial stratification and their implications for exposure to infectious disease were evident. Despite the recent dismantling of apartheid, this conjunction of factors leading to sickness and early death persists. The South African example thus brings into high relief the *multiplicity of factors*, activated by a sharply stratified social order, that ultimately compromise health. We conclude this section with a summary of research on tuberculosis in a racially diverse U.S. sample. Of interest is the racial/ethnic distribution of tuberculosis in a society characterized by racial diversity but lacking the officially sanctioned discrimination of apartheid.

Tuberculosis and U.S. Minorities

Seven out of ten tuberculosis sufferers in the United States are minorities, and during the last decade the increase in the number of tuberculosis cases has been concentrated among minorities.[15] Efforts to explicate the linkages between minority status and disease outcomes have regularly implicated environmental factors. A large body of literature on "environmental racism," in fact, provides evidence that in the United States, the location of environmental hazards is positively associated with the presence of minorities.[16,17] Historically, it has been argued that during the first decades of this century substandard and overcrowded housing was "the most lasting problem" facing blacks who migrated to urban centers in the Northeast and the Midwest. More recently, a study of the incidence of tuberculosis among migrant farm workers in North Carolina (most of whom are black or Hispanic) revealed that harsh living conditions in farm labor camps explain the unusually high tuberculosis infection and active tuberculosis rates.[18]

Using data from the state of New Jersey, Acevedo-Garcia examined the residential distribution of cases of tuberculosis from 1985 to 1992. The targeted minority groups were African Americans, Hispanics, and Asians, and the key residential factors (derived from 1990 socioeconomic census information) were degree of minority isolation and density as well as exposure to poverty, crowding, and immigration. The research demonstrated that communities with large numbers of tuberculosis cases were sharply distinct from the rest; that is, the prevalence of any single risk factor for tuberculosis (e.g., poverty, overcrowded housing, a large fraction of multi-family housing units, immigration, proximity to New York, and high population density) was higher in zip code areas with high rates of tuberculosis. Moreover, the co-occurrence of two or more risk factors was significantly more common in these communities.

With regard to minority variation, the study showed that the tuberculosis problems among blacks are different than among Hispanics and Asians. Specifically, immigration plays a more important role in the epidemiology of tuberculosis among the latter two groups.

Furthermore, segregation appears to have a direct influence on the epidemiology of tuberculosis among African Americans. Blacks were shown to be the only racial/ethnic group experiencing high isolation levels, often combined with high density levels. Thus, many African Americans live in environments conducive to transmission of tuberculosis (and other infectious diseases), especially *among blacks themselves*.

Juxtaposed with the South African example, this U.S. study points to a parallel conjunction of environmental factors underlying the disproportionate incidence of tuberculosis among American minorities. U.S. minorities do not show the same dramatic hierarchical arrangement as seen in racial subgroups in South Africa, although segregation appears to play a strong role in tuberculosis transmission among black Americans, while immigration is more influential for Hispanics and Asians. High exposure to poverty and crowded housing is central to the epidemiological tale of all three U.S. minorities. Thus, in two dramatically distinct countries, tuberculosis is differentially distributed across racial/ethnic lines, and the life situations of individuals in these groups are recurrently characterized by negative environmental factors.

What needs systematic documentation in both contexts is the wider array of intervening mechanisms and processes, such as racial differences in access to health care or proper nutrition. Psychosocial factors (e.g., stress, coping, social supports) are notably missing from current empirical accounts. These are particularly critical for their well-established connections to mental health indicators, which, in turn, are implicated in physiological processes tied to physical health.[19] In short, the preceding literatures provide yet more examples of racial differences in disease; they illustrate the role of intervening environmental factors tied to racism and discrimination; and importantly, they call for future studies of intervening psychosocial mechanisms.

A similar set of questions about intervening mechanisms and processes needs to be explored before conclusions about racial differences in cognitive ability or success in life can be reached. *The Bell Curve* does not explore such alternatives.

A Life-History Approach to Health Outcomes: The Wisconsin Longitudinal Study

Racial and ethnic inequalities, operationalized by economic, educational, and political power differentials, define the hierarchical system that reflects, at each point in a life-history, a person's relative advantage or disadvantage as well as the potential for future exposure to adversity, and, importantly, the availability of restorative strategies that enable resilience and recovery. We propose that disease and positive health outcomes are directly influenced by, and reciprocally influence, the individual's relationship with his or her own multilevel ecosystem. The preceding examples elaborate these ecosystems, differentiated by multiple levels of interrelated physical, biological, and social components, which define the environments and experiences comprising individual lives. Life-history approaches are required to explicate the pathways of cumulating experiences in these diverse life domains and how they progressively undermine, or sustain, health and well-being. The primary challenge faced by those who advocate life-history agendas is the management of complexity, that is, how to organize the dynamic, multilevel factors that comprise individual lives.

We illustrate the life-history approach with research based on analyses of data from the Wisconsin Longitudinal Study (WLS).[20–22]

This study was begun in 1957 with a random sample of over 10,000 Wisconsin high school graduates. Data were collected on respondents' family background, starting resources, academic abilities, youthful aspirations, and in subsequent waves, on educational and occupational achievements, work events and conditions, family events, social support and relationships, social comparisons, and physical and mental health. Our analytical strategy provides an explicit example of how data spanning multiple time periods and multiple life domains have been organized and linked to health outcomes, and it informs the call to implement life-history analyses in studies of racial and ethnic inequalities in health.[23] The strength of this approach, we emphasize, is not the identification of new mechanisms underlying racial (and socioeconomic) differences in health; like others,[4,24] we draw on extensive intervening factors (e.g., work conditions, family life, medical care, health behaviors, environmental exposure, early life conditions, economic resources, personality, social supports, coping strategies). The distinguishing feature of the life-history approach, rather, is putting these separate pieces *together*. So doing entails an essential shift away from single (or dual or triple) variable explanations toward person-oriented accounts that tell a full life story.

Organizing Principles

Essential to the life-history approach are organizing principles, which provide conceptual guides through extensive multidomain, temporally expansive data. (These principles are elaborated in Ref. 23). Briefly reiterated here, they are stated in terms of hypothesized consequences for health outcomes.

(1) Adversity and its cumulation over time has negative health consequences. The idea that negative events and chronic life conditions contribute to health problems (physical and mental) has longstanding presence in prior health research, particularly literatures (human and animal) on stress.[25] The life-history approach calls for the tracking of negative experience in multiple life domains, thereby capturing patterns of "pileup" (cumulative and often co-occurring experience of problems in work, family, neighborhood, etc.). This emphasis on cumulation also points to enduring, persistent features of life strains, which are key to understanding the physiological substrates involved in illness and disease.

(2) Advantage and its cumulation over time has positive health consequences. Because we are interested in understanding not just states of illness, but also positive health, our life-history analyses emphasize the positive features of individual lives (see Ref. 19). These may come in multiple forms: starting resources (e.g., intact family, parents with high socioeconomic status [SES]), personal capacities and abilities (e.g., high IQ, optimism), positive events (e.g., job promotions, birth of desired children). Although adversity is believed to compromise health, experiences of advantage are hypothesized to provide protective and buffering resources.

(3) Reactions to adversity and advantage can exacerbate or ameliorate the impact of life experiences. Consistent with the extensive prior research on the importance of how individuals perceive and cope with stressful life experiences,[26–28] we see reactions to and interpretations of what happens to people as central to understanding mechanisms that affect health. However, because our life histories, include not only unexpected life events, but also chronic conditions, normative life transitions, and general life evaluations, we broaden the scope of what is typically examined under the heading of "reactive responses."

(4) Position in social hierarchies across life domains has health consequences. Positions in social hierarchies are seen to influence a broad range of outcomes,[29] and they are viewed as ubiquitous features of human and animal life.[30,31] We concur with this view, but expand the meaning of hierarchy to encompass

more than traditional SES classifications (i.e., education, income, occupational status). Although class and race constitute prominent dimensions of social stratification, individuals are also differentiated according to hierarchies of ability, positions of power and influence in the family and community, degree of autonomy, and authority in the workplace. We propose that negative health results from the cumulative effects of low social standing across diverse domains, and alternatively, that high social standing in stable hierarchies and its cumulation over time has positive health consequences. We also emphasize the need to assess hierarchies at both social structural and psychological levels. The former captures actual position in a larger distribution, while the latter involves social comparison processes in which individuals evaluate their lot in life relative to others.

(5) Social relationships can exacerbate or ameliorate the impact of life experience and enduring conditions. This principle converges with the vast literature on social supports and their role in reactivity to life stress.[32,33] Following this tradition, we examine the buffering effects of quality relationships in the face of difficult life experiences but, in addition, consider the role of quality ties with significant others in facilitating processes (psychological, emotional, physiological) that protect and maintain the integrity of the organism (see Ref. 19, for expanded discussion).

These organizing principles can be applied to multiple data sources, including structured longitudinal survey data (such as the WLS), focused interviews with supplemental aggregate-level sociodemographic information,[34] and ethnographies of individual lives, which are particularly informative for their "thick description."[35] The analytic task across these different sources is that of "thinning" the rich detail of individual lives down to generalizable features that characterize more than a single person, and yet retain sufficient complexity and detail to differentiate diverse classes of life histories that can subsequently be linked with health outcomes.

In the WLS, we use these principles to understand the *mental health* of the respondents as assessed in midlife. Detailed description of the steps followed to generate diverse life histories and link them to distinct mental health groups are described elsewhere (see Ref. 23). Without elaborating details, we note that the process begins with the writing of randomly selected individual biographies and culminates with tests of whether distinct histories can successfully discriminate the mental health groups. Other presentations of ideas that underpin our life-history approach include Ragin[36] and Tilly.[37]

Resilient Women in the Wisconsin Longitudinal Study (WLS)

To illustrate products from the approach, we briefly describe findings for one select group of respondents: *resilient women*. To be classified as mentally resilient, respondents had to have reported a prior episode (or episodes) of serious depression (assessed with a subset of items from the Composite International Diagnostic Interview, CIDI), but also reported high levels of psychological well-being at the time of the most recent data collection (1992-93).[38,39] There were 168 women in the WLS fitting this description. (The other mental health groups included in our analyses include the *healthy*—those with high well-being and no past history of depression; the *depressed*—those with past depression and low well-being; and the *vulnerable*—those with no past history of depression but low well-being).

Beginning with detailed biographies, the life histories of resilient women were progressively thinned (from the 250 variables with which the analysis began) to their responses on each of 17 vectors, which covered multiple aspects of their early life resources (e.g., parental education, intact family, IQ, high school grades); chronic conditions (e.g., alcohol problem at home, stressful work conditions); the age intervals over which the conditions occurred, acute events, and the age of respondent at the time of occurrence (e.g., death of parent, occupational advancement); quality of social relationships and social integration; and social comparisons with parents and sibs. Certain of these variables were comprised of *complex Boolean statements*, which are logical AND and OR statements that combine complex strings of life characteristics or conditions. Although rich in longitudinal information about chronic conditions and acute events across multiple life domains and position in SES hierarchies, the WLS is relatively weak in assessment of reactions to life experiences and quality relationship indicators. New data collection, on a subset of WLS respondents, designed to alleviate these deficiencies is currently under way.

The larger objective is to retain the richness and texture of individual lives and, at the same time, to simplify the variability into more parsimonious summaries. These procedures resulted in the differentiation of four life-history pathways of resilience among WLS women. Briefly summarized here (see Ref. 23, for detailed descriptions), the first pathway (subgroup H_1) was comprised of women with generally positive beginnings (e.g., high starting abilities, no alcoholism in childhood home) and subsequently experienced upward job mobility. They also perceived that their achievements in life compared favorably with their parents and siblings. Despite these advantages, all of these women had experienced the death of one parent, most had participated in caregiving for an ill person, and approximately half had two or more chronic health conditions. Thus, their lives involved multiple difficulties of particular acute or chronic adversities that were offset by positive work experiences, good beginnings, and favorable self-evaluations.

The second subgroup, H_2, was comprised of women for whom the primary early-life adversity was growing up with alcohol problems in the childhood home. All women in the subgroup met this condition. In addition, many (65%) of these women had experienced 3 or more major acute events (e.g., death of parent, child, spouse; divorce; involuntary job loss). However, the women had important advantages involving social relationships and social participation, early employment with stable or upward occupational status, and positive comparative evaluations. The latter are presumably implicated in their high well-being in midlife.

The third subgroup, H_3, showed primarily advantage in early life: all had parents who were both high school graduates, no alcohol problems existed in the childhood home, and the women had high starting abilities (high school grades, IQ). Later, however, they confronted various forms of adversity (e.g., poor social relationships, downward occupational mobility, job loss, divorce, single parenthood, caregiving). Thus, their lives were characterized by various forms of family adversity occurring largely in adulthood, but they began their life journeys with important strengths that likely facilitated recovery from these adverse experiences.

The final subgroup, H_4, were women whose early lives showed mixed advantages (intact families, no alcoholism) and disadvantage (all had one parent with less than a high school diploma). As life unfolded, the women confronted an array of adversities: job loss, downward mobility, living with an alcohol problem in the home, divorce/single parenthood, high profiles of major acute events. This array of negatives, combined with their less than uniformly positive beginnings, makes difficult the explanation for their resilience. As such, this subgroup underscores the need for additional information pertaining, for example, to their *reactions* to differing life challenges, or the quality of their *significant social relationships*.

Overall, the analyses underscore the diversity in what was bad and good in these women's lives: difficulties occurred across multiple life domains; some were chronic and enduring, others acute; some occurred early in life, others in adulthood. Their advantages and resources also varied across life domains and as to when they occurred. From this variety emerged differing tales of why the women may have succumbed to depression and what their routes out of it were. Final analyses documented that their diverse life histories were, in fact, distinguishable from the lives of women in the other mental health groups (depressed, healthy, vulnerable; see Ref. 23).

Resilience in South Africa

We conclude this section with application of the life-history approach to the story of a young black man growing up under the adverse conditions of apartheid in South Africa. We include this example for three purposes. First, it illustrates our prior point that life- history analyses are not restricted to survey studies but are richly applicable to case studies, biographies, autobiographies, and narrative sources of data. Second, this particular life is a breathtaking story of resilience that is powerfully pertinent to the study of racial and ethnic inequalities in health, and more specifically, to the question of why some ethnic/minority individuals stay well in the face of overwhelming adversity. Third, the account puts into human terms the meaning of apartheid for individual lives and families. It is, in short, a "thick description" of the personal significance of racial inequalities and their consequences for health.

Mark Mathabane's recounting of his first eighteen years of life is entitled *Kaffir Boy*.[40] A word of arabic origin, *Kaffir* is used in South Africa to refer disparagingly to blacks; it is the equivalent of *nigger*. His story begins with acknowledgement of the pervasive sense of hierarchy in his country of origin—"In South Africa there is a saying that to be black is to be at the end of the line when anything of significance is to be had" (p. 4). Living at the bottom of this social order translated to a life of unending adversity, beginning with terrifying accounts of late-night police raids when he was only 5 years of age and left to care for his younger brother and sister, as his parents were forced to escape lest they be arrested for not having their identification passes in order. These raids created a life of constant uncertainty, and children were frequently brutalized for trying to protect or hide their parents. Parents, particularly fathers, were subjected to repeated humilitations. Describing the latter, he conveys the psychological and emotional trauma of apartheid:

> My father forced a face smile. It was not a spontaneous smile—my father never smiled. It was a begging smile, a passive acceptance of the policeman's authority. After smiling my father again dropped his eyes to the floor. He seemed uncharacteristically powerless and contrite, a far cry from the tough, resolute and absolute ruler of the house I knew him to be, the father whose words were law. I felt sorry for him. The policeman, still brandishing the bulky black book, leaned into my father's ear and whispered something.
>
> The other policeman meantime was still at the doorjamb, reveling at the sight of my father being humiliated. The emotional and physical nakedness of my father somehow made me see him in a different light—he seemed a stranger, a total alien. Watching him made tears surge into my eyes, but I fought desperately to keep them from flowing. I cannot cry, I told myself, I would not cry, I should not cry in front of these black beasts. For the first time in my life I felt hate and anger rage with furious intensity inside me. What I felt was no ordinary hate or anger; it was something much deeper, much darker, frightening, something even I couldn't understand.... I watched impassively as they led him through the front door, his head bowed, his hands manacled, his self-esteem drained, his manhood sapped. (pp. 22–23)

The raids by black and white policemen were a tormenting presence in life; they came frequently, always unannounced. For a child barely 6 years old, they meant constant terror: "They haunted me in real life and in my dreams, to the extent that I would often wake up dreaming in the middle of the night, claiming that the police were after me with dogs and flashlights, trying to shoot me down" (p. 28).

His father's arrests and absences, some over lengthy periods, created still worse difficulties—wages for food and rent were lost. This meant days with nothing to eat and constant threats from the landlord about eviction from their dingy shack. The malnutrition, in turn, led to illnesses in the children, which could not be treated because there was no money for the clinic. In their desperation, Mark, his siblings, and their mother dug for food at the nearby garbage dump and searched for diseased chickens and rejected eggs from the nearby chicken factory.

> When it seemed that no help was forthcoming, we resigned ourselves to the inevitable: eviction and starvation. Luck of some sort came when my maternal grandmother—who had been away in the Shangaan Bantustan attending a ceremony to exorcise evil spirits from a raving mad relative—came back unexpectedly. My mother told her of our plight. Granny had some money to spare. She paid our rent a week before we were to be evicted; bought us bread, sugar, and mealie meal; and gave my mother one hundred cents to take George and Florah to the clinic. (p. 43)

Other significant strengths were present in these lives of enormous deprivation. As children, the Mathabanes, of course, had no storybooks or nursery rhymes, but Mark recalls his mother's stories serving as a kind of library, "a golden fountain of knowledge where we children learned about right and wrong, about good and evil" (p. 79).

> She was such a mesmerizing storyteller that once she began telling a tale, we children would remain so quiet and transfixed, like mannequins, our eager and receptive minds under her hypnotic voice, that we would often hear ourselves breathing. Whenever she ended a particular story, saying that it was past our bedtime, we would implore her to tell another one—a request she always heartily granted—until either my father screamed from the bedroom that the candle be snuffed, or we children simply dozed off into faraway worlds, our minds pregnant with fantastic yarns we wished never to forget. (p. 78)

Amid the endless hardship—including the constant strain of inadequate food and shelter; numerous pregnancies and births (7 children in all); a husband who, from bitterness and frustration about his inability to find work and earn a living, descended into alcoholism and gambling; and her own desperate pursuit for work of any kind—Mathabane's mother maintained an astonishing sense of hope and belief in her children's future. Her commitment to education was unwaivering, even when it meant beatings from her husband, who objected to the children going to school.

> "Your father didn't go to school," she continued, dabbing her puffed eyes to reduce the swelling with a piece of cloth dipped in warm water, "that's why he's doing some of the bad things he's doing. Things like drinking, gambling, and neglecting his family. He didn't learn how to read and write; therefore, he can't find a decent job. Lack of any education has narrowly

> focused his life.... I [also] can't find a job because I don't have papers, and I can't get papers because white people mainly want to register people who can read and write. But I want things to be different for you, child. For you and your brothers and sisters. I want you to go to school, because I believe that an education is the key you need to open up a new world and a new life for yourself, a world and life different from either your father's or mine.... Education will open doors where none seem to exist. It'll make people talk to you, listen to you and help you; people who otherwise wouldn't bother. It will make you soar, like a bird lifting up in to the endless blue sky, and leave poverty, hunger and suffering behind." (pp. 133–134)

School became critical for more than the knowledge it imparted; it kept Mathabane from joining the youth gangs who roamed the filthy streets of Alexandra in search of food and adventure.

> These boys had long left their homes and were living in various neighborhood junkyards, making it on their own. They slept in abandoned cars, smoked glue and benzene, ate pilchards and brown bread, sneaked into the white world to caddy and, if unsuccessful, came back to the township to steal beer and soda bottles from shebeens, or goods from the Indian traders on First Avenue. (p. 123)

Many were lured into lives of prostitution and crime; many ended up in penitentiaries or in the grave. Again, Mark's mother played a critical role in keeping him from the pressure of these gangs. When he began skipping school to return to the gangs, his mother intervened, sanctioning a severe beating delivered by the principal and male teachers.

> "Whip him good," she impassively gave the order ... The teachers descended on me like starved vultures out of the sky. They commenced the savage beating, taking turns whenever one teacher's hand got tired. I fainted. They revived me, only to whip me some more. I spent an entire week bedridden, unable to sit up or sleep. For the rest of my primary school years I seldom, if ever, cut school for any reason. Even when I was gravely ill, I would crawl to school, and the teachers would send me back home." (p. 161)

School also provided an important source of success in a world where failure and hopelessness was everywhere:

> That night, my mother, upon hearing the news I had come out number one in my class, hugged and kissed me so many times and made me so many promises. Before the night was over, the yard was buzzing with news of a teacher-to-be at the Mathabane's household. Granny was told the good news, and she promised me many things. I felt happy and proud. I even began to like school. What was so remarkable about my first-year performance was that I had come out number one despite chronically lacking books, having had to resort to paying more attention during class lectures, borrowing other children's books to do homework and relying on a picture-book memory during exams. (p. 143)

Juxtaposed with these positive occasions were continuing adversities from the surrounding world of crime and violence. At age 10, Mathabane witnessed a grisly murder. A few months later, he attempted suicide.

> I was weary of being hungry all the time, weary of being beaten all the time: at school, at home and in the streets. I felt that somehow the whole world was against me. I felt that the courage, the resiliency and the unswerving, fanatical will to survive, to dream of a bright future, to accomplish, to conquer, of early years had deserted me. (p. 167)

It was, again, his remarkable mother who delivered Mark from despair, first, by conveying to him how much he would be missed by his younger brothers and sisters, and most importantly, by how she would feel were he to die.

> She held me tighter and said, "I would miss you more than anyone else. I too would want to die if you were to die. You're the only hope I have. I love you very much. . . . " For years afterward, I was to think of that suicide attempt in the following terms: whenever the troubles of the world seemed too much, it helps to have someone loving and understanding to share those troubles with; and life takes its true meaning in proportion to one's daily battles with suffering." (pp. 169–170)

It was Granny, the gardener for white people, who exposed Mathabane to the world outside his closed environment, first, by bringing comic books from the son of the Smiths for whom she worked, and second, by taking Mark into that world. His first observations of white school children—their snow-white shirts, blazers, badges, shiny black shoes; their large red-brick school house with its big windows and flower beds, the athletic fields, swings, and merry-go-rounds—left him stunned. At the home of the Smiths, he was confronted with Clyde, a same-aged son, who did not want a "pickaninny" in his house but was required by his mother to show Mark around.

> I followed him around as he showed me all the things his parents regularly bought him: toys, bicycles, go-carts, pinball machines, Ping-Pong tables, electric trains. I only half-listened: my mind was preoccupied with comparing my situation with his. I couldn't understand why he and his people have all the luxuries money can buy, while I and my people lived in abject poverty. Was it because they were whites and we were blacks? Were they better than we? I could not find the answers; yet I felt there was something wrong about white people having everything, and black people nothing.
>
> We finally came to Clyde's playroom. The room was roughly the size of our house, and was elaborately decorated with posters, pennants of various white soccer and cricket teams, rock stars and photographs of Clyde in various stages of development. But what arrested my attention were the stacks of comic books on the floor, and the shelves and shelves of books. Never had I seen that many books in my life; even our school, with a student population of over two thousand, did not have half as many books. I was dazed. (p. 191)

Because Mark could not read English, or more precisely, Shakespeare, Clyde announced he was retarded, and like other Kaffirs unable to learn such things, because he had a smaller brain. Curiously this comment fueled a driving future passion—to learn to express his thoughts and feelings effectively in English.

> The remark that black people had smaller brains and were thus incapable of reading, speaking or writing English like white people had so wounded my ego that I vowed that, whatever the cost, I would master English, that I would not rest until I could read, write and speak it just like any white man, if not better. Finally, I had something to aspire to. (p. 192)

Mathabane's "passport to freedom" came not only from his excellence in school, but also from tennis. He won a government scholarship for secondary schooling and chose a school with a tennis team—a decision he would later view as the most important of his life. Tennis brought him into contact with Arthur Ashe, who was a great source of inspiration—"I marveled at how proudly he walked. I had never seen a black man walk that proudly among whites. He appeared calm, cool and collected, even though he was surrounded by a sea of white faces" (p. 236). Ashe, to him, revealed the hidden truth about the need to *rise above one's own suffering*, to conquer one's own fears. It also underscored his critical need for role models out of the all-encompassing adversity. Tennis brought him into contact with other German and South African players who taught him about the ANC liberation movement. He discovered, via these developing friendships, that not all whites in South Africa were the same.

> He [Wilfred] had now become my unofficial tennis sponsor; from time to time he gave me money to enter tournaments and travel to them, gave me tennis clothes, rackets and balls; and once in a while would buy me a schoolbook, or help my family through hard times. And he did it all out of the goodness of his heart, without any trace of condescension or paternalism. It was the knowledge that white people like him existed that reinforced my belief that though whites in South Africa were generally prejudiced toward blacks, and in many cases treated us as less than dogs, there were, among them, those handful who would bend over backward to help black people—not as servants, but as equals. (p. 237)

It was through the American tennis player, Stanley Smith, that possibilities for Mathabane to attend college on a tennis scholarship in the United States materialized. Like all prior opportunities, he used this chance to improve his life to the fullest and progressed eventually toward the writing of his own remarkable story. Mathabane's writings put real faces, real lives of apartheid before the rest of the world. In so doing, he became a voice for those who suffered, communicating the wretchedness of their daily lives. Speaking to his tennis colleagues, who could not understand the riots and anarchy in the townships because the blacks they knew said conditions were improving, he corrected their misperceptions.

> "You've just described the elite black middle class," I said, "a hopelessly small minority whose aspirations are no different from those of the white man. Many of them make money any way they can so they can live comfortably. The fact of the matter is that the majority of blacks are peasants. They live in shacks and mud houses, they walk barefoot in dirt streets strewn with rocks and broken bottles, their infants die of malnutrition, their children's growth is stunted, for them education and medical treatment are not free, they live in constant fear of running afoul of the degrading pass laws, of being deported to impoverished tribal reserves." (p. 275)

Along the way, Mathabane articulated not just the deplorable living conditions, but what they did to the core of the person, their essential humanity. His description of the thousands of black migrant workers in Alexandra forced to live hundreds of miles from their families because of the Influx Control laws revealed a penetrating understanding.

> Housed mostly in sterile single-sex barracks, they were prey to prostitution, alcoholism, robbery, and senseless violence; they existed under such stress and absorbed so much emotional pain that tears, grief, fear, hope and sadness had become alien to most of them. There were the walking dead.
>
> Stripped of their manhood, they hated the white man with every fiber of their being. Anger would leap into their eyes each time the words "white man" were uttered. Rage would heave their chest each time something or someone reminded them that it was the white man who kept their families away from them....
>
> There is a death far worse than physical death, and that is death of the mind and the soul, when, despite toiling night and day, under sweltering heat, torrential rain, blistering winds, you still cannot make enough to clothe, shelter, and feed your loved ones, suffering miles away, forcibly separated from you. (p. 181)

The overriding question is how did he do this? How could a child confronted with endless adversity involving violation of the most basic of human needs as well as relentless feedback that he had no status in a strictly hierarchical world, become a superb scholar, talented athlete, and, above all, a powerful spokesman about the horrors of apartheid? Clearly, the most profound advantage was his remarkable mother, herself a *sine qua non* of human resilience.

> Since I was four years old I had held dear my mother's protracted and impassioned urges to keep on fighting, to keep on trying regardless, to never succumb to poverty, fear, pain and suffering, to study hard and to keep on knocking at the doors of opportunity and patiently wait for them to open eventually. "Stay in school, study hard, and keep on trying, child," my mother had always told me. "And things will get better." (p. 181)

But there were others as well, his grandmother, aunts, and uncles, and the tennis coaches and friends, all of whom provided sustaining and supportive social relationships. Then there were his own reactions to the adversity, his capacity to turn the negative into powerful positive forces, such as his realization of the strength that comes from rising above one's own suffering, and even his passion to learn English, having been told his brain was too small for such a quest. Mark Mathabane clarifies that resilience follows from numerous strengths and advantages—quality relationships, occasions of success, unswerving aspirations, inspiring role models, and, paradoxically, the awfulness of the adversity itself, from which he discovered his own inner resources. He is, of course, a dramatic exception to the usual life outcomes following from severe poverty and discrimination, but as such, his history offers dramatic insight about the forces that enable health and well-being in the face of staggering obstacles.

Physiological Substrates: The Character of the Current Knowledge Base

Overall there appears to be greater cumulated psychosocial adversity among U.S. blacks relative to whites and, correlatively, less cumulated advantageous and compensating experience. In the case of South Africa under apartheid there was, in effect, an officially mandated ordering of cumulating psychosocial adversity according to the racial/ethnic sequence White, Asian (Indian), coloured, African. Thus, psychosocial life-history pathways, analogous to those constructed for the resilient women in the WLS, and illustrated with the autobiography of Mark Mathabane, are heavily weighted with diverse chronic adversities among the more disadvantaged groups.

It would be highly desirable to have directly comparable life-history data at the level of richness of the WLS or the British National Survey of Health and Development[41] that covers multiple racial and ethnic groups. The extant evidence supporting racial and ethnic disparities across multiple life domains in cumulating adversity, advantage, mobility in social hierarchies, and quality of relationships is an agglomeration of segments of life histories from multiple surveys and biographical sources, in each of which data are heavily dominated by one racial/ethnic group.

The multiple psychosocial pathways—where an ordering on cumulated adversity induces a racial/ethnic ordering—are one among several components of explanations for both disease and positive health outcomes. Genetic, nutritional, developmental, and maturational factors interact with the psychosocial pathways in a complex manner that has yet to be fully delineated.[42] With the exception of gross physical injury resulting from natural or human-made disasters, and leading either to recovery, disability, or death, experiences perceived as stressful (or threatening) do not on their own produce disease. They do, however, interact with the above-listed factors to facilitate the onset of disease.

For example, African Americans who, more than whites, are exposed to violence and police brutality, who live in crowded conditions in areas of a city marked by social disorganization and economic deprivation, and whose personal and marital lives are disrupted, have higher blood pressure levels than whites living in middle class neighborhoods in which these social conditions do not prevail.[43] This, in turn, leads to much higher hypertension rates among older African Americans relative to whites (Table 5.2) (see Ref. 5).

Relevant to our discussion of tuberculosis, it is important to observe that infection with the *Mycobacterium* of tuberculosis does not specify a particular disease outcome. Complex interactions involving multiple host factors and the *Mycobacterium tuberculosis* agent define a multiplicity of pathways that either place an individual at high risk for disease or are protective and, therefore, endow resistance to it.[44] The central point is that even from infection—following environmental exposure—there are multiple subsequent pathways. In principle, the methodology used on the psychosocial histories in the WLS are equally applicable to biological measurements that would delineate alternative chains of events from infection to disease or resistance. Full understanding of pathways to disease outcomes requires, in fact, the weaving together of the psychosocial with the biological/biochemical histories. It is this synthesis that comprises the research frontier of life histories and health.

Despite the current unavailability of sufficiently precise biological information to allow for specification of the chains of events along multiple pathogenetic routes to disease, there is strong associational evidence that provides requisite clues to guide ongoing and future research on disease pathogenesis. Examples of such evidence are as follows:

1. The mental interpretation of an experience as threatening initiates a cascade of hormone secretion driven by the hypothalamus in the brain, signaling a release of corticotropin releasing factor (CRF), which, in turn, stimulates the pituitary gland to release corticotropin, and this stimulates the adrenal glands to release cortisol into the blood.[25,45] Activation of the sympathetic nervous system, another consequence of the interpretation of experience(s) as stressful (threatening), leads to the release of catecholamines (i.e., norepinephrine and epinephrine) from the inner portion (medulla) of the adrenal gland to the blood. Immune cell function is altered by the action of these hormones and transmitters.[46] In combination, these processes can damage regulatory feedback systems and reduce immune system capability.
2. McEwen and Stellar introduced the notion of "allostatic load" (meaning the strain on multiple organs and tissues resulting from repeated fluctuation in physiological response to perceived threat), in which high levels are associated with organ-system breakdown, compromised immune response, elevated cortisol and insulin secretion, and ultimately disease.[47] Allostatic load measures the cumulative physiologic effects of multiple forms of

adversity and is predictive of a diverse range of organ-system breakdown and disease incidence.

There are two central features to the accumulation of allostatic load. One reflects the wear and tear associated with *acute shifts* (generally elevations) in physiologic activity in response to specific stimuli; this is the process frequently referred to as *physiologic reactivity.* The second contributing factor to allostatic load is chronic elevations in physiologic activity outside of basal operating ranges. Conditions such as hypertension and diabetes are examples of such *chronic elevations.* These states of chronic elevation in physiologic activity presumably represent levels of system operations that follow from repeated and cumulative adversity experiences viewed by the organism as threatening.

Allostatic load for an individual has been operationalized as the number of indicators from the list in Table 5.3 for which an individual's assessed value satisfies the stated inequality.[48] As measures of possible physiological system impairment, high systolic and diastolic blood pressure are indices of cardiovascular activity; waist-hip ratio is an index of metabolism and adipose tissue deposition; serum HDL and total cholesterol are indices of atherosclerotic risk; blood plasma levels of glycosylated hemoglobin indicate glucose metabolism; 12-hour integrated measure of urinary cortisol excretion is an indicator of hypothalamic-pituitary-adrenal (HPA) axis activity; and urinary norepinephrine and epinephrine excretion levels are indices of sympathetic nervous system (SNS) activity.

In a study of elderly persons in the United States,[49] it has been shown (Ref. 48) that higher allostatic load scores among persons with no reported cardiovascular disease (CVD; i.e., myocardial infarction, stroke, diabetes, or high blood pressure) at baseline in 1988 predicted a subsequent decline in cognitive function (particularly memory decline), a decline in physical

Table 5.3. Indicators of high allostatic load

Indicator
Systolic blood pressure ≥148 mm. Hg
Diastolic blood pressure ≥83 mm. Hg
Waist-hip ratio ≥ 0.94
Ratio total/HDL cholesterol ≥ 5.9
Glycosylated hemoglobin ≥ 7.1%
Urinary cortisol ≥ 25.7 g/g creatine
Urinary norepinephrine ≥48 g/g creatine
Urinary epinephrine ≥5 g/g creatine
HDL cholesterol ≤56 ug/dl
DHEA-S ≤91 mg/dl

Abbreviations: HDL, high-density lipoprotein; DHEA-S, dihydroepiandrosterone sulfate.

performance, an increase in mortality, and an increase in CVD incidence 2.5 years later. Thus, the measure of allostatic load based on counts of the number of indices (in Table 5.3) for which a person's assessment exceeds the specified thresholds served as a good summary of an individual's level of physiological activity across a range of regulatory systems known to have important influences on disease risks.

3. There is a growing body of evidence indicating a direct relationship between naturalistic stressors and immune system impairment. As an example of the role of persistent adversity, residents of the area surrounding the Three Mile Island nuclear power plant were assessed 10 years after the serious accident of 1979. They were found to have more antibody to herpes viruses than in demographically matched control-group residents, suggesting lower cellular immune competence.[50] Kiecolt-Glaser et al. found that caregiving was associated with distress and reduced response to an influenza vaccine challenge.[51] Kiecolt-Glaser et al. found that 16 separated and divorced women had higher levels of herpes antibody, a lower percentage of NK cells, and a lower lymphocyte proliferative response to phytohemagglutinin (PHA) and concanavalin A (Con A) than a comparison group of 16 married women.[52] Perceived availability of social support has also been associated with immune function. In a study of 256 elderly adults, it was found that blood samples from persons reporting they had confiding relationships proliferated more in response to PHA than samples from those without confiding relationships.[53] (For more examples and a superb overview of empirical studies relating psychological factors to physical disease, see Ref. 54).

Three key messages follow from the above-mentioned examples relating the social environment and behavior to physiological substrates. The first is that racial and ethnic inequalities in opportunity, and accompanying increments in adversity, map correspondingly onto disparities in impaired immune function, and HPA axis and SNS reactivity. Delineation of these impairments, working interactively with the psychosocial factors, is critical to comprehensive understanding of the pathogenetic pathways to racial and ethnic equalities in incidence of infectious and chronic disease. In short, essential biological mechanisms must be elaborated and incorporated into prevailing accounts of environmental and psychosocial mechanisms hypothesized to underlie racial differences in health (e.g., Ref. 4).

A second point is that the life-history approach, which brings together multiple domains of experience unfolding through time, as well as complex physiological underpinnings, is essentially an *aggregation enterprise*. That is, the analytic focus is not on single types of adversity (e.g., poor job conditions) or single biochemical factors (e.g., cortisol), but rather on the cumulation of multiple experiences through time and the combination of physiologic responses summarized by allostatic load. This approach stands in sharp contrast to more delimited and focused scientific agendas that track a single, or, a small subset of, intervening mechanisms. When contrasted with such research, the present agenda may appear excessively expansive, if not foolhardy. We assert, however, that the narrower agendas are fundamentally short sighted in their capacity to explicate pathways to health outcomes.

Finally, returning to our recurring theme of positive health, we note that all of the preceding examples of physiological substrates pertain to pathways of pathogenesis—routes to disease. Equal attention must be paid to pathways of salubrity—promotion of well-being. This is essentially the *preventive research agenda*, wherein life-history studies can contribute vital knowledge regarding the factors (psychosocial or physiological) that prevent cumulative spirals of negative experience and/or related cascades of harmful physiological reactivity. Our focus on resilience points to the salubrious qualities of supportive social relationships, positive ways of framing and reacting to adversity, and opportunities for advancement in social hierarchies. We contend that these pluses, and more generally the experience of flourishing, has its own physiological substrates, many of which may be protective of the organism, such as by keeping allostatic load habitually low and maximizing immune competence (see Ref. 19). In short, the call to connect the external world of experience with the internal workings of the body need not be, despite the preponderance of existing evidence, an exclusively disease-oriented agenda.

Discussion

Because of the overwhelming emphasis we have given to environmental factors in discussing the propensity for diverse disease and health outcomes, it seems appropriate to clarify why we have ignored genetics to such an extent. First, blacks and whites have, as an initial approximation, over 99.9% of their nucleotides in common.[55] Thus, the general expectation is that there is similarity, not differences, between racial groups in the vast majority of genetic response to disease producing situations.

Second, most of the major diseases differing in frequency between Caucasoids and Negroids, for example, essential hypertension in the United States, appear to be diseases of complex etiology with a multifactorial genetic basis and a strong environmental component in the determination of how the inherited susceptibility is expressed. Identification of gene complexes responsible for varying degrees of susceptibility is a subject still very much in its infancy. In addition, even when an allele–disease association is clearly established for one racial/ethnic group, it cannot be automatically assumed that the same relationship holds in a different ethnic group. In the presence of a strong allele–disease association, only a minority of carriers develop disease. Actual realization of the potential disease relationship depends on further modifying genetic factors—see Lifton for an illustration regarding hypertension—and environmental triggers.[56]

Focusing on hypertension as a generic example, we first observe that numerous studies have established its familial nature.[57] However, the correlation between adult sibs usually varies between 0.2 and 0.3 for both systolic and diastolic blood pressures, and for parent–offspring it is the same or somewhat lower. For monozygous twins, the correlations between systolic and diastolic blood pressures are, respectively, 0.55 and 0.58, whereas for dizygous twins the corresponding figures are 0.25 and 0.27. Although in the extant data these values are all statistically different from 0, they are well below what one would expect of a continuously distributed trait that is *completely* genetically determined. It is the gene × environment interaction that is fundamental.

The susceptibility genes for hypertension—to the extent that they are presently known—appear to be relatively common and widespread. This raises the possibility that we could regard susceptibility genes as having been present in humans prior to civilization, that is, in hunter-gatherer and traditional agrarian societies, serving a useful function but kept from going to fixation by a balance with opposing selective forces.[58]

Essential hypertension is genetically heterogeneous, with attempts to tease out subtypes representing a currently very active research area. The disease manifestations are highly diverse; one example that pertains to our focus on racial inequalities is the fact that blacks have a lower risk for coronary artery disease but a higher risk for stroke and renal failures than whites. The incidence of blacks with end-stage renal disease is 3 to 4 times higher than whites at all age groups.[59] A possible partial explanation is the longer duration (i.e., earlier onset) of disease in blacks relative to whites. However, full answers to the question, How does this outcome come about? require elaboration of life-history pathways and, in particular, the environmental conditions that activate genes responsible for hypertension.

A final point concerning our focus on environmental factors is a concern about the character of the alleged interface between race and genetics. Race, as we have used the term, is a somewhat arbitrary social construct (see Ref. 4). Whether we are comparing chronic disease incidence rates, mortality rates, or gene frequencies, black/white or Negroid/Caucasoid differentials represent comparisons across social categories. There has been more than a century of attempts to provide a genetically based classification scheme for the notion of race. This effort led to the contemporary consensus among geneticists that, due to the enormous heterogeneity within culturally similar groups, the concept of race is so vague that it cannot be given a clear meaning.[60] A future molecular genetic–based classification of human populations in terms of susceptibility to the onset of particular diseases and the character of environmentally induced gene expression is likely to have scant connection to the current social constructs of race and ethnicity.

Given these multiple qualifiers to genetic explanations for racial differences in health, this chapter has dealt primarily with features of the environment. What we have set forth is

1. A characterization of the psychosocial and physical environmental life histories/pathways that provide the beginning of a comprehensive, multiple-domain explanation for the racial and ethnic differentials in both positive health and disease outcomes. The multiple pathways that we have exhibited and implicated are a reflection of the enormous breadth of plasticity in the life histories of vertebrates generally.[61] Limits on the possible extent of such plasticity derive from both constitutional features of the human body plan and the particular genes that an individual carries.
2. The structure of pathways defining resilience to adversity, thereby laying the groundwork for more informed health promotion and disease prevention interventions. The features of positive interpersonal relationships, strong encouragement to pursue education, and participation in supportive community organizations (e.g., church groups) that underlie multiple resilient pathways form the basis for understanding the potential to increase the number of persons who begin life in disadvantaged environments but nevertheless succeed to high levels of well-being in middle and later life.
3. A selection of physiological substrates that interact with psychosocial experience in a complex feedback system that should ultimately lead to explanations for racial and ethnic inequalities in health outcomes in terms of tightly integrated representations of mind–body relationships. Reactions to challenge in the social environment stimulate cascades of neuroendocrine, sympathetic nervous system, and immune system events, and these, in turn, set in motion biochemical processes that are templates for future behaviors and environmental exposures.[61,62] A broad-based ordering of racial and ethnic groups by the extent of cumulative psychosocial adversity, with delineation of biological substrates, provides the beginning of an explanation for the coarse associational evidence of racial/ethnic-group ordering by mortality and morbidity rates.

We believe that the life-histories approach presented in this chapter has far broader applicability than the regression model approach adopted by Herrnstein and Murray in *The Bell Curve*. Success in life has, we believe, complex determinants, including biologic ones, but these are not necessarily genetically based.

References

1. Herrnstein, R., and Murray, C. (1994), *The Bell Curve: Intelligence and Class Structure in American Life*, The Free Press, New York.
2. WHO, (1983), *Apartheid and Health*, Geneva, World Health Organization.

3. Williams, D. (1990), "Socioeconomic Differentials in Health: A Review and Redirection," *Social Psychology Quarterly*, 53, 81–99.
4. Williams, D., and Collins, C. (1995), "U.S. Socioeconomic and Racial Differences in Health: Patterns and Explanations," *Annual Review of Sociology*, 21, 349–386.
5. Berkman, L., Singer, B., and Manton, K.G. (1989), "Black/White Differences in Health Status and Mortality Among the Elderly," *Demography*, 26, 661–678.
6. Huntley, J.C., Brock, D.B., Ostfeld, A.M., Taylo, J.O., and Wallace, R.B. (1986), *Established Populations for Epidemiologic Studies of the Elderly: Resource Data Book*, U.S. Government Printing Office (NIH Pub. No. 86-2443), Washington, D.C.
7. Ware, J.E. (1986), "The Assessment of Health Status," in L.H.Aiken and D. Mechanic (Eds.), *Applications of Social Science to Clinical Medicine and Health Policy*, 9th ed., Rutgers University Press, New Brunswick, NJ.
8. Collins, T.F.B. (1982), "The History of Southern Africa's First Tuberculosis Epidemic," *South African Medical Journal*, 62, 780–788.
9. Dubos, R.J., and Dubos, J. (1953), *The White Plague: Tuberculosis, Man, and Society*, Victor Gollancz Ltd., London.
10. Packard, R.M. (1989), *White Plague, Black Labor: Tuberculosis and the Political Economy of Health and Disease in South Africa*, University of California Press, Berkeley.
11. City of Capetown, (1979), *Annual Report of the Medical Officer for Health*, Capetown, South Africa.
12. Legassick, M., and deClerq, F. (1978), Capitalism and Migrant Labor in Southern Africa: The Origin and Nature of the System, ILO-ECA Conference on Migrant Labor in Southern Africa, April.
13. Wilson, F. (1980), "Suggested Directions for the Future," in SAIRR, *Towards Economic and Political Justice in South Africa*, South African Institute of Race Relations, Johannesburg.
14. SAIRR. (1977), *Survey of Race Relations*, South African Institute of Race Relations, Johannesburg.
15. Acevedo-Garcia, D. (1996), *Has Residential Segregation Shaped the Epidemiology of Tuberculosis among US Minorities? The Case of New Jersey, 1985–1992*. Doctoral Dissertation, Princeton University, Princeton, NJ.
16. Commission on Racial Justice, United Church of Christ. (1987), *Toxic Wastes and Race in the United States*, Public Data Access, Inc., New York.
17. Mohai, P., and Bryant, B. (1992), "Environmental Racism: Reviewing the Evidence," in B. Bryant and P. Mohai (Eds.), *Race and the Incidence of Environmental Hazards: A Time for Discourse*, Westview Press, Boulder, CO.
18. Ciesielski, S., Esposito, D., Protiva, J., and Piehl, M. (1994), "The Incidence of Tuberculosis among North Carolina Migrant Farm Workers, 1991," *American Journal of Public Health*, 84, 1836–1838.
19. Ryff, C.D., and Singer, B. (1997), *The Contours of Positive Human Health,Psychological Inquiry* (in press)
20. Hauser, R.M., Sewell, W.H., Logan, J.A., Hauser, T, Ryff, C., Caspi, A., and McDonald, M. (1992), *The Wisconsin Longitudinal Study: Adults as Parents and Children at Age 50*, CDE Working Paper 92-2, University of Wisconsin, Center for Demography and Ecology.
21. Sewell, W.H, and Hauser, R.M. (1980), "The Wisconsin Longitudinal Study of Social and Psychological Factors in Aspirations and Achievements," in A.C. Kerckhoff (Ed.), *Research in Sociology and Education*, 1, JAI Press, Greenwich, CT, pp. 59–99.

22. Sewell, W.H., and Shah, V.P. (1967), "Socioeconomic Status, Intelligence, and the Attainment of Higher Education," *Sociology of Education*, 40 (Winter), 1–23.
23. Carr, D., Ryff, C.D., Singer, B., and Magee, W. (1996), *Life Histories and Mental Health*, Technical Report, Center for Demography and Ecology, University of Wisconsin-Madison.
24. Adler, N.E., Boyce, T., Chesney, M., Folkman, S., and Syme, L. (1993), "Socioeconomic Inequalities in Health: No Easy Solution," *Journal of American Medical Association*, 269, 3140–3145. Adler, N.E., Boyce, T., Chesney, M., Cohen, S., Folkman, S., Kahn, R.L., and Syme, L. (1994), "Socioeconomic Status and Health: The Challenge of the Gradient," *American Psychologist*, 49, 15–24.
25. McEwen, B.S., and Schmeck, H.M., Jr. (1994), *The Hostage Brain*, Rockefeller University Press, New York.
26. Lazarus, R.S., and Folkman, S. (1984), *Stress, Appraisal, and Coping*, Springer-Verlag, New York.
27. Pearlin, L., and Schooler, C. (1978), "The Structure of Coping," *Journal of Health and Social Behavior*, 19, 2–21.
28. Thoits, P.A. (1994), "Stresses and Problem-Solving: The Individual as a Psychological Activist," *Journal of Health and Social Behavior*, 34, 143–159.
29. House, J., Kessler, R.C., Herzog, A.R., Mero, R.P., Kinney, A.M., and Breslow, M.J. (1992), "Social Stratification, Age, and Health," in K.W Schaie,.D. Blazer, and J.S. House (Eds.),*Aging, Health Behaviors, and Health Outcomes*, Lawrence Erlbaum, Hillsdale, NJ, pp. 1–31.
30. Weisfeld, G.E. (1980), "Social Dominance and Human Motivation," in D.R. Omark, F.F. Strayer, and D.G. Freedman (Eds.), *Dominance Relations: An Ethological View of Human Conflict and Social Interaction*, Garland Press, New York.
31. Wright, R. (1994), *The Moral Animal: The New Science of Evolutionary Psychology*, Pantheon Books, New York.
32. Wethington, E., and Kessler, R.C. (1986), "Perceived Support, Received Support and Adjustment to Stressful Life Events," *Journal of Health and Social Behavior*, 24, 208–229.
33. Wheaton, B. (1985), "Models for the Stress-Buffering Functions of Coping Resources," *Journal of Health and Social Behavior*, 26, 352–364.
34. Newman, K. (1988), *Falling from Grace: The Experience of Downward Mobility in the American Middle Class*, New York: Basic Books
35. Geertz, C. (1973) *The Interpretation of Cultures*, New York: Basic Books.
36. Ragin, C. (1987), *The Comparative Method: Moving Beyond Qualitative and Quantitative Strategies*, University of California Press, Berkeley.
37. Tilly, C. (1984), *Big Structures, Large Processes, Huge Comparisons*, Russell Sage Foundation, New York.
38. Ryff, C.D., (1989), "Happiness Is Everything, or Is It?: Explorations on the Meaning of Psychological Well-Being," *Journal of Personality and Social Psychology*, 57, 1069–1081.
39. Ryff, C.D., and Keyes, C.L.M. (1995), "The Structure of Psychological Well-Being Revisited," *Journal of Personality and Social Psychology*, 69, 719–727.
40. Mathabane, M. (1986), *Kaffir Boy*, Plume, Penguin Books, New York.
41. Wadsworth, M.E.J. (1986), "Serious Illness in Childhood and its Association with Later-Life Achievement," in R.G.Wilkinson (Ed.), *Class and Health: Research and Longitudinal Data*, Tavistock Publications, London, pp. 50–74.

42. Weiner, H. (1992), *Perturbing the Organism: The Biology of Stressful Experience*, University of Chicago Press, Chicago.
43. Harburg, E., Erfurt, J.C. Hauenstein, L.S., Chape, C., Schull, W.J., and Schork, M.A. (1973), "Socio-ecological Stress, Suppressed Hostility, Skin Color, and Black-White Male Blood Pressure: Detroit," *Psychosomatic Medicine*, 35, 276–296.
44. Copeland, D.D. (1977), "Concepts of Disease and Diagnosis," *Perspectives on Biology and Medicine*, 20, 528–538.
45. Sapolsky, R.M. (1994), *Why Zebras Don't Get Ulcers: A Guide to Stress-Related Diseases and Coping*, W.H. Freeman, New York.
46. Maier, S.F. Watkins, L.R., and Fleshner, M. (1994), "Psychoneuroimmunology: The Interface Between Behavior, Brain, and Immunity," *American Psychologist*, 49, 1004–1017.
47. McEwen, B.S., and Stellar, E. (1993), "Stress and the Individual." *Archives of Internal Medicine*, 153, 2093–2101.
48. Seeman, T., Singer, B., Horwitz, R.I., and McEwen, B. (1996), *Operationalizing Allostatic Load*, Manuscript, Andrus Gerontology Center, University of Southern California.
49. Seeman, T., Charpentier, P., Berkman, L., Tinetti, M., Guralnik, J., Albert, M., Blazer, D., and Rowe, J. (1994), "Predicting Changes in Physical Performance in a High Functioning Elderly Cohort," MacArthur Studies of Successful Aging, *Journal of Gerontology*, 49, M97–M108.
50. McKinnon, W., Weisse, C.S., Reynolds, C.P., Bowles, C.A., and Baum, A. (1989), "Chronic Stress, Leukocyte Sub-populations, and Humoral Response to Latent Viruses," *Health Psychology*, 8, 389–402.
51. Kielcolt-Glaser, J.K., Glaser, R., Gravenstein, S., Malarkey, W.B., and Sheridan, J. (1996), "Chronic Stress Alters the Immune Response to Influenza Virus Vaccine in Older Adults," *Proceedings of the National Academy of Sciences USA*, 93, 3043–3047.
52. Kiecolt-Glaser, J.K., Fisher, L.D., Ogrocki, P., Stout, J.C., Speicher, C.E., and Glaser, R. (1987), "Marital Quality, Marital Disruption, and Immune Function," *Psychosomatic Medicine*, 49, 13–25.
53. Thomas, P.D., Goodwin, J.M., and Goodwin, J.S. (1985), "Effect of Social Support on Stress-Related Changes in Cholesterol Level, Uric Acid Level, and Immune Function in an Elderly Sample," *American Journal of Psychiatry*, 142, 735–737.
54. Cohen, S., and Herbert, T. (1996), "Health Psychology: Psychological Factors and Physical Disease from the Perspective of Human Psychoneuroimmunology," *Annual Reviews of Psychology*, 47, 113–142.
55. Neel, J. (1994), Are Genetic Factors Involved in Ethnic Differences in Late-Life Health? Paper presented for Workshop on Racial and Ethnic Differences in Health in Late Life in the US, December 12–13, 1994, Committee on Population, National Academy of Sciences, Washington, DC.
56. Lifton, R.P. (1995), "Genetic Determinants of Human Hypertension," *Proceedings of National Academy of Sciences USA*, 92, 8545–8551.
57. Burke, W. and Motulsky, A.G. (1990), "Hypertension," in R.A. King, J.I. Rotter, and A.G. Motulsky, (Eds.), *The Genetic Basis of Common Diseases*, Oxford University Press, New York, pp. 170–191.
58. Julius, S., and Jammerson, K. (1994), "Sympathetics, Insulin Resistance and Coronary Risk in Hyptertension: The 'Chicken and Egg' Question," *Journal of Hypertension*, 12, 495–502.

59. Lopes, A.A.S., Port, F.K., James, S.A., and Agodoa, L. (1993), "The Excess Risk of Treated End-Stage Renal Disease in Blacks in the United States," *Journal of the American Society of Nephrology*, 3, 1961–1971. Lopes, A.A.S., Hornbuckle, K., James, S.A., and Port, F.K. (1994), "The Joint Effects of Race and Age on the Risk of End-Stage Renal Disease Attributed to Hypertension," *American Journal of Kidney Diseases*, 24, 554–560.
60. Cavalli-Sforza, L.L., and Cavalli-Sforza, F. (1995), *The Great Human Diasporas*, Addison-Wesley, Reading, MA.
61. Finch, C.E. (1996), "Biological Bases for Plasticity During Aging of Individual Life Histories," in D. Magnusson (Ed.), *The Lifespan Development of Individuals: Biological and Psychosocial Perspectives, a Synthesis*, Cambridge University Press, Cambridge, pp. 488–512.
62. Finch, C.E., and Rose, M.R. (1995), "Hormones and the Physiological Architecture of Life-history Evolution," *Quarterly Review of Biology*, 70, 1–52.

PART III

INTELLIGENCE AND THE MEASUREMENT OF IQ

What is intelligence? Is it one-dimensional? Can it be described with a single score? Is it simply what IQ tests measure? By making IQ the centerpiece of their study of inequality in American society, Herrnstein and Murray have raised a number of questions about the nature and measurement of intelligence. They resolve these questions, however, by an appeal to the authority of measurement psychologists rather than through an independent analysis of the components of intelligence or of the tests that measure it. The existence of "such a thing as a general factor of cognitive ability on which humans differ," they argue, is "by now beyond technical dispute" (p. 22).

They are quite right, Carroll argues, that measurement psychologists can largely agree that there is such a general factor (*g*), that it accounts for a lot of the successful performance on related tests, and that it is the main component in IQ. There is also quite a long tradition, going back to Spearman, of representing it unidimensionally. Factor analysis has been a valuable tool in explaining the patterns of correlation in psychometric data. Although many different kinds of intelligence can be profiled in test data, a general ability construct continues to explain most of the correlations. Carroll offers a detailed history of the development and use of factor analysis, along with a review of important datasets. His analysis of the ASVAB dataset and its operational and conceptual representation of *g* is of particular interest, since ASVAB scores were heavily used by Herrnstein and Murray as an indicator of IQ.

Are test scores the only source for our knowledge about the forms and operation of general cognitive ability? While Hunt also sees test results and factor analysis as important, he brings to us findings about learning that cognitive scientists have gathered through laboratory experiments. Laboratory data reinforce the views of Horn, also discussed by Carroll, that intelligence can be seen in three dimensions—fluid, crystallized, and visual-spatial. This analysis has important implications for learning in our society.

Fluid intelligence can be described as the ability to learn how to learn or through the metaphor of number-crunching capacity. It tends to wane as people age, but crystallized intelligence, which involves command of problem-solving strategies and cultural knowledge, can develop with experience. Hunt finds that those with lots of fluid intelligence have an easier time learning problem-solving strategies. But great gains can be made independently in field knowledge without calling for more fluid intelligence. Hunt agrees with Herrnstein and Murray that intelligence is increasingly important to workplace performance, but having defined intelligence differently, remains more optimistic about ways of meeting the growing demand.

chapter six

Theoretical and Technical Issues in Identifying a Factor of General Intelligence

JOHN B. CARROLL

A main thrust of Herrnstein and Murray's *The Bell Curve*[1] was to describe the role of intelligence in American society, or, in particular, the role of the famous *g* factor originally postulated by Spearman.[2,3] Early on in their book, they made several assertions about *g* that they believed were "by now beyond significant technical dispute" (p. 22), including the proposition that "[t]here is such a thing as a general factor of cognitive ability on which human beings differ." It is understandable that they felt able to make such an assertion, in that at least some of the more prominent current experts in psychometrics were on record as holding the view that *g* exists.[4–13] Moreover, they mentioned, or could have mentioned, various social scientists who assume the existence of *g*.[14–16] In citing Snyderman and Rothman's (1988) survey of opinions in a large sample of educational psychologists and other scholars, they were correct in inferring that most of the respondents in that survey believed that a general factor of intelligence can be identified.[17]

Nevertheless, as Sternberg has pointed out, psychologists who have considered the matter are not unanimous in the opinion that there exists a general factor of intelligence.[18] For example, the late J.P. Guilford explicitly rejected this idea, preferring a so-called structure of intellect model whereby intelligence is composed of many separate components.[19,20] Horn has claimed that a general factor of intelligence cannot presently be supported by evidence, even though certain broad factors of ability (G_f, fluid intelligence, and G_c, crystallized intelligence) have a resemblance to a general factor.[21–23] Other critics of the notion of general intelligence include Ceci[24] and Gardner.[25]

There was thus some warrant for Gould's complaint that Herrnstein and Murray gave little attention to justifying the notion of *g* and spelling out its support in theoretical and empirical writings on the subject.[26] However, Gould is hardly to be trusted to provide an appropriate comment. Long before the publication of *The Bell Curve*, Gould put forth his opinion in his book *The Mismeasure of Man*, that there is no such thing as *general intelligence* or *g* and implied that "experts" in psychometric research were guilty of false or unscientific thinking.[27] Though he did not explicitly criticize psychological testing, many of his readers took him to intend such a criticism. In a paper published elsewhere, I have pointed out that an entire generation of readers (or "public intellectuals") seems to have been persuaded to adopt Gould's negative, arguably incorrect views.[28]

There have been two recent attempts by psychologists to issue balanced statements about the nature of intelligence and the existence of *g*. One of these was published in the *Wall Street Journal* and included the assertion that "[i]ntelligence is a very general mental capability that, among other things, involves the ability to reason, plan, solve problems, think abstractly, comprehend complex ideas, learn quickly and learn from experience."[29] Another was a report by a task force of psychologists appointed by the American Psychological Association,[30] intended to communicate to the public what is known and not known about intelligence and its many aspects and components. It states (p. 81) that

> [w]hile some psychologists today still regard *g* as the most fundamental measure of intelligence [e.g., Ref. 13], others prefer to emphasize the distinctive profile of strengths and weaknesses present in each person's performance. A recently published review identifies over 70 different abilities that can be distinguished by currently available tests [Ref. 4]. One way to represent this structure is in terms of a hierarchical arrangement with a general intelligence factor at the apex and various more specialized abilities arrayed below it. Such a summary merely acknowledges that performance levels on different tests are correlated; it is consistent with, but does not prove, the hypothesis that a common factor such as *g* underlies those correlations. Different specialized abilities might also be correlated for other reasons, such as the effects of education. Thus while the *g*-based factor hierarchy is the most widely accepted current view of the structure of abilities, some theorists regard it as misleading [see Ref. 24].

Even the latter report, however, does not contain the details needed to judge the acceptability of a *g*-based hierarchy. My intention here is to discuss the methodologies employed in analyzing data from cognitive ability tests (principally, factor analysis and the theory of mental tests) and the bases for drawing various possible conclusions about the dimensionality of intelligence. I offer information about the possible existence and composition of *g* (and related concepts) in sufficient depth to give statisticians outside psychometrics, or anybody who is willing to follow a fairly technical presentation, a sense of the basis for experts' debate about the nature of intelligence, or cognitive ability, and the extent to which Herrnstein and Murray's use of the concept of *g* can be accepted.

Is There a *g*? A Brief History

As far as we know, most languages of the world have words that refer to individual variations in cognitive ability—words that can be translated as "wise," "clever," "bright," "stupid," "dull," and so on. From this standpoint, intelligence can be viewed as a universal societal concept, treated in some of the writings of Plato and Aristotle. In modern times it was more directly formulated by the late nineteenth century British polymath Francis Galton, who was concerned to demonstrate that "eminence" (or "genius") in various lines of endeavor, such as law, statesmanship, science, and literature, runs in families and could therefore be attributed in part to heredity.[31] He was one of the first to characterize the distribution of intellectual ability in the population, apparently assuming that intellectual ability has only one dimension. He even attempted to devise tests for such an ability, but his tests consisted mainly of simple tasks involving reaction time, sensory discrimination, and the like. (See Ref. 32 for analyses of some of his results.) Galton also was one of the originators of the concept of correlation between variables, and one of his close associates, Karl Pearson, provided a mathematical formulation for it.[33] (It is interesting that the concept of correlation arose chiefly in the context of studying heredity and individual differences.) Using Pearsonian correlation coefficients applied to small correlation matrices of tests and school marks, Charles Spearman was the first to provide what he regarded as evidence for the notion that variations in intelligence could be summarized in a single number, in effect, by scores on a factor Spearman called *g*, for *general intelligence*. Spearman was one of the founders of the discipline now called *psychometrics*, making seminal contributions both to the theory of mental tests (in his studies of what is now called the *reliability* of measures) and to the technique now called *factor analysis* (see Refs. 2 and 3).

In the meantime, other early psychologists began to investigate mental abilities and their development. The French psychologist Alfred Binet addressed the problem of developing tests for determining children's prospects of succeeding in school.[34] The Binet scale comprised a series of age-scaled tests to determine the mental age of a child and assumed, in effect, that intelligence could be measured as a single dimension of ability. The German psychologist Wilhelm Stern proposed that if the mental age is divided by the child's chronological age (and multiplied by 100), producing the Intelligence Quotient, IQ, this would constitute a measure of the child's intelligence relative to the child's age.[35]

Thus, practically all research by early investigators was based on the assumption that intelligence is essentially unidimensional. But the issue of whether a single number or scale was sufficient to describe variations in cognitive ability has been a preoccupation of psychometricians from the earliest days of mental testing. The answer depended on whether, in fact, all measures of mental ability and achievement could be found to be correlated in such a way as to permit the conclusion that a single factor, g, could account for them. Spearman recognized that a score on a given test could reflect not only a *general* factor of intelligence, but also a *specific* factor of ability unique to that test, s. In fact Spearman's theory has often been called a "two-factor theory," the two factors being g and s. The s-loading of a test, however, would not contribute to the correlation between that test and another test because it pertains only to that test.

The model of intelligence espoused by Spearman can be specified as a factor matrix with n rows and $n + 1$ columns, where n is the number of tests or variables. The first column of such a matrix contains values, $a_i(i = 1, 2, \ldots, j, \ldots, n)$, specifying the weight or loading of a general factor in producing the score on test i. The remaining columns contain weights of the specific factor, s_i, for each test. The correlation of any test i with another test, j, is expected to be $r_{ij} = a_i a_j$. If the total matrix is denoted $\mathbf{F}$, the matrix of correlations is predicted to be $\mathbf{FF'}$, that is, the matrix $\mathbf{F}$ multiplied by its transpose. Obviously the values s_i do not contribute to the correlations. Spearman provided methods for estimating the values of a_i for such a matrix. He also formulated a method (now only of historical interest) for determining whether a matrix of observed correlations could be regarded as properly fitted by a single factor g—the method of *tetrad differences* whereby the expressions $r_{ik}r_{jl} - r_{jk}r_{il}$ for all possible combinations of values of i and j (for rows of the matrix) and k and l (for columns) would be distributed around zero with a variance that depended only on the number of cases, N, in the sample. In work that he summarized in a volume published in 1927, he endeavored to show that many actual matrices of observed correlations among certain types of mental tests could be well fitted by assuming the operation of a single factor g. In fact, the object was to find what kinds of tests displayed correlations that were fitted by the general-factor model; by examining the nature of these tests, he sought to define the nature of the g factor. In this way he claimed to demonstrate that the *g*-factor could be described as calling on three types of mental operations or functions, which he called the apprehension of experience, the eduction of relations, and the eduction of correlates.

Even in the first several decades of the twentieth century, critics pointed out that many matrices of mental test correlations could not be well fitted by the assumption of a single *g*-factor. There was a period around 1927 to 1935 in which Spearman, colleagues, and students further investigated the extent to which a single *g*-factor could account for correlations among mental tests, and eventually Spearman (see Ref. 36, a work published after Spearman's death in 1945) had to acknowledge that several other factors (called *group factors*) of mental ability must be assumed to exist. The American educational psychologist Holzinger, among others who had worked with Spearman, developed more complex models for explaining correlations, including a "bifactor" method that found group factors as well as a general factor.[37] For various reasons, the bifactor method was never widely accepted. It is notable, however, that this method assumed that test scores could be modeled as linear combinations of scores on a general factor and one or more group factors.

The American psychologist L.L. Thurstone, working independently of the Spearman group, further explored the idea that intelligence is composed of multiple factors.[38] He developed methods of factor analysis that were in some ways more sophisticated than those of Spearman and his colleagues. In particular, he developed a reasonably efficient method (the centroid method) of factoring a correlation matrix and a method of rotating the axes of factor matrices to "simple structure" in order to facilitate interpretation of factors. In a large, 57-variable matrix of correlations among a variety of mental ability tests, he claimed to find at least seven interpretable group factors of intelligence but no *g*-factor. This result was disputed by Spearman[39] and Eysenck,[40] who reworked Thurstone's data and found a *g*-factor, as well as a series of group factors. The difference in results can now be explained as being due to technical problems that had not been completely resolved by Thurstone at the time of his 1938 publication. In later publications, Thurstone acknowledged the possible existence of a general factor of intelligence.[41,42] In these publications, Thurstone computed loadings on a "second-order" factor *g* derived from the correlations of "first-order" factors, such as verbal ability, memory ability, spatial ability, and so on. In 1957, Schmid and Leiman published a method that generalized the concept of *factor order* to any number of orders[43]; for example, using this method Carroll found examples of matrices in the factorial literature that required at least three orders of factors to explain the correlations satisfactorily, including a third-order factor that might be interpreted as *g*.[4,44]

The next major effort to explore the idea of multiple factors was that of Guilford in a series of investigations conducted at the University of Southern California, inspired by Guilford's work as the director of psychological testing for the selection of aircrew personnel in World War II.[19,45] In Guilford's 1967 first book-length report of these investigations, he discussed Spearman's theory of *g* and *s*, as well as several hierarchical models (those of Burt[46,47] and Vernon[48]) that incorporated higher order factors, including *g*, but argued against accepting such factors. His objection was based mainly on his opinion that "we have as yet no good way of estimating intercorrelations of factors at any level," because he had "never been able to accept the locations of oblique factor vectors and their angles of separation as the basis for estimating factor intercorrelations" (p. 59). (In effect, he rejected oblique rotation of axes.) As a consequence, Guilford's well-known Structure-of-Intellect (SOI) model contained no provision for a general factor; instead, it could provide for as many as 120 orthogonal factors, organized into a 3-dimensional structure in such a way that each factor was regarded as embodying one of 5 "operations," 4 "contents," and 6 "products." Guilford claimed that the "fundaments" of Spearman's *g*-factor were taken care of in several of the factors incorporated in his SOI model. However, Guilford's SOI model has had its critics.[5,49–52] Guilford (1981) eventually modified his model to provide for higher order factors, but still never acknowledged the existence of a general *g*-factor of intelligence.[53] Carroll reanalyzed many of Guilford's datasets and found numerous higher-order factors, including factors that could be regarded as similar to Spearman's *g* (see Refs. 4 and 44).

Guilford's factorial methods were highly idiosyncratic and were generally not accepted by other investigators of cognitive abilities. These other investigators further developed Thurstone's methods, relying on extraction of factors according to various rules for deciding how many factors to accept, oblique rotation of factorial axes to meet criteria of "simple structure" as described by Thurstone and others, and further factor analysis of correlations among factors to compute higher-order factor loadings. These methods of what has come to be known as *exploratory factor analysis* are described, for example, in treatises on factorial methodology by Harman[54] and Cattell.[55] They are the basis, for example, for factor analyses of test data provided by Hakstian and Cattell[56,57] and for the extensive reanalyses of factorial data (including those of Guilford) by Carroll (see Ref. 4), who made available his final results in the form of a series of hierarchical orthogonalized factor matrices (see Ref. 44). Such hierarchical matrices frequently included test loadings on one or more higher-order factors, including factors that could be considered for interpretation as examples of a general factor *g*.

More recently, techniques of what has come to be known as *confirmatory factor analysis* or *structural factor analysis*[58] were developed by a group headed by the Swedish statistician Karl Jöreskog.[59] These methods, incorporated in widely used computer programs such as LISREL[60] and EQS,[61] make it possible to test for the presence of a putative general factor in a set of data.

Gustafsson, for example, has used these methods to argue for a hierarchical model of cognitive abilities, including a general factor *g* at the third order, several broad abilities at the second order, and a series of narrower abilities at the first order (see Ref. 8–10).

There are fundamental differences between exploratory and confirmatory factor analysis; the methods are actually complementary. The former is concerned with analyzing correlational data to suggest a satisfactory model for describing those data. The latter is concerned with appraising the probability that any given proposed model, even one that might seem quite unlikely, could generate the observed data. Exploratory factor analysis is essentially descriptive, while confirmatory factor analysis appeals to statistical significance testing. Confirmatory factor analysis cannot proceed until one proposes a model that can be tested. One source of such a model is a model produced by exploratory analysis, but this is not the only source; a model could be suggested by a psychological theory.

Partly because of the difficult logistics of administering large batteries of tests or obtaining other types of relevant information (such as ratings) on substantial numbers of individuals, data are not yet available to help decide on the true structure of higher order abilities. The *g*-factors thus far isolated are based largely on batteries of traditional academically oriented paper-and-pencil tests and may not represent all the higher-order variance that might be present in batteries that contain not only traditional tests but also, for example, various physiological measures and tests of "practical intelligence."[62] Further research on the higher-order structure of cognitive abilities is urgently needed.

Concrete Example of a Factorial Dataset

In order to discuss, in limited space, several factorial methods and the considerations involved in identifying a possible general factor, I present a small set of data drawn from a study conducted many years ago[63] with a group of 145 seventh- and eighth-grade children and used as an illustration in a manual on LISREL analysis (Ref. 60, pp. 135–137). The dataset analyzed in this example should not be taken as typical of datasets appropriate for advanced research in the structure of cognitive abilities; the number and variety of variables are very limited. Also, I must omit discussion of many problems that arise in actual analysis.

Virtually all factorial methods start with the analysis of a correlation matrix. There are problems, which I discuss elsewhere, with deciding what types of correlations are most appropriate in given cases.[64] In the example presented here, the correlations were Pearsonian, which were probably most appropriate in view of the fact that the variables were psychological test scores with distributions that can be assumed to approximate the normal curve. The correlations are presented in Table 6.1. It may be noted that this dataset was drawn from a somewhat larger dataset containing 24 variables (treated in a subsequent section). The nine variables were chosen because they had been found to measure three first-order factors, interpreted by the authors as "verbal," "visual," and "speed." Each of these factors was represented by three tests; for technical reasons, it is desirable to have at least three tests for defining a given factor, although many datasets in the factorial literature are found to contain factors defined by only two variables.

At this point it is useful to emphasize that the goal of all factorial methods, as applied to a given correlation matrix **R**, is to specify a factor matrix, **F**, that reproduces, to some satisfactory degree, the given matrix **R**, when **F** is used to generate the predicted matrix $\mathbf{R}_p$. (If **F** is orthogonal, the "reproduced" matrix is $\mathbf{FF'} = \mathbf{R}_p$; readers unfamiliar with matrix algebra can compute any entry in $\mathbf{FF'}$ by computing the sum of the products of corresponding values in the corresponding two rows of the matrix. If **F** is not orthogonal, the computations are more complicated, involving the correlations among the factors.) Of course, if matrix **R** (for n variables) is positive definite, a Principal Component matrix, with n factors, can always be found that will reproduce **R** exactly. But factor analysis seeks to determine a factor matrix with the least number of factors, m, that will satisfactorily reproduce the given **R**. Satisfactory reproduction can be defined in several ways, for example, the extent to which m factors appear to account for the common factor variance (the communalities of the variables), or the extent to which the residuals (the differences between the observed correlations and the reproduced correlations) are close to zero. If confirmatory factor analysis methods are used, satisfactory reproduction can be measured and tested statistically.

However, because there is a virtual infinity of factor matrices that will satisfactorily reproduce the given **R**, reproductive accuracy is not the only

Table 6.1. Correlation matrix for nine psychological variables

	1	2	3	4	5	6	7	8	9
1 Visual perception	1.000								
2 Cubes	.318	1.000							
3 Lozenges	.436	.419	1.000						
4 Paragraph comprehension	.335	.234	.323	1.000					
5 Sentence completion	.304	.157	.283	.722	1.000				
6 Word meaning	.326	.195	.350	.714	.685	1.000			
7 Addition	.116	.057	.056	.203	.246	.170	1.000		
8 Counting dots	.314	.145	.229	.095	.181	.113	.585	1.000	
9 Straight curved capitals	.489	.239	.361	.309	.345	.280	.408	.512	1.000

Source: Ref. 60. $N = 145$.

criterion for arriving at a satisfactory factor matrix. There is also the criterion of *meaningfulness*, embodied (partly) in the principle of *simple structure* outlined by Thurstone, and (in confirmatory factor analysis) in the process of specifying patterns of zero and nonzero weights in the factor matrix (see Refs. 38 and 41). The meaning of this concept can best be illustrated in practical examples, as will be discussed in the following material.

Exploratory Factor Analysis of the Sample Dataset

First, we conduct an exploratory factor analysis of the correlation matrix, using procedures that have been developed over the years since Thurstone and others first outlined them (see Ref. 38). The results are shown in Table 6.2. (For greater clarity, names of matrices are capitalized.)

Initially, we compute a Principal Component Matrix, but only for the purpose of obtaining certain kinds of evidence (the latent roots of the correlation matrix, shown below the Principal Component Matrix) concerning how many factors should be retained as significant, namely, the scree test of Cattell (Ref. 55), as well as the so-called Kaiser–Guttman test.[65] The scree test looks for an "elbow" in the plot of the roots against their order number; the number of factors to retain is where the elbow occurs, before the plotted curve tends to flatten out. The Kaiser–Guttman test states that the number of factors to retain is the number of roots that are greater than one. Neither of these tests is definitive in every case, but here they are consistent in reporting that the number of factors to retain is three.

Also shown with the Principal Component Matrix is a column (h^2) reporting the communalities, or sums of squares of the loadings for each row, attained with three factors. However, these communalities are inflated because they include specific variance as well as common factor variance. For most purposes in exploratory factor analysis, principal component analysis is unacceptable.

Next, we compute a Principal Factor Matrix, using squared multiple correlations (SMCs, shown as the last column in this part of the table) as the initial estimates of communalities, but with computations iterated through sufficient cycles (in this case 12) to ensure that no difference between successive communality estimates exceeds a small quantity, .0005. I find that when principal factor computations are iterated to this strict criterion, the results are generally equivalent (at least within an orthogonal transformation) to a Maximum Likelihood solution, often regarded as the best type of solution, but somewhat more difficult to compute.[66] The resulting Principal Factor Matrix, as well as the communalities, is shown in the table. Also shown in the table are data for evaluating the number of factors to retain according to the Montanelli and Humphreys parallel analysis criterion.[67] That is, three latent roots of the correlation matrix (SMC roots), when lower bound communalities are used as initial estimates of the true communalities, are found to be greater

Table 6.2. Exploratory factor analysis of the correlation matrix in Table 6.1

A. PRINCIPAL COMPONENT MATRIX (first 3 factors)

	Factor			
Test	**1**	**2**	**3**	h^2
1 Visual perception	.648	.101	.379	.574
2 Cubes	.449	−.008	.611	.576
3 Lozenges	.608	−.044	.524	.647
4 Paragraph comprehension	.748	−.454	−.235	.821
5 Sentence completion	.744	−.366	−.341	.804
6 Word meaning	.730	−.459	−.223	.793
7 Addition	.455	.579	−.469	.762
8 Counting dots	.510	.712	−.128	.783
9 Straight-curved caps	.691	.413	.045	.649
			Sum =	6.430
Latent Roots (1–3)	3.582	1.576	1.252	
. . . (4–6)	.710	.538	.433	
. . . (7–9)	.354	.307	.255	

B. PRINCIPAL FACTOR MATRIX (after 12 iterations for communalities)

Test	**1**	**2**	**3**	h^2	**SMCs**
1 Visual perception	.574	.107	.333	.453	.368
2 Cubes	.373	.022	.368	.275	.208
3 Lozenges	.548	.003	.459	.511	.342
4 Paragraph comprehension	.749	−.413	−.164	.759	.631
5 Sentence completion	.735	−.318	−.254	.706	.587
6 Word meaning	.713	−.389	−.130	.677	.584
7 Addition	.426	.484	−.402	.577	.414
8 Counting dots	.486	.663	−.118	.689	.472
9 Straight-curved caps	.628	.353	.080	.525	.443
			Sum =	5.172	
SMC roots (1–4)	3.073	1.079	.620	.030	
Random data roots	.494	.335	.235	.139	
Converged roots	3.192	1.232	.748	.030	

than the corresponding roots (random data roots) of a matrix of correlations of random data. The converged roots are the roots found after the iterations for communalities have converged—slightly larger than the SMC roots.

The Principal Factor Matrix gives the coordinates of the variables on three arbitrary orthogonal axes. It is desirable to rotate these axes so that the

Table 6.2. (continued)

C. VARIMAX-ROTATED MATRIX

Test	1 (Verbal)	2 (Speed)	3 (Visual)	h^2
1 Visual perception	.209	.210	.604	.453
2 Cubes	.105	.037	.512	.275
3 Lozenges	.203	.065	.682	.511
4 Paragraph comprehension	.836	.077	.234	.759
5 Sentence completion	.804	.183	.161	.706
6 Word meaning	.783	.065	.244	.677
7 Addition	.167	.739	−.046	.577
8 Counting dots	−.009	.795	.241	.689
9 Straight-curved caps	.199	.532	.450	.525
Sum of squares	2.122	1.555	1.495	5.172

D. OBLIQUE ROTATED MATRIX (by Tucker-Finkbeiner DAPPFR)

	Factor		
Test	1 Verbal	2 Speed	3 Visual
1 Visual perception	.017	.104	.548
2 Cubes	−.026	−.043	.466
3 Lozenges	.023	−.049	.605
4 Paragraph comprehension	.722	−.037	.009
5 Sentence completion	.692	.080	−.038
6 Word meaning	.676	−.044	.031
7 Addition	.020	.719	.010
8 Counting dots	−.244	.751	.339
9 Straight-curved caps	−.022	.444	.447

E. CORRELATIONS AMONG FACTORS

				(R matrix reproduced by factor matrix below)		
	1	2	3	1	2	3
1 Verbal	1.000	.283	.513	1.000	.231	.519
2 Speed	.283	1.000	.123	.231	1.000	.189
3 Visual	.513	.123	1.000	.519	.189	1.000

Table 6.2. (continued)

F. FACTOR MATRIX FOR CORRELATIONS AMONG FACTORS

	G?
1 Verbal	.796
2 Speed	.290
3 Visual	.652

G. HIERARCHICAL ORTHOGONALIZED FACTOR MATRIX

	Factor				
	1	2	3	4	
Test	**G?**	**Verbal**	**Speed**	**Visual**	h^2
1 Visual perception	.465	.013	.104	.484	.462
2 Cubes	.316	−.019	−.043	.411	.271
3 Lozenges	.467	.017	−.049	.535	.506
4 Paragraph comprehension	.693	.530	−.036	.008	.762
5 Sentence completion	.660	.505	.080	−.034	.698
6 Word meaning	.660	.493	−.044	.028	.681
7 Addition	.245	.015	.718	.009	.576
8 Counting dots	.270	−.163	.750	.299	.751
9 Straight-curved caps	.454	−.016	.443	.395	.559
					5.267
Sum of squares	2.224	.807	1.228	.938	

coordinates will be on more meaningful scales, that is, closely related variables will receive high positive weights (or "loadings") on a given factor, and variables much less related to those closely related variables will receive loadings that are lower, nearer to zero. A commonly used method for doing this is the so-called Varimax rotation developed by Kaiser.[68] Next shown in the table is the Varimax-Rotated Matrix. Varimax rotation preserves the orthogonality of the rotated axes, and thus the communalities can still be computed as the sums of squares of the entries for a given variable; they are identical to those computed from the Principal Factor Matrix. It is evident that the Varimax Matrix shows an interesting pattern in the loadings. The columns are ordered in terms of the decreasing sizes of the sums of squares of the loadings for given columns. The highest loadings for factor 1 are for variables 4–6, all uniquely "verbal" tests. Similarly, the highest loadings for variables 7–9 are for "speed" tests, and the highest loadings for variables 1–3 are for "visual" tests.

Next, it is desirable to determine whether the points defined by the Varimax matrix are structured in such a way as to suggest that the distribution of loadings on a given factor would be improved by further rotation of axes to non-orthogonal, oblique positions, that is, to suggest that the factors are correlated. One way of doing this is to plot the points on orthogonal axes and to inspect their structure. Figure 6.1 shows three pairwise plots of these points (axis 1 against 2, 1 against 3, and 2 against 3). It is evident that the structure of the points would be improved by further, oblique rotation of axes, in the sense that the projections of the points on rotated axes would make for a greater contrast between high and low (or "vanishing") loadings for a given factor.

In the early days of exploratory factor analysis, rotations of axes were made by *graphical rotation*, that is, by inspecting plots like those in Figure 6.1. Because there was a considerable degree of subjectivity in this process, methodologists have sought ways to make the process more objective. One such procedure is the so-called Promax rotation, involving raising Varimax loadings to a certain power while preserving their signs, and then finding the rotations that produce the best fit to the powered loadings.[69] This procedure is still subjective to the extent that it requires a decision as to the power to which loadings should be raised. Instead of the Promax procedure, here I use what I regard as a better procedure developed by Tucker and Finkbeiner[70] but further modified.[71] I have found that this DAPPFR (Direct Artificial Probability Function Rotation) procedure generally achieves satisfactory approximations of the results of the "best" graphical rotations. The procedure can also be justified in terms of its overall accuracy in recovering known, hypothetical factor patterns.

The Oblique Rotation Matrix for the sample data is next shown in Table 6.2. Some of the loadings in each column of the table are highly positive, while most of the remaining loadings are near zero or "vanishing." For example, in column 1, variables 4 (PARCMP), 5 (SENCMP), and 6 (WDMEAN) have

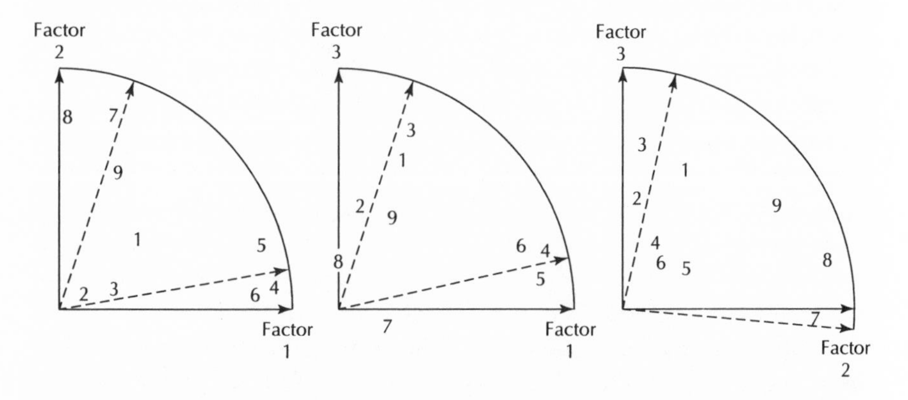

Figure 6.1. Pairwise orthogonal plots of factor loadings from the Varimax-rotated matrix of Table 6.2C. The oblique vectors (broken lines) are approximately those implied by the Oblique Rotated Matrix of Table 6.2D.

loadings .722, .692, and .676, respectively, while the other variables have largely vanishing loadings, suggesting that rotated factor 1 can be interpreted, at least tentatively, as uniquely measuring "verbal ability." Overall, the Oblique Rotation Matrix exhibits good "simple structure."

Because the axes underlying the Oblique Rotated Factor Matrix are not orthogonal, the correlations among the factors are generally other than zero. In the present case the intercorrelations are all positive, as shown in the next matrix in the table. The entries in the Oblique Rotated Factor Matrix can be shown to predict quite accurately the intercorrelations of the variables shown in Table 6.1, but only when the intercorrelations of the factors are taken into account.

The final step we take is to employ the technique developed by Schmid and Leiman to produce a Hierarchical Orthogonalized Factor Matrix, on the presumption that underlying the correlations among factors there is a further variable or factor that might be interpreted as a factor that is general to all the nine tests of this battery (see Ref. 43). Thus, the next step is to factor-analyze the matrix of correlations among factors. In doing so, we follow the same steps that were used in the factor-analysis of the correlations among the nine variables. However, there being only three variables (i.e., the three first-order factors), our analysis must be confined to estimating only one higher order factor. (For n variables, the maximum number of factors that can be defined is given by the expression $[(2n + 1) - \sqrt{8n + 1}]/2$; see Ref. 41, p. 293).

The next matrix in the table is a single-column "second-order" factor matrix that *approximately* reproduces the matrix of correlations among the factors. The "verbal" and "visual" factors have high loadings, while the "speed" factor has a low but still nonvanishing loading.

Finally, we find a Hierarchical Orthogonalized Factor Matrix, which is next shown in the table. In effect, it redistributes the variance to higher- and lower-order factors, with the factor labeled G? being the higher-order factor. This matrix approximately reproduces the original correlation matrix (and also its estimated communalities), but not as well as the previous nine-variable factor matrices because the correlations among the original factors failed to be reproduced exactly by one factor. Nevertheless, factor 1 in this matrix is a general factor in the sense that it has substantial loadings on all nine of the original variables. The pattern of loadings on the remaining factors was preserved; that is, salient loadings for the VERBAL factor occurred for variables 4–6, for the SPEED factor for variables 7–9, and for the VISUAL factor for variables 1–3. The loadings are generally lower than corresponding ones in the Oblique Rotated Matrix because part of the variance for each of the three factors in the original matrix has been absorbed into the "general" factor. Also, communalities computed from this matrix are slightly different from those previously computed, because of the approximation involved in computing the second-order factor matrix. It should be noted, furthermore, that the orthogonalized matrix is only of rank (dimensionality) three, because the loadings on factor 1 are completely predictable from the loadings on the remaining three factors. It is characteristic of the Schmid–Leiman technique that it does not increase the rank of the matrix on which it is based.

Confirmatory Factor Analysis of the Sample Dataset

The results of this "exploratory" factor analysis can be regarded as constituting, in a sense, a set of hypotheses about the structure of the original correlation matrix (as shown in Table 6.1). We can therefore proceed to a "confirmatory" or "structural" factor analysis matrix, using the LISREL program developed by Jöreskog and Sörbom. Several results from this analysis are shown in Table 6.3.

First we test the hypothesis that the structure is similar to what was given by the Hierarchical Orthogonalized Factor Matrix at the end of Table 6.2. The purpose of this analysis (as of any factor analysis using LISREL) is to estimate the weights in a factor matrix that produce the best fit to the sample data, given the specified factor pattern, and then to appraise the likelihood that the actual data could be generated (on a statistical sampling basis for a given number of cases) on the assumption that the estimated weights apply in the population. If the actual data fit this pattern, on the basis of various statistical tests, the specified pattern can be at least tentatively accepted as properly characterizing the data. Even if the data do well fit a specified pattern, there is, however, no guarantee that this is the only pattern that can be shown to fit well. A general strategy in using LISREL in factor analysis is to find the solution that fits the data most satisfactorily while avoiding the temptation to overfit the data.[72]

In the present case, we test the Initial Pattern Matrix (shown at Table 6.3A), which is derived from the Hierarchical Orthogonalized Factor Matrix produced by exploratory factor analysis. The pattern contains values of either 1 or 0 for each entry. The entry "1" means that the value in the matrix to be computed is "freed" to be other than zero (even possibly negative), while the entry "0" means that the value in the matrix to be computed is "fixed" to be zero. The principle by which patterns of loadings are specified in this way is the analogue of the exploratory factor analysis concept of simple structure. In our test, we specify four factors and constrain the correlations among the factors to be zeroes. (An alternative test would have been to specify only three factors (factors 2, 3, and 4 of the initial pattern) and allow the correlations among the factors to be freed. When this test was run, the value of chi-square was found to be 24.86 with 22 degrees of freedom ($p = .304$). The factor correlations were found to be .510 (VISUAL/VERBAL), .169 (VISUAL/SPEED), and .214 (VERBAL/SPEED), similar to those found for our exploratory factor analysis model, and apparently justifying testing a model that incorporates a general factor).

The resulting factor matrix, estimated with a maximum likelihood solution, is shown at Table 6.3B. Apparently it was not possible to solve for the pattern satisfactorily because iterations did not converge, and the estimated communality for variable 5 exceeded unity. Nevertheless, χ^2 was 14.49 with 16 degrees of freedom ($p = .562$), indicating a highly satisfactory fit, and the presence of a "general" factor seemed to be confirmed. The pattern of loadings found for the general factor was roughly the same as that found in the Hierarchical Orthogonalized Factor Matrix in Table 6.2.

In view of the inadequacy of this first pattern, a reduced pattern, whereby estimates for the verbal factor were absorbed into the general factor, was tested. This three-factor pattern, shown at Table 6.3C, was tested under the constraint that the factors were to be orthogonal. The matrix of LISREL estimates is shown at Table 6.3D. Fit of the data to this pattern was excellent, as shown by the χ^2 test [$\chi^2(19) = 14.70, p = .742$]. Unfortunately, the pattern tested was somewhat arbitrary, except that the fit was better than when the pattern tested combined *g* with the visual factor [$\chi^2(21) = 25.90, p < .210$] or with the speed factor [$\chi^2(21) = 58.89, p < .001$] instead of the verbal factor. Without further investigation I cannot be sure why the convergence problem arose; possibly it was because there were not enough variables to define a general factor separately from the three first-order factors. In any event, presence of a valid "general" factor is justified by the fact that all loadings on this factor are substantial, most of them highly significant. (I place the word *general* in quotes because this factor's identification as a truly general factor is only provisional at this point.) Moreover, if the values for variables 1, 2, 3, 7, 8, and 9 on this factor are fixed at zero (thus, in effect, eliminating the *g*-factor), the fit of the pattern to the data is significantly (and drastically) degraded, as detailed in a final note to Table 6.3.

With the presentation of two factorial analyses of a small set of actual data, one "exploratory" and the other "confirmatory," we can proceed to consideration of what the results might mean in terms of a supposed general factor.

Justifying a General Factor

What would a "general factor of intelligence" be? How could we recognize one? What would it mean?

Basically, there are two ways of defining a general factor: (1) conceptually and (2) operationally. A conceptual definition would draw on what theory and research experience might indicate about the nature of a general factor. An operational definition would provide criteria for deciding whether a general

Table 6.3. Confirmatory factor analysis of the correlation matrix in Table 6.1

A. INITIAL PATTERN MATRIX

Test	LISREL factor			
	G?	Verbal	Speed	Visual
1 Visual perception	1	0	0	1
2 Cubes	1	0	0	1
3 Lozenges	1	0	0	1
4 Paragraph comprehension	1	1	0	0
5 Sentence completion	1	1	0	0
6 Word meaning	1	1	0	0
7 Addition	1	0	1	0
8 Counting dots	1	0	1	1
9 Straight-curved caps	1	0	1	1

B. LISREL ESTIMATES FROM THE ABOVE PATTERN MATRIX

Test	LISREL factor				
	G?	Verbal	Speed	Visual	h^2
1 Visual prception	.388*	---	---	.612 **	.525
2 Cubes	.250*	---	---	.403 **	.225
3 Lozenges	.384*	---	---	.519 **	.416
4 Paragraph comprehension	.874 **	.045	---	---	.766
5 Sentence completion	.787 **	.746	---	---	1.176
6 Word meaning	.814 **	.059	---	---	.666
7 Addition	.240*	---	.802 **	---	.701
8 Counting dots	.141	---	.688 **	.376 **	.634
9 Straight-curved caps	.367*	---	.400 **	.494 **	.538

$\chi^2(16) = 14.49$; $p = .562$. Goodness of Fit Index = .977. Adjusted Goodness of Fit Index = .937. Root Mean Square Residual = .031. *Warning*: Number of Iterations exceeded 87. *t*-values for values in column 2 (Verbal) were vanishing. By *t*-test, **p* < .05; ***p* < .01. The LISREL factors are orthogonal.

factor can be shown to exist in a particular set of data, or more generally, by inference, across all conceivable datasets in the cognitive domain.

In formulating either conceptual or operational definitions, it is necessary to adopt several premises: (1) *g* must be defined in the context of the possibility of specifying a large space of "cognitive variables," that is, variables signifying individual differences in performing cognitive tasks. (2) It must be possible to measure these variables with significant reliability. (3) For consistency, the variables must be scored in such a way that numerically higher numerical

Table 6.3. (continued)

C. REVISED PATTERN MATRIX

Test	LISREL factor		
	G	**Speed**	**Visual**
1 Visual perception	1	0	1
2 Cubes	1	0	1
3 Lozenges	1	0	1
4 Paragraph comprehension	1	0	0
5 Sentence completion	1	0	0
6 Word meaning	1	0	0
7 Addition	1	1	0
8 Counting docs	1	1	1
9 Straight-curved caps	1	1	1

D. LISREL ESTIMATES FROM THE ABOVE REVISED PATTERN

Test	LISREL factor			
	G	**Speed**	**Visual**	h^2
1 Visual prception	.382 **	---	.615 **	.525
2 Cubes	.240 **	---	.411 **	.227
3 Lozenges	.376 **	---	.525 **	.417
4 Paragraph comprehension	.871 **	---	---	.759
5 Sentence completion	.829 **	---	---	.687
6 Word meaning	.823 **	---	---	.677
7 Addition	.244*	.796 **	---	.693
8 Counting dots	.149	.690 **	.369 **	.633
9 Straight-curved caps	.368 **	.400 **	.492 **	.538

$\chi^2(19) = 14.70$ $p = .742$. Goodness of Fit Index = .977. Adjusted Goodness of Fit Index = .946. Root Mean Square Residual = .030. By *t*-test: * $p < .05$; ** $p < .01$. The LISREL factors are orthogonal. When variables 1–3 and 7–9 on factor G are fixed to zero, $\chi^2(25) = 56.53$, $p < .0005$, thus χ^2 for the difference is $\chi^2(6) = 41.53$, $p < .0005$.

values indicate greater ability. (4) These variables are to be analyzed with an appropriate procedure of factor analysis as applied to representive samples of the populations to which the results apply. In particular, the populations must include individuals at different points in wide ranges of whatever ability is assumed to be defined as a general factor of intelligence. Lack of a sufficient range of ability in the sample could lead to failure to confirm a general factor, even when it actually exists.

It is necessary to reject two proposals that have sometimes been made to support the notion of a general intelligence factor: (1) *The finding of uniformly positive intercorrelations of cognitive variables as an indicator of the presence of a general factor.* The mere fact that cognitive variables are positively correlated does not adequately validate the presence of a *single* general factor. It might indicate the presence of multiple general factors. (2) *General factors identified as first principal components or principal factors in the factor analysis of a dataset.* The first principal component, or eigenvector, is that vector that produces (under appropriate constraints) the maximal variance obtainable from a linear combination of the variables. The major problem with this proposal is that even if no general factor underlies the variables, the size of the first eigenroot is necessarily still relatively large, as compared with other eigenroots. The first principal component derived from a matrix of randomly generated correlations is necessarily larger than the remaining components. The first principal component is therefore not a valid indicator of the presence of a general factor. The same applies to the first principal factor (computed in such a way as to estimate the communalities of the variables).

In view of these considerations, only the complete factor analysis (either exploratory or confirmatory, or both) of a set of variables should be used to judge the presence of a general factor.[73] We now give tentative conceptual and operational definitions of a general factor.

A *conceptual* definition: A general factor of intelligence is one that would exhibit significantly positive loadings on all, or nearly all, the individual difference variables that could be selected or devised in the total domain of cognitive abilities, whether or not these loadings are accompanied by loadings on lower-order factors in this domain. Further, it is required that the general factor is independent of any lower-order factors and constitutes a "true ability," as could be defined by relevant operations. (The meaning of this statement will be discussed later.) It can be seen that this conceptual definition of a general factor requires a theory or knowledge of what the cognitive domain consists of, and of its scope. Our present knowledge of the cognitive domain is surely limited. In my book on cognitive abilities, I outlined our present knowledge of this domain, treating it in subdomains such as language and communication, reasoning, memory and learning, visual and auditory perception, idea production, and cognitive speed (see Ref. 4). I posited a hierarchical, three-stratum structure of abilities, with a general factor at the apex or third stratum, several broad abilities at the second stratum, and a series of narrower abilities at the first stratum. I also posited that all these abilities can be expressed in such a way that they are independent of each other, although phenotypically they would often be found to be correlated (because any phenotypic measurement could include variance from different strata). Nevertheless, I would suppose that further research could readily lead to further important domains or to modifications in the total structure.

An *operational* definition: If the dataset is found to measure only one factor (by an appropriate analysis), to the extent that all variables in the dataset exhibit significantly positive loadings, it is a general factor, but only if the variables are all drawn from different parts of the cognitive domain. This definition would not apply, for example, if the variables are drawn from a limited portion of the cognitive domain, because they might then serve to define only a single first-stratum or second-stratum factor. If, however, it is found that the correlations in the dataset must be described with more than one factor, a general factor is one that exhibits significantly positive loadings on at least a large majority, and preferably all, of the variables in the dataset, regardless of the loadings on other factors that may be identified in the dataset. Ideally, there should be at least three lower-order factors (at stratum II) in order adequately to define a general factor at stratum III on the basis of the correlations among the lower-order stratum II factors. In any case, it would be desirable to show also that a general factor so identified constitutes a true ability, independent of lower-order factors, rather than being merely a measure of associations among those lower-order factors that might be due, for example, to the effects of common learnings.

In terms of this operational definition, it could be said that a general factor is to be found in the small set of data treated in Tables 6.2 and 6.3, at least to the extent that a general factor is found to have significant loadings on all or most of the nine variables in the dataset, in which three lower-order factors from different parts of the cognitive domain are also found. At the very least, the general factor represents covariance in the dataset that is not accounted for by the lower-order factors or variables. But from the correlational data as they exist, it is impossible to tell whether the general factor represents a "true ability."

The remaining problem, both for a conceptual definition and an operational definition, is to demonstrate that a "general" factor is a "true ability." Factor analysis, as such, cannot be used to make any such demonstration because it is concerned only with analyzing covariances among variables. To demonstrate that factors are measures of ability, one must appeal to other techniques or models offered by theories of measurement. In the case of test scores that are factor analyzed, item response theory (IRT) provides some help in the sense that it offers a testable model for the relation of item responses to a continuum of ability.[74] Suppose, however, that we are interested in demonstrating that a first-order factor, derived from a matrix of individuals' scores on a series of tests that all have significant loadings on that factor—and only on that factor—represents a true ability as described by IRT. Presumably this could be done by analyzing, using IRT, the pool of items represented by all the tests that load on the factor. I am not aware that this problem has ever been addressed seriously by psychometricians, but conventional wisdom and logic would indicate that this suggested course of action could be practicable and successful. The procedure would become problematic when a factor embraces tests that have multiple sources of variance, as would often happen, particularly if one is trying to demonstrate that a second-order factor, or worse still, a third-order factor, is a "true ability" independent of lower-order factors or variables, rather than, say, merely covariance with extraneous, non-ability variables such as gender or amount of education. This is partly because items selected from measures of a higher-order factor are likely to entail confounding with variance from lower-order variables. It should be remembered that the Schmid–Leiman technique, producing loadings on independent higher-order factors, does not produce a rank of the hierarchical matrix that is greater than the rank of the lower-order matrices on which it is based. The problem is less serious in testing the presence of a general factor with confirmatory techniques because the estimated factor matrix has no constraints on its rank similar to those entailed in the Schmid–Leiman technique used in exploratory factor analysis. At this point, I can only state my belief that the problem has not been adequately studied, and that satisfactory conclusions cannot be drawn at this time. Later, however, I will voice a proposal whereby something like a general factor can be taken to characterize the population.

The reader will have observed that our attribution of "generality" to a factor would have to depend upon demonstrations that the factor applies to the "total cognitive domain." This entails the possibly insoluble problem of defining what the "total cognitive domain" is. Even upon tentatively defining such a domain and finding some "general" factor that applies to it, it would always be possible that someone could propose some other part of this domain to which the factor would not apply, thus destroying the "generality" of this factor. *Generality* is thus a relative term.

Evidence from Selected Datasets

In my book on human cognitive abilities (Ref. 4; see in particular Chapter 15 and Table 15.4), I was able to identify and analyze a considerable number of datasets in the factorial literature, spanning major portions of the cognitive domain, for which it was possible to isolate a presumed general factor by exploratory factor analysis techniques. These "general factors" were necessarily somewhat different because they were based on somewhat different assemblages of variables. It must be recognized that the composition of an operationally defined general factor can be somewhat skewed or biased by the particular selection of variables used for the dataset. For example, many datasets emphasize multiple-choice, paper-and-pencil tests involving abilities in dealing with printed language, while others emphasize "nonverbal" tests, even though both verbal and nonverbal tests may be present in the dataset. In the former type of study, the general factor may be biased toward verbal abilities, while in the latter type of study, it may be biased toward spatial or reasoning abilities. Nevertheless, the presumption is that underlying either type of "general" factor there could be a truly general latent trait or factor. If it were possible to compute correlations among different types of general factors, one would expect that the intercorrelations would be universally high.

In view of the limited space available, for present purposes it seems reasonable to examine just three datasets (out of several dozen available) that present plausible information about a possible general factor. In contrast to the exploratory factor analyses performed in preparing my book, however, I present confirmatory factor analyses of these datasets in order to avoid the possible pitfalls of exploratory analyses.

The Holzinger and Swineford Dataset

First, it is useful to look at the complete dataset, analyzed by Holzinger and Swineford (see Ref. 63), from which the small nine-variable dataset that was examined in Tables 6.1 to 6.3 was drawn. Table 6.4 presents a LISREL-tested factor matrix for that dataset. I arrived at this by the following steps: (1) Exploratory factor analysis of the original correlation matrix yielded a five-factor rotated matrix with correlated factors. (2) Factor correlations were factor-analyzed to yield one second-order factor, and the data were then further transformed to a six-factor hierarchical orthogonalized factor matrix by the Schmid–Leiman technique (see Ref. 43). (3) The salient loadings in the latter matrix yielded an initial pattern for LISREL analysis, giving LISREL estimates yielding $\chi^2(227) = 333.42, p < .0005$, indicating that the initial pattern matrix provided an unsatisfactory fit to the data. (4) Modification indices provided by the LISREL program were used selectively to modify the pattern in a series of runs until the results shown in Table 6.4 were attained, with $\chi^2(214) = 234.94$, $p = .156$, considered to indicate a satisfactory fit to the data. (Except in the final stages of these runs, suggested changes in the pattern matrix were accepted only if the estimated change would yield a positive LISREL estimate.)

The LISREL analysis confirmed the presence of a second-order general factor, labeled 2G in the table. The factor is general, that is, in that it has significant ($p < .01$ by t-test) loadings for all psychological test variables in the dataset. These loadings are accompanied by one or more loadings on first-order factors found in this dataset, covering verbal, spatial, reasoning, cognitive speed, and memory abilities, abilities that may be regarded as comprising the majority of the abilities tested by standard intelligence tests. It is of interest that one of the variables (variable 19) was found to load *only* on the general

Table 6.4. LISREL-estimated structure of 24 psychological tests studied by Holzinger and Swineford (1939): $N = 145$

	LISREL factor						
	1	2	3	4	5	6	
	2G	Verb.	Space/	Speed	Mem.	?	
Test			Reason.				h^2
1 VisPer	.466	—	.546	—	—	.184	.551
2 Cubes	.320	—	.336	—	—	—	.215
3 PapFmBd	.328	—	.469	-.168*	—	—	.361
4 Flags	.394	—	.428	—	—	—	.339
5 GenInfo	.656	.459	—	—	—	—	.641
6 ParaComp	.671	.472	—	—	—	—	.673
7 SenComp	.614	.602	—	—	—	—	.739
8 Word Class	.569	.385	.228	—	—	—	.523
9 WordMean	.726	.442	—	-.127*	—	—	.736
10 Addition	.350	—	—	.795	—	—	.754
11 Code	.461	—	—	.412	.252	.456^e	.659
12 CountDots	.270	—	.318	.641	—	—	.572
13 S-cCaps	.353	.190*	.474	.381	—	.456^e	.733
14 WordRec	.401	—	—	—	.466	—	.378
15 NumRec	.355	—	—	—	.434	—	.314
16 FigureRec	.412	—	.325	—	.435	—	.469
17 ObNum	.413	—	—	.180*	.432	—	.391
18 NumFig	.401	—	.247	.252	.292	—	.371
19 FigWord	.451	—	—	—	—	—	.204
20 Deduction	.611	—	.262	—	—	-.212*	.486
21 NumPuzz	.508	—	.287	.296	—	—	.423
22 ProbReas	.658	—	.155*	—	—	—	.456
23 SerComp	.665	—	.332	—	—	—	.553
24 ArithProb	.639	—	—	.376	—	—	.549

$\chi^2(214) = 234.94, p = .156$, Goodness of Fit Index = .890; Adjusted Goodness of Fit Index = .846; Root Mean Square Residual = .045. All values indicated as "—" were fixed to zero, ^{e} Values equated within column. * By t-test, $p > .01$. For all other nonzero values, $p < .01$. Descriptions of variables are to be found in Ref. 63. The LISREL factors are orthogonal.

factor. Also, note that because the six factors are orthogonal and independently estimated, the rank of the matrix is six; thus, one cannot complain that the general factor is merely a linear composite of the other factors. However, this dataset is limited to two orders of factors. The computed general factor may, therefore, cover several stratum-II factors as well as possibly a true general factor.

The Hakstian and Cattell Dataset

Hakstian and Cattell studied correlations among 57 mental ability tests selected to represent 19 of the most frequently reported primary, first-stratum factors that had been demonstrated, or at least proposed at the time of their study (see Ref. 56). Scores on these tests, which they regarded as spanning the full range of the ability domain, were obtained for a highly heterogeneous sample of 343 young adults and were factor-analyzed, yielding 19 correlated primary abilities. Their publication presented only the "oblique primary-factor pattern" achieved through use of Cattell's factor rotation procedures, and the correlations among the 19 factors thereby obtained. (The original 57-variable correlation matrix was not available in published form.) I analyzed their factor correlation matrix, continuing the analysis to the third order, finding a general factor and five second-order factors (see Ref. 4). In a subsequent analysis, I was able to find eight second-order factors as well as a general factor. These results were then subjected to the Schmid–Leiman procedure to produce a hierarchical orthogonalized factor matrix for the second and third orders. The latter matrix was the basis for establishing a conservative initial pattern matrix for confirmatory factor analysis, which when tested yielded a value of χ^2 equal to 459.82 with 129 degrees of freedom, $p < .0005$. I then proceeded through a number of LISREL runs, using modification indices as guides in the attempt to find a more satisfactory fit for the data. I was unable to find a solution that yielded a truly satisfactory fit; possibly further work (e.g., by positing a different number of second-order factors) could produce a better solution. However, the value of χ^2 finally achieved, 219.48 with 109 degrees of freedom, was only about twice its number of degrees of freedom; some authorities would regard this as indicating satisfactory fit. The final estimated structure, shown in Table 6.5, produced a third-order general factor that had highly significant loadings for nearly all variables in the study.

The Wothke et al. Dataset

Another large dataset, involving all 10 subtests of the Armed Services Vocational Aptitude Battery (ASVAB) and 46 tests (presumably measuring 23 different cognitive factors) from the Educational Testing Service Kit of Factor-Referenced Tests,[75] was investigated by Wothke et al.,[76] principally in order to determine whether the ASVAB measures a sufficient range of domains of cognitive ability. Sets of tests from this battery were administered to 6,900 Air Force recruits in small groups (an average of about 230 each) in such a way that it was possible to estimate values for the total correlation matrix. (However, scores on all ASVAB subtests were available for all subjects.) The sample was probably somewhat restricted in range, both at the bottom and at the top, because the recruits had already been selected on the ASVAB, and there were

Table 6.5. LISREL-estimated structure of 19 first-order factors studied by Hakstian and Cattell: $N = 343$

First-order factor	LISREL factor									
	3*g*?	**2F**	**2R**	**2C**	**2Y**	**2V?**	**2B**	**Esth.J**	**2S?**	h^2
V:	.56	—	.27	.30	—	—	.42	—	—	.64
N:	.48	.43	.20	.31	—	—	.16	—	—	.55
SR:	.41	.61	—	—	—	.18	—	—	—	.57
P:	.39	.43	—	—	—	—	—	—	—	.34
CS:	.65	.28	—	.28	—	—	—	.26	.44	.82
I:	.66	.47	—	—	—	—	—	—	—	.66
MA:	.60	—	—	—	.41	—	—	—	—	.53
MK:	.11*	.30	—	—	—	—	.80	.22	—	.82
CF:	.50	.35	—	—	—	.34	—	—	.16	.52
MS:	.35	.33	—	.12	—	—	—	—	—	.23
SG:	.47	—	—	.91	—	—	—	—	—	1.05
EJ:	.73	—	—	—	-.29	—	—	.30	—	.71
MM:	.62	—	—	—	.49	—	—	—	—	.63
FO:	.70	.09*	.27	—	—	—	—	-.34	.12*	.70
FI:	.53	—	.54	.14	—	—	—	—	—	.60
WF:	.51	—	.21	.38	—	—	—	—	.38	.60
FOALT:	.44	—	.33	—	—	—	.26	—	.30	.47
AI:	.37	—	—	—	.30	.26	—	—	—	.29
REPDR:	.57	—	—	—	—	.74	—	—	—	.88

$\chi^2(109) = 219.48, p < .0005$. Goodness of Fit Index = .942. Adjusted Goodness of Fit Index = .89. Root Mean Square Residual = .040. All values indicated as "—" were fixed to zero. * By *t*-test, $p > .05$; for all other nonzero values, $p < .01$. 2F, fluid intelligence; 2R, broad retrieval ability; 2C, fluid intelligence; 2Y, broad memory ability; 2V, broad visual ability; 2B, broad information ability; Esth. J: esthetic judgement; 2S, broad cognitive speed; LISREL factors are orthogonal. V, verbal, N, numerical; SR, space; P, perceptual speed; CS, closure speed; I, induction; MA, associative memory; MK, mechanical knowledge; CF, flexibility of closure; MS, memory span; SG, spelling; EJ, esthetic judgment; MM, meaningful memory; FO, originality; FI, ideational fluency; WF, word fluency; FOALT, originality (Alt.); AI, aiming; REPDR, representaional drawing.

fewer college-trained members than would be expected in the general population. Nevertheless, this dataset has the advantage that it better applies to the adult population as considered by Herrnstein and Murray as opposed to the younger populations involved in the other datasets considered here. At the same time the dataset involves several problems of design, including the fact that only two tests of each first-order factor were included from the ETS kit, thus likely limiting the number of factors that could be recognized by the analysis. I reanalyzed these data, extracting more factors than the original investigators did, but using more adequate rotational procedures, and found eleven first-order factors, four second-order factors, and one third-order factor (see Ref. 44). I characterized the one third-order factor as a "general factor" on the basis of its content—involvement of verbal and reasoning processes—even though some of the first-order and second-order factors had vanishing loadings on it. I have now conducted a confirmatory factor analysis at the second and third orders, using the second- and third-order portion of the hierarchical orthogonalized factor matrix (from the exploratory analysis) as the basis for an initial pattern matrix and testing the LISREL pattern against the matrix of correlations among the eleven first-order factors. As with the other datasets, I conducted what Jöreskog and Sörbom term a *specification search* to find a plausible well-fitting model. The results are presented in Table 6.6. (For lack of time and resources, I deemed it impractical even to attempt conducting a confirmatory factor analysis for the original 56-variable matrix. I have not investigated whether the LISREL program that I use with a personal computer would be sufficient, in terms of memory requirements, to conduct such an analysis.)

As may be seen, LISREL confirmed a model with a general factor that had positive and highly significant loadings on nearly all the first-order factors identified in this dataset. Ordered by magnitude of positive loading, these included quantitative reasoning (RQ), memory span (MS), ideational fluency (FI), associative memory (MA), perceptual speed (P), word fluency (WF), and lexical knowledge (VL), covering a broad span of the cognitive domain. Actually, at the first order the factor somewhat arbitrarily named quantitative

Table 6.6. LISREL-estimated structure of eleven first-order factors in data studied by Wolthke et al. (Ref. 76) and reanalyzed by Carroll (Ref. 44): $N = 230$

	LISREL factor				
	1	2	3	4	
First-order factor	*g*?	2F?	2R?	?	h^2
SR: *Space*	.196	.705**	—	.364**	.666
VL: *Lexical Knowledge*	.288**	.302**	—	—	.174
P: *Perceptual Speed*	.386**	– .324**	.227*	—	.312
FI: *Ideational Fluency*	.482**	.166	.384**	—	.403
RQ: *Quantitative Reasoning*	.673**	.489**	—	—	.693
MA: *Associative Memory*	.393**	—	—	.197*	.194
MS: *Memory Span*	.483**	—	—	—	.233
VU: *Verbal Closure*	.137	—	—	.313**	.117
FF: *Figural Fluency*	– .181*	—	.728*	—	.563
MK: *Technical Knowledge*	– .020	.604**	—	—	.365
WF: *Word Fluency*	.307**	—	—	.651**	.518

$\chi^2(13) = 44.39, p = .056$. Goodness of Fit Index = .956. Adjusted Goodness of Fit Index = .925. Root Mean Square Residual = .042. All values indicated as "—" were fixed to zero. By *t*-test: $*p < .05$; $**p < .01$ 2F, Fluid intelligence; 2R, broad retrieval ability.

reasoning covered not only a number of mathematical tests but also several verbal tests, such as the Paragraph Comprehension test of the ASVAB and the inductive Letter Sets test of the ETS kit.

The small loadings on *g* are also relevant to interpreting this third-order factor. The loading of the factor figural fluency (FF) was −.181, but as I characterize this factor in my book (Ref. 4, p. 436), it may be "nothing more than speed of doodling." It cannot be regarded as one requiring any substantial degree of mental ability. Somewhat unexpectedly, the technical knowledge (MK) factor had a vanishing loading of only −.020 on *g*. At the first order, this covers such ASVAB tests as Auto/Shop Information, Electronics Information, and Mechanical Comprehension, measuring specialized information that is apparently acquired by only a small proportion of the population, unrelated to their status on the *g* factor. Finally, note that the spatial relations (SR) factor had a low, insignificant loading on *g*; in contrast, it had a high loading (.705) on the second-order factor, labeled, somewhat hesitatingly, 2F (to suggest its possible identification as the fluid intelligence factor G_f). (The evidence from this analysis, in fact, suggests that factor *g* is *not* identical to factor G_f, as suggested by Gustafsson in Ref. 8.)

Conclusions and Final Comment

Evidence assembled here and elsewhere suggests that a third-stratum factor *g* can be *operationally* identified in psychological datasets that are sufficiently well designed to permit such an identification (Ref. 4 and 44). In that it tends to cover the major portion of the domain of cognitive abilities, it tends also to conform to a *conceptual* definition of *g*, corresponding to the statement cited earlier from the *Wall Street Journal* which was expounded by Gottfredson[77]: "[i]ntelligence is a very general mental capability that, among other things, involves the ability to reason, plan, solve problems, think abstractly, comprehend complex ideas, learn quickly and learn from experience" (Ref. 29).

Furthermore, the evidence presented here tends to substantiate the claim I have made elsewhere (see Refs. 4 and 28) that a general factor contributes a substantial portion of the variance of most cognitive variables. Omitting a general factor in factor patterns submitted to confirmatory factor analysis procedures drastically downgrades the fit of the pattern to the data.

What has not been adequately demonstrated and proven at this time is that *g* is a "true ability" independent of more specific cognitive abilities defined by various types of psychological tests and observations. An operationally defined *g* could be partly a function of non-ability variance arising from societal and environmental effects, such as differential amounts, patterns, and qualities of parenting and of schooling, migratory trends, and even lead pollution in the air and elsewhere. Nevertheless, the fact that wide variations in general ability occur even among individuals who appear to be approximately equal in their experiences and backgrounds suggests that environmental and societal effects cannot be the sole sources of variations in *g*. Genetic factors are other sources that are being widely investigated, although these investigations thus far may not have adequately attended to the different traits that may underlie *g*.[78]

Horn may be correct in concluding that the existence of *g* cannot be properly supported in factorial investigations, although the confirmatory factor analyses presented here would seem to contradict Horn's view (see Refs. 21–23). I would draw attention, however, to the typically substantial correlations between fluid and crystallized intelligences, suggesting that a latent trait of intelligence underlies them in some way—call it *g* or what you will. Even if such a trait does not exist, the linear combination of measures of fluid and crystallized intelligences should nevertheless yield an index of overall mental ability that will be highly correlated with standard measures of intelligence such as the IQ and that will produce wide variance in the population.

For the time being, therefore, my conclusion is that however one views the analyses and interpretations put forth in their book, Herrnstein and Murray were justified in presuming that "[t]here is such a thing as a general factor of cognitive ability on which human beings differ" (p. 22).

References

1. Herrnstein, R.J., and Murray, C. (1994), *The Bell Curve: Intelligence and Class Structure in American Life*, The Free Press, New York.
2. Spearman, C. (1904). "'General Intelligence', Objectively Determined and Measured," *American Journal of Psychology*, 15, 201–293.
3. Spearman, C. (1927), *The Abilities of Man: Their Nature and Measurement*, Macmillan, New York.
4. Carroll, J.B. (1993a), *Human Cognitive Abilities: A Survey of Factor-Analytic Studies*, Cambridge University Press, New York.
5. Eysenck, H.J. (1967), "Intellectual Assessment: A Theoretical and Experimental Approach," *British Journal of Educational Psychology*, 37, 81–98.
6. Eysenck, H.J. (Ed.) (1982), *A Model for Intelligence*, Springer-Verlag, Berlin, Germany.
7. Eysenck, H.J. (1988), Editorial: "The Concept of 'Intelligence': Useful or Useless?" *Intelligence*, 12, 1–16.
8. Gustafsson, J.-E. (1984), "A Unifying Model for the Structure of Intellectual Abilities," *Intelligence*, 8, 179–203.
9. Gustafsson, J.-E. (1988), "Hierarchical Models of Individual Differences in Cognitive Abilities," in R.J. Sternberg (Ed.), *Advances in the Psychology of Human Intelligence*, Vol. 4, Erlbaum, Hillsdale, NJ, pp. 35–71.
10. Gustafsson, J.-E. (1989), "Broad and Narrow Abilities in Research on Learning and Instruction," in R. Kanfer, P.L. Ackerman, and R. Cudeck (Eds.), *Abilities,*

Motivation, and Methodology: The Minnesota Symposium on Learning and Individual Differences, Erlbaum, Hillsdale, NJ, pp. 203–237.

11. Humphreys, L.G. (1979), "The Construct of General Intelligence," *Intelligence*, 3, 105–120.
12. Humphreys, L.G. (1982), "The Hierarchial Factor Model and General Intelligence," in N. Hirschberg and L. G. Humphreys (Eds.), *Multivariate Applications in the Social Sciences*, Erlbaum, Hillsdale, NJ., pp. 223–239.
13. Jensen, A.R. (1980), *Bias in Mental Testing*, The Free Press, New York.
14. Gottfredson, L.S. (1986), "Societal Consequences of the g Factor in Employment," *Journal of Vocational Behavior*, 29, 379–410.
15. Hunt, E. (1980), "Intelligence as an Information-Processing Concept," *British Journal of Psychology*, 71, 449–474.
16. Plomin, R., and DeFries, J.C. (1980), "Genetics and Intelligence: Recent Data," *Intelligence*, 4, 15–24.
17. Snyderman, M., and Rothman, S. (1988), *The IQ Controversy, The Media and Public Policy*," Transaction Books, New Brunswick, NJ.
18. Sternberg, R.J. (1995), "For Whom the Bell Curve Tolls: A Review of *The Bell Curve*," *Psychological Science*, 6, 257–261.
19. Guilford, J.P. (1967), *The Nature of Human Intelligence*, McGraw-Hill, New York.
20. Guilford, J.P. (1985), "The Structure-of-Intellect Model," in B.B. Wolman (Ed.), *Handbook of Intelligence: Theories, Measurements, and Applications*, Wiley, New York, pp. 225–266.
21. Horn, J.L. (1985), "Remodelling Old Models of Intelligence," in B.B. Wolman (Ed.), *Handbook of Intelligence: Theories, Measurements, and Applications*, Wiley, New York, pp. 267–300.
22. Horn, J.L. (1988), "Thinking About Human Abilities," in J.R. Nesselroade and R.B. Cattell (Eds.), *Handbook of Multivariate Experimental Psychology*, 2nd ed., Plenum, New York, pp. 645–685.
23. Horn, J.L., and Noll, J. (1994), "A System for Understanding Cognitive Capabilities: A Theory and the Evidence on Which it Is Based," in D.K. Detterman (Ed.), *Current Topics in Human Intelligence*, Vol. 4, *Theories of Intelligence*, Ablex, Norwood, NJ., pp. 151–203.
24. Ceci, S.J. (1990), *On Intelligence . . . More or Less: A Bioecological Treatise on Intellectual Development*, Prentice-Hall, Englewood Cliffs, NJ.
25. Gardner, H. (1983), *Frames of Mind: The Theory of Multiple Intelligences*, Basic Books, New York.
26. Gould, S.J. (1994, November 28), "Curveball," [Review of R.J. Herrnstein and C. Murray, *The Bell Curve* (The Free Press, New York, 1994)] *The New Yorker*, pp. 139–149,
27. Gould, S.J. (1981), *The Mismeasure of Man*, Norton, New York.
28. Carroll, J.B. (1995a), Editorial: "Reflections on Stephen Jay Gould's *The Mismeasure of Man* (1981): A Retrospective Review," *Intelligence*, 21, 121–134.
29. *Wall Street Journal*, December 13, 1994, p. A18.
30. Neisser, U., Boodoo, G., Bouchard, T.J., Jr., Boykin, A.W., Brody, N., Ceci, S.J., Halpern, D.F., Loehlin, J.C., Perloff, R., Sternberg, R.J., and Urbina, S. (1996), "Intelligence: Knowns and Unknowns," *American Psychologist*, 51, 77–101.
31. Galton, F. (1869), *Hereditary Genius: An Inquiry Into its Laws and Consequences*, Collins, London, England.
32. Johnson, R.C., McClearn, G.E., Yuen, S., Nagoshi, C.T., Ahern, F.M., and Cole, R.E. (1985), "Galton's Data a Century Later," *American Psychologist*, 40, 875–892.

33. Pearson, K. (1895), "Regression, Heredity and Panmixia (Abstract)," *Proceedings of the Royal Society*, 59, 69–71.
34. Binet, A., and Simon, T. (1905), "Méthodes Nouvelles pour le Diagnostic du niveau Intellectuel des Anormaux" [New Methods for Diagnosing the Intellectual Level of Abnormal Persons], *Année Psychologique*, 11, 191–336.
35. Stern, W. (1916), "Der Intelligenz Quotient als Masz der kindlichen Intelligenz, inbesondere der Unternormal" [The Intelligence Quotient as Measure of Children's Intelligence, Especially for those Below Normal]," *Zeitschrift für angewandte Psychologie*, 11, 1–18.
36. Spearman, C., and Wynn Jones, L. (1950), *Human Ability: A Continuation of "The Abilities of Man,"* Macmillan, London, England.
37. Holzinger, K.J., and Harman, H.H. (1938), "Comparison of Two Factorial Analyses," *Psychometrika*, 3, 45–60.
38. Thurstone, L.L. (1938), "Primary Mental Abilities." *Psychometric Monographs*, No. 1.
39. Spearman, C. (1939), "Thurstone's Work Reworked," *Journal of Educational Psychology*, 30, 1–16.
40. Eysenck, H.J. (1939), Review of Thurstone's *Primary Mental Abilities*, 1938, *British Journal of Educational Psychology*, 9, 270–275.
41. Thurstone, L.L. (1947), *Multiple Factor Analysis: A Development and Expansion of* The Vectors of Mind, University of Chicago Press, Chicago.
42. Thurstone, L.L., and Thurstone, T.G. (1941), "Factorial Studies of Intelligence," *Psychometric Monographs*, No. 2.
43. Schmid, J., and Leiman, J.M. (1957), "The Development of Hierarchical Factor Solutions," *Psychometrika*, 22, 53–61.
44. Carroll, J.B. (1993b), *Human Cognitive Abilities: A Survey of Factor-Analytic Studies*, Appendix B: *Hierarchical Factor Matrix Files* (Diskettes), Cambridge University Press, New York.
45. Guilford, J.P., and Hoepfner, R. (1971), *The Analysis of Intelligence*, McGraw–Hill, New York.
46. Burt, C. (1949), "The Structure of Mind: A Review of the Results of Factor Analysis," *British Journal of Educational Psychology*, 19, 100–111, 176–199.
47. Burt, C. (1955), "The Evidence for the Concept of Intelligence," *British Journal of Educational Psychology*, 25, 158–177.
48. Vernon, P. E. (1950), *The Structure of Human Abilities*, Methuen, London, England.
49. Carroll, J.B. (1968), Review of J.P. Guilford's *The nature of human intelligence*, (New York: McGraw-Hill, 1967), *American Educational Research Journal*, 5, 249–256.
50. Carroll, J.B. (1972), "Stalking the Wayward Factors," review of J.P. Guilford and R. Hoepfner, *The Analysis of Intelligence* (McGraw-Hill, New York 1971), *Contemporary Psychology*, 17, 321–324.
51. Cronbach, L.J., and Snow, R.E. (1977), *Aptitudes and Instructional Methods: A Handbook for Research on Interactions*, Irvington, N.Y.
52. Horn, J.L., and Knapp, J.R. (1973), "On the Subjective Character of the Empirical Base of Guilford's Structure-of-Intellect Model," *Psychological Bulletin*, 80, 33–43.
53. Guilford, J.P. (1981), "Higher-Order Structure-of-Intellect Abilities," *Multivariate Behavioral Research*, 16, 411–435.
54. Harman, H.H. (1976), *Modern Factor Analysis*, 3rd ed., revised, University of Chicago Press, Chicago.

55. Cattell, R.B. (1978), *The Scientific Use of Factor Analysis in Behavioral and Life Sciences*, Plenum, New York.
56. Hakstian, A.R., and Cattell, R.B. (1974), "The Checking of Primary Abilities Structure on a Broader Basis of Performances," *British Journal of Educational Psychology*, 44, 140–154.
57. Hakstian, A.R., and Cattell, R.B. (1978), "Higher-Stratum Ability Structures on a Basis of Twenty Primary Abilities," *Journal of Educational Psychology*, 70, 657–669.
58. McArdle, J.J. (1996), "Current Directions in Structural Factor Analysis," *Current Directions in Behavioral Science*, 5, 11–18.
59. Jöreskog, K.G. (1978), "Structural Analysis of Covariance and Correlation Matrices," *Psychometrika*, 43, 443–477.
60. Jöreskog, K.G., and Sörbom, D. (1989), *LISREL 7 User's Reference Guide*, Scientific Software, Mooresville, IN.
61. Bentler, P. M. (1985), *Theory and Implementation of EQS, a Structural Equations Program*, BMDP Statistical Software, Los Angeles.
62. Sternberg, R.J., Wagner, R.K., Williams, W.M., and Horvath, J.A. (1995), "Testing Common Sense," *American Psychologist*, 50, 912–927.
63. Holzinger, K., and Swineford, F. (1939), *A Study in Factor Analysis: The Stability of a Bifactor Solution*, Supplementary Educational Monographs, No. 48, University of Chicago Press, Chicago.
64. Carroll, J.B. (1961), "The Nature of the Data, or How to Choose a Correlation Coefficient," *Psychometrika*, 26, 347–372.
65. Kaiser, H.F. (1960), "The Application of Electronic Computers to Factor Analysis," *Educational and Psychological Measurement*, 20, 141–151.
66. Carroll, J.B. (1995b), "On Methodology in the Study of Cognitive Abilities," *Multivariate Behavioral Research*, 30, 429–452.
67. Montanelli, R.G., Jr., and Humphreys, L.G. (1976), "Latent Roots of Random Data Correlation Matrices with Squared Multiple Correlations on the Diagonal: A Monte Carlo Study," *Psychometrika*, 41, 341–348.
68. Kaiser, H.F. (1958). "The Varimax Criterion for Analytic Rotation in Factor Analysis," *Psychometrika*, 23, 187–200.
69. Hendrickson, A.E., and White, P.O. (1964), "PROMAX: A Quick Method for Rotation to Oblique Simple Structure," *British Journal of Statistical Psychology*, 17, 65–70.
70. Tucker, L.R. and Finkbeiner, C.T. (1981), "Transformation of Factors by Artificial Personal Probability Functions," Educational Testing Service Research Report RR-81–58, Princeton, NJ.
71. Tucker, L.R. (1984), personal communication.
72. MacCallum, R. (1986), "Specification Searches in Covariance Structure Modeling," *Psychological Bulletin*, 100, 107–120.
73. Jensen, A.R., and Weng, L.-J. (1994), "What is a Good *g*?," *Intelligence*, 18, 231–258.
74. Lord, F.M., and Novick, M. R. (1968), *Statistical Theories of Mental Test Scores*, Addison-Wesley, Reading, MA. With contributions by Allan Birnbaum.
75. Ekstrom, R.B., French, J.W., and Harman, H.H. (1976), *Manual for Kit of Factor-Referenced Cognitive Tests, 1976*, Educational Testing Service, Princeton, NJ.
76. Wothke, W., Bock, R.D., Curran, L.T., Fairbank, B.A., Augustin, J.W., Gillet, A.H., and Guerrero, C., Jr. (1991), *Factor Analytic Examination of the Armed Services Vocational Aptitude Battery (ASVAB) and the Kit of Factor-Referenced Tests*, Air Force Human Resources Laboratory, Brooks Air Force Base, TX.

77. Gottfredson, L.S. (1997), "Mainstream Science on Intelligence: An Editorial with 52 Signatories, History, and Bibliography," *Intelligence*, 24, 13–23.
78. Plomin, R., and Petrill, S.A. (1997), "Genetics and Intelligence: What's New?," *Intelligence*, 24, 52–76.

chapter seven

The Concept and Utility of Intelligence

EARL HUNT

Debates over intelligence have recurred through history. The idea that some of us are simply smarter than others touches a raw nerve in a society that tries to combine an Athenian commitment to democracy with a capitalist commitment to reward according to product. What happens when we are confronted with evidence that socially important talents are not distributed equally over society, and even more frighteningly, that there are powerful biological forces working to continue the differential distribution of ability across generations? This argument was made in 1869 by Francis Galton in his work *Hereditary Genius*,[1] and again more than one-hundred years later by Richard Herrnstein and Charles Murray in *The Bell Curve*.[2] The intervening years have witnessed a great deal of point and counterpoint on the issue, battles fought with more than the normal academic bickering because they expose the conflict between our ideal of a democratic society of equals and the practice of rewarding the best individual effort. We would be more comfortable if we could keep these beliefs as separate as we keep our feelings about baby lambs and lamb chops.

Unfortunately the debate is often cast in extremes, in spite of massive evidence for an intermediate position. The debate following the publication of Herrnstein and Murray's book revolved around three points, that "intelligence" (1) exists, monodimensionally and without qualification; (2) is very important for success in our society; and (3) is differentially distributed across ethnic groups, with whites and Asians on the top, and blacks on the bottom. Understandably, the claims about racial inequalities in intelligence generated the most heat, seconded by the remarks about the importance of intelligence in society. The assertion that intelligence exists as a unitary dimension was ignored completely. Yet the assertions about racial differences and the importance of intelligence would have to be much qualified if the claims about the nature of intelligence are incorrect.

This point has been virtually ignored in the public debate. Herrnstein and Murray acknowledged that some people believe that intelligence is a multidimensional rather than unidimensional attribute of the mind, and that others feel that we should move away from trying to document the attributes along which mental competence varies to deal with individual differences in the mental processes that underlie intelligent action. Herrnstein and Murray referred to people who hold these positions as "splitters" and "revisionists," and having acknowledged their presence, ignored their ideas. Media representatives and learned columnists who commented on Herrnstein and Murray also ignored these points, either because they were not aware of the alternative positions, because they regarded them as academic squabbles, or because it simply made better TV to cast the argument in simplistic terms.

I shall argue that these quibbles and qualifications are crucial, both for scientific accuracy and for a possible resolution of the economic, philosophical, and moral dilemmas inherent in our society's simultaneous pursuit of equality and reward of individual accomplishments. The argument will be presented in four steps. First, I shall review the concept of general intelligence and the evidence that it is a quality that we use to achieve success in our society. (If it is not, why the debate?) Second, I shall take a closer look at the splitter argument that there are different types of intelligence. Third, I will present the revisionist position, by summarizing studies of individual differences based on laboratory research rather than conventional intelligence and/or aptitude tests. I will argue that the splitter and revisionist theories have converged to present a picture of human intelligence that is strongly supported by several different sources of evidence. Most people who evaluate psychological theories regard convergent evidence of this sort as persuasive, because in psychology the crucial single experiment is unlikely in the extreme. Fourth, I shall discuss the implications of the convergent view on some of the social issues that Herrnstein and Murray raise.

A Very Brief History of Intelligence Tests

Many of the debates about the use of mental tests begin with the assumption that the reader and writer have some agreed upon idea about how intelligence is tested, and proceed to argue about whether or not the results of the test matter. When testing procedures are discussed, example items are often presented in isolation, to appeal to the reader's sense of whether or not a particular item is "fair" or "fair to group X." However, it is a good idea to look at the actual testing procedures in their entirety. My presentation must, of necessity, be brief. A fuller treatment can be found in Brody,[3] Hunt,[4] or Sternberg.[5]

The modern intelligence test dates from the turn of the century work of Alfred Binet, who developed tests that were to be used to identify French school children who (to use the pseudo-medical terminology that is popular with today's educationists) were at risk of experiencing failure in their school work. To this day, psychometric tests that are the direct descendants of Binet's procedures are used in many school systems to identify children who may best be served by special education programs.*

Binet developed his tests by asking educators what children of particular ages can do. Based on what he was told, Binet then developed trial tasks that could calibrate a child's progress relative to the norm for an appropriate reference population. Examples of tasks are counting to a set level, displaying one's vocabulary, or explaining something that is presumed to be common knowledge in the relevant population. The result provides a relative rather than absolute measurement of intellectual capacity.

The mental tests developed by Binet and his successors measure intelligence in the same sense that the decathlon measures athletic ability. They both do so indirectly, by providing standard forums in which various underlying abilities can be demonstrated. For that reason, David Wechsler has argued that intelligence itself is a collection of special skills.[7] Individually administered tests, group intelligence tests, and most paper-and-pencil tests employed in scholastic, military, and industrial screening consist of batteries of subtests measuring skills such as paragraph comprehension, elementary mathematical reasoning, and simple inductive and deductive problem solving. The decathlon analogy certainly applies to the three most widely used tests: the Armed Services Vocational Aptitude Battery (ASVAB), which is used to screen recruits

*This particular use of tests has led to a great deal of controversy, culminating in a variety of legal actions that either have or have not upheld the use of tests to make educational decisions about individual children (Ref. 6). To tip off my own political position at the outset, I assert that the tests, themselves, are politically neutral. They provide information about children that, used in an appropriate context and in conjunction with other information, can be of considerable assistance in planning an appropriate education for a child. They can also be misused to label children inappropriately and to force them into bad educational experiences. Does this mean that school authorities should or should not use tests? The answer depends upon how one feels about the competence and intentions of those authorities. The debate over the use of IQ tests in the schools is a political question about the authorities, not an issue about the tests.

wishing to enlist in the U.S. Armed Services,* the Scholastic Achievement Test (SAT) (previously known as the Scholastic Aptitude Test) developed by the Educational Testing Service (ETS), and the General Aptitude Test Battery (GATB) used by the Department of Labor until the late 1980s. These tests are often referred to as "batteries" (in the artillery rather than electrical sense) because they contain subtests that fire at their target from different directions. For instance, the Scholastic Aptitude Test contains tests of word knowledge, paragraph comprehension, and mathematical knowledge. No one of these exercises defines intelligence, but most of us would agree that they all tap some sort of mental competence.

In theory, the different mental skills tapped by different parts of a test battery or by different tests might be independently distributed in the population. In fact, they are not. The most striking example is the high correlation between group tests designed for educational or psychological screening and the individual (Binet-type) tests that are now primarily used in clinical and counseling situations. Although adequate scholarship is lacking, it is generally believed that the correlations between the most widely used group and individual tests are in the .8 range.[8] Furthermore, performances on the different subtests of a test battery are also highly correlated. These correlates are the essence of the case for *g*. A great many tests have been designed to measure different aspects of mental competence. Virtually all of these tests are positively correlated, often to a substantial degree. At the least, a reasonable argument can be made to the effect that all these tests are, imperfectly, tapping into a single mental construct, general intelligence.

The fact that statistics based on test and subtest correlations indicate the existence of general intelligence is of interest only if the tests themselves are correlated with socially important indices of mental performance, such as accomplishments in education or the workplace. Virtually every knowledgeable observer agrees that some such correlation exists, but there is considerable debate over whether or not the correlation is large enough to be important. Here the problem is more complex because the appropriate statistic depends upon the precise question that you wish to ask.

To make the issues concrete, consider the problem faced by the ETS when it evaluates its educational testing program. The ETS denies that the SAT measures intelligence, in the sense of some innate, immutable property of the individual. However, the ETS does claim that the SAT is an *imperfect* measure of "cognitive readiness for college," because it samples the various cognitive skills (reading, paragraph comprehension, mathematical skills) that a college

*Officer candidates are screened using an analogous procedure, but with different tests.

student is going to require. By the same token, no rational professor would claim that grades are a perfect measure of the amount a student has learned in class. Most do claim, in their actions if not in their words, that grades are positively related to the amount of learning.

Acknowledging these imperfections, it is still reasonable to attempt to predict the relationship between SAT scores and college grades. Certainly if there is no relationship there is no excuse for continuing the test! In fact, the correlation (r) between SAT scores and college grades is somewhere in the .25 to .35 range, depending upon the institution and year that you are talking about. (For instance, for the University of Washington the correlation is slightly above .3).* This figure has held remarkably constant over years and institutions. In a sense this correlation speaks for itself because it can be used to provide a reliable prediction of grade point average given knowledge of the test score. However the prediction is far from perfect. Critics of test use object that predicted scores will "account for" only r^2 of the percent of variance of the actual scores, in this case, somewhere between 5% and 10% of the variance. To the critics, this prediction seems low enough to be virtually useless.

This argument misses two important points. The first is that "percent of variance" is a statistical concept that, in spite of its name, does not exactly capture the intuitive concept of variation. (This is not to say that the concept is not useful. It is useful when properly used by statisticians who understand it.) A better test of the importance of a correlation coefficient can be obtained by looking at the correlation coeffecients for variables that we understand. The correlation between height and weight in adult males is about .5, and the correlation between chief executive officer income and stockholder's return on investment is close to zero. While we cannot predict grades from SATs as well as we can predict height from weight, boards of directors of our leading companies would be delighted if someone developed a test that was as accurate a selector of CEOs as the SAT is a selector of college students.

The second point is that the admissions officer has to make decisions. Therefore, providing that there is some reliable relationship between a predictor and grade points, the admissions officer's real question is not How high is the predictor? but Is there a better predictor? The answer to this question is no. With all their faults, SAT scores and high school grades are the best predictors we have, and the combination of the two provides a better prediction than either predictor used alone.

Now suppose that we change the argument slightly. Instead of viewing the SAT–grade point average (GPA) correlation from the viewpoint of the college

---

*Based on a sample of 2,400 University of Washington students graduating in the 1993–1994 period. Data courtesy of the Educational Assessment Center, University of Washington.

selection office, let us look at it from the viewpoint of a psychologist who is interested in knowing whether the abilities required to do well on a test are related to the abilities required to do well in college. This question is subtly different from the question about predicting grades, because the issue now is not the relation between observed scores and grades but the relation between the abilities that gave rise, imperfectly, to the observations. Nevertheless, the question can be answered, using a somewhat involved statistical argument.[9] It turns out that the correlation between underlying abilities is about .5. As this example shows, the correlation that ought to be used varies substantially depending upon whether you are interested in prediction of scores or in understanding the relationship between abilities.

It has frequently been said that intelligence tests predict "academic" rather than "on-the-job" intelligence. In support of this point, there are a number of studies of "on-the-job" situations in which one can demonstrate unarguably intelligent performance by people who do not have high test scores.[10] All these demonstrations show is that intelligence is not all that is important on the job, and no one ever said that it was. The studies showing failures of intelligence as a predictor of performance have been so small as to be almost anecdotes. Massively larger studies of the correlations between various aptitude tests and measures of workplace performance have shown that the correlations between test and measure are only slightly, if at all, lower than the correlations found in academic situations, such as the SAT-GPA example.[11–13] Furthermore, the findings go beyond studies that simply compute correlation coeffecients. During the years when the American Telephone and Telegraph (AT&T) had a virtual monopoly on telephone services in the United States, the company conducted a longitudinal study in which candidate managers were interviewed and tested early in their careers, and then followed for more than 15 years.[14] A test much like the SAT, given at the outset of the executives' careers, was the best single predictor of eventual level of management achieved. However, the correlation was only slightly below .4, and personality tests added more to predictability. This does not mean that the personality tests were better predictors than the intelligence tests. They were not. It means that the combination of intelligence and personality test scores provides a better prediction than either test score alone. This issue is not whether "intelligence" or personality is more important to success. They both are.

Intelligence tests of various sorts clearly measure something that is related to both academic and workplace performance. However, intelligence clearly is not the sole determinant of either academic or workplace performance. Cognitive psychologists such as myself (Ref. 4), specialists in industrial-organizational psychology,[15] specialists in personality,[16] and educators[17] have all argued that there is room for both intelligence and personality measures ("emotional intelligence") as predictors of success. Going a bit further, it would be quite unrealistic to expect any measure of an individual's motivation or ability to be a perfect predictor of social outcomes simply because social rewards are not solely determined by the merits of the individual receiving them. Nevertheless, intelligence is clearly one of the largest of many factors that determine success. It cannot be disregarded.

One of the most unfortunate polarizations in the public debate has been between people who argue that intelligence is virtually the only individual trait that determines success and those who argue that because other traits can be shown to predict success, intelligence is not important. Neither position is correct.

What Is Intelligence? The Psychometric Evidence

If intelligence is important it makes sense to ask, in more detail, just what intelligence is. There are two sorts of evidence: evidence based on an examination of the distribution of test scores and evidence based on experimental studies of the process of thought. I will refer to these bodies of data as *psychometric* and *cognitive* evidence. Because the psychometric evidence has dominated the discussion, it is examined first.

The argument for general intelligence has already been stated: virtually all measures of cognitive competence are positively correlated. Technically, this phenomenon is known as *positive manifold.* Charles Spearman, a British psychologist who figured prominently in the early development of tests and test theory, concluded that positive manifold occurs because each individual test score is produced by a person's general intelligence, *g*, plus specialized skills required for that test.[18,19] Subsequently, Spearman and others developed a technique known as *factor analysis* to investigate this phenomenon. Spearman (see Ref. 19), Jensen,[20] Herrnstein and Murray, and other general theorists have concluded that factor analyses of intelligence test results support the concept of general intelligence. Other eminent psychologists, notably Thurstone,[21] Cattell,[22] and Horn,[23] have concluded that it does not! While this sort of disagreement between scientists is frustrating to policy makers, it is a fact of life. Furthermore, it is an important fact of life that needs to be understood in order to carry on a sophisticated discussion about the nature of intelligence.

Part of the problem arises from the nature of factor analysis. It is a sophisticated technique for determining how many underlying traits are required to account for the observed correlations between a battery of tests. Early factor analytic procedures derived results using purely mathematical criteria. Instead of applying a mathematical/statistical criteria to evaluate a psychological theory, the early factor analysts used mathematical criteria to define the theory to be evaluated. Not surprisingly, the mathematical criteria used had an inordinate influence on the psychological theory that was developed. In the 1970s, new factor analytic techniques were developed that allowed people to test the extent to which a structure of abilities *defined by a psychological theory* could account for the observed correlations. This technique is a more normal use of statistics in science and has resulted in unifying what had been a very scattered and confusing literature.

Following on ideas first developed by Raymond Cattell, John Horn and his colleagues have conducted this sort of analysis in a large number of investigations, extending over almost 30 years.[23–25] A somewhat similar and massively heroic analysis was conducted by John Carroll,[26] who used modern factor analysis to see if the same simple structure could fit a large number of the classic data sets. While there are some differences between Carroll's and Horn's approaches, the similarities in their findings vastly outweigh the differences. Therefore, I shall adopt Horn and Cattell's terminology, because I believe that it provides the best intuitive picture of what factor analysis has told us.

Skipping over some details, Horn and Cattell divide human intellectual competence along three dimensions, fluid intelligence (Gf), crystallized intelligence (Gc), and visual-spatial reasoning (Gv). They describe the three dimensions as follows:

> *Fluid intelligence* is the ability to develop techniques for solving problems that are new and unusual, from the perspective of the problem solver.
>
> *Crystallized intelligence* is the ability to bring previously acquired, often culturally defined, problem-solving methods to bear on the current problem, implying both that the problem solver knows the methods and recognizes that they are relevant in the current situation.
>
> *Visual-spatial reasoning* is a somewhat specialized ability to use visual images and visual relationships in problem solving; for instance, to construct in your mind a picture of the sort of mental space that I described earlier, in discussing factor analytic studies.

Crystallized and fluid intelligence measures are substantially correlated. For instance, Horn reported a study in which Gf and Gc measures were extracted from an analysis of the Wechsler Adult Intelligence Scale (WAIS), possibly the most widely used intelligence test (see Ref. 23). The correlation between factors was .61. This is the sort of evidence that leads *g* theorists to postulate a single intelligence factor. Gf–Gc theorists view the correlation in another way. The *g* advocate begins with the assumption that general intelligence exists, and is expressed with varying purity, in both tests of Gc and Gf. The Gc–Gf theorist assumes that crystallized intelligence is developed by the interaction between fluid intelligence and (useful) cultural experiences. Therefore, Gc and Gf will be correlated because one's Gc, at time t, will be an increasing function of Gf at time $t - 1$.

Although this discussion is a bit oversimplified, it is enough to show two things. First, the fact that there is a correlation between tests intended to draw out pure reasoning ability and tests intended to evaluate cultural knowledge does not distinguish between *g* and Gc–Gf theories. Second, the Gc–Gf theory is essentially a "rich get richer" theory. If two individuals start with different amounts of fluid intelligence and have equivalent experiences, the one with the higher Gf will better utilize these experiences and come out of them with more Gc than the less talented person.

The discussion also illustrates an important point. There is really no way that one can distinguish between *g* and correlated Gf–Gc factors on the basis of psychometric evidence alone. Either analysis can provide a summary of the data. What summary one prefers depends on either theoretical predilections or the purpose of the summary. On the other hand, the distinction between Gc and Gf theories can be addressed by stepping outside of factor analysis and looking at how Gf and Gc measures respond to different manipulations. It turns out that they respond differently.

The most striking example is aging. It has been very well documented that measures of Gf generally decrease from early adulthood, whereas Gc measures increase throughout most of the working years (see Ref. 23). This is not surprising. Experience counts; most of the key leadership positions in our society are held by people over 40. On the other hand, it can be shown that middle aged and older people do take longer than younger people to understand new problem-solving methods appropriate for unfamiliar problems. The fact that Gf and Gc measures respond differently to outside influences is sufficient to show that a general intelligence theory is not correct.

The distinction between general intelligence and the Gc–Gf model is not simply an academic nicety. I shall argue later that it has major implications for social policy. First, though, we want to look at what Herrnstein and Murray call the *revisionist* approach to intelligence.

The Revisionist Position: An Alternative Approach to Intelligence

Herrnstein and Murray, like many authors, present the revisionist approach as an alternative to the psychometric approach. A more correct view is that the revisionist approach and the psychometric approach are complementary ways of looking at the complex issues of individual differences in human mental competence (see Ref. 5). I will now try to present the revisionist, or more formally, the *cognitive science* approach, and to show that in many ways it converges with the Gc–Gf analysis just developed.

The gist of the cognitive science approach is that IQ tests, including various aptitude tests that evolved to solve practical problems in applied psychology, should not be viewed as the sanctified theoretical definition of intellectual ability. Cognitive scientists are even more vehement that statistical abstractions from multiple tests, such as the general intelligence factor, should not be reified. They argue that the definition of individual differences in mental competence should flow from a general theory of cognition. Therefore, in order to understand the approach it is necessary to understand what theory is being espoused.

Researchers in cognitive science (e.g., Ref. 27) who look at thought processes in general but often disregard individual differences, describe thinking as the process of creating a mental representation of the current problem, retrieving information that appears relevant, and manipulating the representation in order to obtain an adequate answer. The problem, its solution, and some of the methods used to solve it are then stored for later reference. The key point in this process is creating the representation, which is assumed to require some sort of working memory capacity so that different ways of looking at a new problem can be tried out. When familiar problems are encountered, the process of building an appropriate representation becomes more efficient because previously acquired information and problem-solving techniques can be used. The demand on working memory is reduced, but not entirely eliminated, unless problem-solving can be reduced to stereotyped stimulus–response sequences.

Note that this account conforms closely to Cattell and Horn's fluid and crystallized intelligence distinction. Both accounts stress the fact that when one embarks on problem-solving in a particular domain, performance will initially depend upon one's ability to develop and maintain alternative ways of looking at the problem (Cattell and Horn's fluid intelligence), but that as experience is developed one shifts from seeking representations to pattern recognition, realizing that past solutions can be applied to present problems. Because the first stage is more demanding of working memory and attention, learning to do an intellectual task will generally be harder than doing it. The theory also implies that people who do well on tests of fluid intelligence should have a large working memory capacity. Indeed, they do.[28]

More generally, the cognitive science approach identifies three different sources of individual variation in mental competence. To understand them it helps to draw an analogy to computing systems. Suppose you were trying to evaluate the "cognitive capacity" of a computing system. (Note that I did not say "computer." We are interested in the whole system.) A computing system can solve problems if it has three things available:

1. Sufficient "number crunching" capacity to attack the problem at hand
2. Programs that are appropriate for solving the problems it faces
3. Access to the data needed to solve a problem

To be more concrete, consider modern accounting practices, which make heavy use of automated book balancing. A computer system can balance the accounts of my local bank only if it can do the arithmetic fast enough to balance all the accounts, if it has a spreadsheet or similar program for book balancing, and only if it has access to the bank's records.

Cognitive scientists draw an analogy between number-crunching capacity, programs and data access, and the cognitive functions of being able to process ideas—any ideas—quickly and accurately, knowing how to solve certain classes of problems, and having access to the data needed to solve particular problems. In psychological terms, human number crunching is a physiological capacity, while knowing how to solve problems and knowing key facts are both the product of learning. Each of these aspects of thought are legitimate parts of intelligence. The physiological capacities are clearly part of Gf, knowing key facts is part of Gc, and having acquired certain problem-solving strategies is a bit of both Gc and Gf. Perhaps most importantly, one function permits the other. You cannot solve problems unless you have the information-processing capacity required by the solution method, and just knowing facts does not help in problem-solving unless you have a "program" that tells you how to use them.

The cognitive science approach has two important implications for the IQ debate. The first is that in defining a person's intellectual capacity you want to distinguish sharply between the ability to learn how to do something and the ability to do it after it has been learned. Viewed this way, it is not surprising that IQ tests, and especially fluid intelligence tests, are associated with academic performance. For many people, high school and college may be the most intellectually demanding stages of life, in the sense that students are continually faced with new challenges. However, the tests' association with learning ability is not restricted to learning in formal school settings. Data from the military have shown that performance on the Armed Forces Qualification Test (AFQT), the test that Herrnstein and Murray use for most of their empirical work, has a strong relationship with performance on the job in the first few months, and then has a progressively reduced, but non-negligible relationship to performance as experience accumulates.[29] Similarly, the Department of Labor's General Aptitude Test Battery (GATB) has been shown to be less valid for older than for younger workers.[30] It has also been established that the extent to which intelligence measures predict performance depends upon the stage of training. The more training, the lower the relation to performance.[31] Experience counts.

Once again, this does not mean that intelligence does not count. If you cannot learn to do something within a reasonable time and at a reasonable cost, then you will not get the chance to do it. However, *if adequate training methods can be developed*, people's intellectual competence in specific domains can be greatly increased, even though their general intelligence cannot be. Some work by one of my own colleagues provides a good example.[32] High school students were given a test of fluid intelligence. They then took a year-long problem-solving oriented course in elementary physics. The IQ test did indeed predict how much physics the students learned. At the end of the year they took an equivalent IQ test. Their IQ scores had not changed a whit. However, all the students learned a great deal of physics, as evidenced by their performance on national examinations. Their intelligence, in the psychometric sense, had not been influenced. Their cognitive competence, in the sense of the problems they could solve, had been increased substantially.

It is important to realize that the cognitive science view of intelligence—the individual differences in the mental "hardware and software" that people bring to bear on a problem—is fundamentally different from the approach used by the psychometricians. This difference is best seen by contrasting the approach of the two disciplines to measurement issues.

People who work in the psychometric tradition begin with the assumption that intelligence is a trait that is normally distributed in the population. They then use the standard deviation unit, which is the deviation of a score from the mean score on a test, divided by the standard deviation, as their unit of measurement. They refer to the standard deviation unit as "like an inch" (their term), because it is a unit of measurement. Cognitive scientists, like myself, counter that the standard deviation unit is not at all like an inch. The standard deviation unit defines a score relative to other scores, whereas an inch or meter defines a length relative to (1) an absolute zero point and (2) a physical standard. To see this it helps to look at what a standard score is. Let x_i be the ith individual's raw score on an achievement test, and let μ and s be the mean and standard deviation of scores in some referent population. The standard score is defined as

$$z_i = (x_i - \mu)/s.$$

Clearly the standard score varies depending upon the referent population used. For example, suppose that we were measuring height. Consider a six foot six inch man, who would have a standard score of about 2 in the North American male population. The same person would have a standard score of approximately 0 with reference to the population of male professional basketball players.

It can be argued that in the case of height, converting to standard scores would not matter for any scientific purpose because the standard score is a linear transformation of physical height. Thus any discussion of the effect of height could use inches, centimeters, or standard scores as the unit of measurement. We can define height independently of the referent population. In the case of the intelligence test, there is no population-independent definition of true mental competence. Therefore, we fall back on a relative definition.

At this point, the revisionists and the psychometricians have their deepest philosophical difference. Cognitive scientists want to define a person's cognitive capacity in terms of that person's absolute ability to do certain carefully selected cognitive tasks. Some of these tasks are designed to measure "hardware" capacities, such as time required to make very simple choices, or to keep track of several things at once. Very recently there has been interest in direct biological measures that can be obtained by examining brain scans as people do various tasks, but this work is in its infancy. In other cases, the absolute measures required are tests of "software," such as knowledge of certain problem-solving methods, often specialized for domains as particularized as chess, physics, or mathematics.

Like Gc–Gf theory, the cognitive science approach shares the important characteristic of being a "rich get richer" theory because people who have greater initial capacities for learning are inevitably going to be able to benefit more from a fixed experience. There is an additional, related implication. Research in cognitive science has shown that as people become more competent in problem-solving, they tend to become more specialized. The theoretical reason for this is simple. Early in training for almost anything, the general intelligence characteristics, such as large working memories and knowledge of general problem-solving skills, exert considerable control over performance. As skill progresses, more specialized abilities, and often the sheer amount of practice that one is willing to do, become the dominating influences on performance. This trend has been shown quite clearly in laboratory studies that examine individual differences in performance over levels of training in a controlled setting.[31,33] It is implied by studies of expert performance[34] and creativity,[35,36] and is consistent with industrial studies that have examined the extent to which test scores can predict job performance when on-the-job experience is varied.

Would we see the same thing in the psychometric data? The answer is that we do. Both *g* theorists and multidimensional intelligence theorists begin with the assumption that the psychometric data are normally distributed. This is not correct. Detterman and Daniel examined the correlations between various measures of mental ability in the "top half" of a multivariate distribution, as defined by a person's position on a general intelligence factor, and the same correlations in the "bottom half" of the distribution.[37]

Figure 7.1 (top panel) shows the results of this analysis applied to the normalization sample used for the Wechsler Intelligence Tests, which are the most widely used individual tests in the United States (see Ref. 3). If the scores were multivariate normally distributed, the correlations in each half should have been the same.* But they were not. The pairwise correlations between measures in the bottom half were higher than those in the top half. The lower panel of Figure 7.1 shows what happened when Derek Chung and I repeated Detterman and Daniel's analysis, using the results for ASVAB scores in the National Longitudinal Study of Youth, the same dataset that Herrnstein and Murray use in many of their analyses. While the absolute numbers are different, which is not surprising considering the difference between the ASVAB and the Wechsler tests, the same pattern appears. As predicted by cognitive theory, but virtually ignored by psychometricians, the data from conventional intelligence tests indicate that lack of "general intelligence" is pervasive, but that having high competence in one field is far from a guarantee of high competence in another.

Implications for Public Policy

I now want to move from a discussion of the scientific evidence for intelligence to a discussion of the social implications of these facts. Any discussion of social implications will, of course, imply a value system because science only tells social leaders what policies are feasible (and at what cost) but cannot tell their leaders what goals policies should seek to achieve. For these reasons, scientists have tended not to make any statement at all about social goals, although some of the conclusions in the social sciences have been presented in a way that at least suggests to me that certain goals are being advocated implicitly. Herrnstein and Murray's *The Bell Curve* was a major departure from that tradition. The authors approached the policy issue openly and made specific policy recommendations. Somewhat surprisingly, they were sometimes criticized for this on the grounds that even if they were right, they had brought up unpleasant facts, especially about racial and ethnic group differences in intelligence, and that it would be best if these things were not discussed. While I do not agree with all of Herrnstein and Murray's analyses or conclusions, I applaud them for bringing into the open a discussion that, after all, was certainly being conducted *sotto voce* in many places.

I have argued that the general intelligence viewpoint adopted by Herrnstein and Murray, although not exactly inaccurate, is simplified in an important

---

*The correlation between two test scores is a function of the regression coefficient between them and the variances of the test scores. If the regression coefficient between two test scores is uniform throughout the range of the scores and if selected populations are defined by subranges with reduced variance on one of the tests (e.g., restricting attention to a specified percentile range), the correlation between tests will be reduced. The amount of the reduction depends on the extent of reduction of the variation. If the population is bivariately normally distributed this effect will be symmetric about the mean of either test because the variance of, say, the lowest decile, will be identical to the variance in the corresponding (i.e., highest) decile on the other side of the mean.

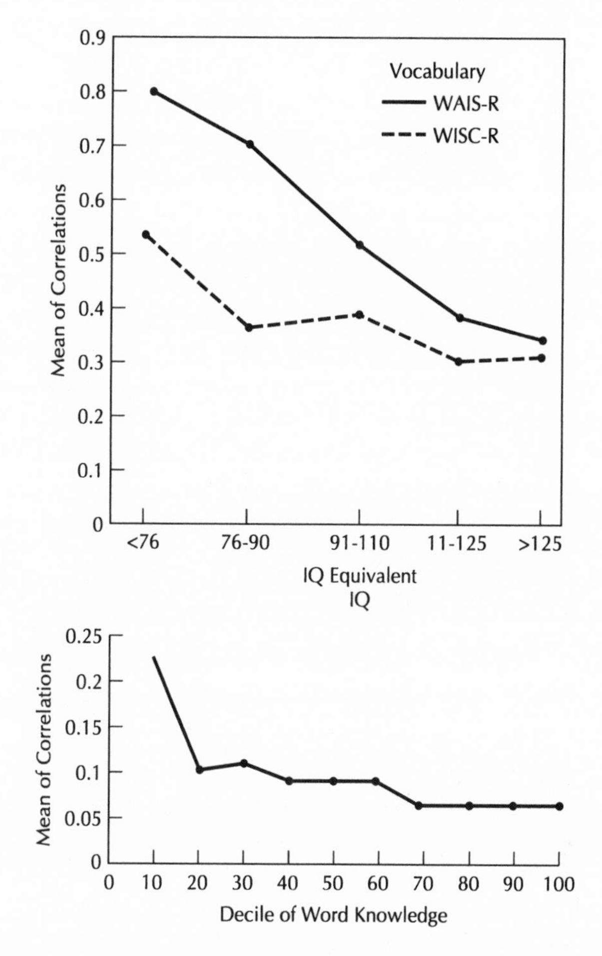

Figure 7.1. Mean correlations between subtests of the Wechsler Intelligence Tests (top panel, after Ref. 37) and the Armed Services Vocational Aptitude Battery (National Longitudinal Survey of Youth database; selected data provided by Charles Murray to match the data reported by Herrnstein and Murray). Correlations are shown by decile of performance on selected subtests.

way. I further argued that the evidence from both psychometrics and cognitive psychology supports a view that distinguishes between the ability to acquire new information and solve new problems, and the ability to apply previously learned problem-solving techniques (problem-solving schema, in cognitive science terms) to the currently encountered problem. To what extent would a more sophisticated view of intelligence alter the policy implications that Herrnstein and Murray drew?

The major premise of *The Bell Curve* is that intelligence matters, and therefore our society must be concerned about how it is developed and utilized. They take the attitude that there is a fixed amount of intelligence in society, and that there is not very much that we can do to change this situation. Therefore, we should utilize the intelligence of those who have it while finding a humane but not economically central role for those who do not. Herrnstein and Murray's analysis proceeds from the overly simplified general intelligence view of human mental competence. To what extent should their social conclusions be modified to accommodate the more complicated view presented here?

Data on the distribution of intelligence test scores have policy implications only if these scores do in fact represent, albeit imperfectly, an important socioeconomic variable. I have already cited the considerable amount of evidence indicating that intelligence counts in the present workplace. There is every reason to believe that there will be an increase in the relation between intelligence and economic/social rewards as we move into the next century. I say this for two reasons. First, there is a well-documented explosion in the social importance of the "knowledge worker," someone who is capable of learning how to manipulate the technology of the information age. Second, there has been a marked increase in the wage differential between those who are able to play lead roles in society and those who play supporting roles. President Clinton's Secretary of Labor, Robert Reich, who is hardly an advocate of conservatism, has pointed out that society is evolving so that the lead roles are increasingly played by those capable of dealing with abstract symbolic, rather than concrete, "in-your-face" economic and social decisions.[38] This theme has been developed by others, including myself.[4,39] There is every reason to believe that the *relative* rewards for being on top of things, as opposed to being just average, are going to increase, and that this differential, imperfectly but substantially, is going to favor those with greater mental competencies.

This issue is more acute because of certain demographic trends. The workforce is aging; within a few years the relative number of over-40 workers in the workforce will reach historic highs. In addition, there has been a very well-publicized shift in the entering workforce from being predominantly white and male to being increasingly nonwhite and just slightly more female. The age trend is shown in Figure 7.2.

Elsewhere (see Ref. 4), I have combined these trends with known facts about the distribution of intelligence test scores by age and ethnic group to project the availability of fluid and crystallized intelligence in the workforce at the start of the twenty-first century. Crystallized intelligence will be increasingly available, whereas fluid intelligence will be in shorter supply than it is now. The driving influence behind this trend is the aging of the workforce, which exerts a much larger effect than the increase in ethnic minorities in the entering workforce. The problem is serious to the extent that the ability to learn new skills, rather than to experience normal ways of doing things, becomes an important part of the workplace. Unfortunately, though, virtually every projection of workplace characteristics assumes that this is precisely what will happen. The need for "learning to learn" in the future workforce is so well accepted that it has even been featured in presidential speeches on the topic. If these widely believed projections are, in fact, correct, the relatively few young workers with high fluid intelligence will be markedly favored over older workers who rely on knowledge of how the workforce used to be, and the large number of young workers who are entering the workplace without substantial workforce skills, including the much sought after "learning to learn."

Herrnstein and Murray take the attitude that not very much can be done about this trend because social and educational programs intended to increase intelligence have generally been ineffective. In part, they are correct. In spite of a great deal of publicity about programs that teach creativity and problem-solving, there is very little evidence that lasting, major gains can be achieved for people of average or above-average intelligence.[40] The only possible exception to this rule is the usefulness of programs that emphasize metacognition, the need for people to be aware of how they are going about solving a problem. On the other hand, there are a number of highly successful programs for improving the teaching of content-specific but highly general problem-solving skills, such as writing and applied mathematics.[41] A single example should suffice. Statistics teachers today routinely teach their students to solve problems that baffled Pascal and Galton. The modern students are in no way the equals in

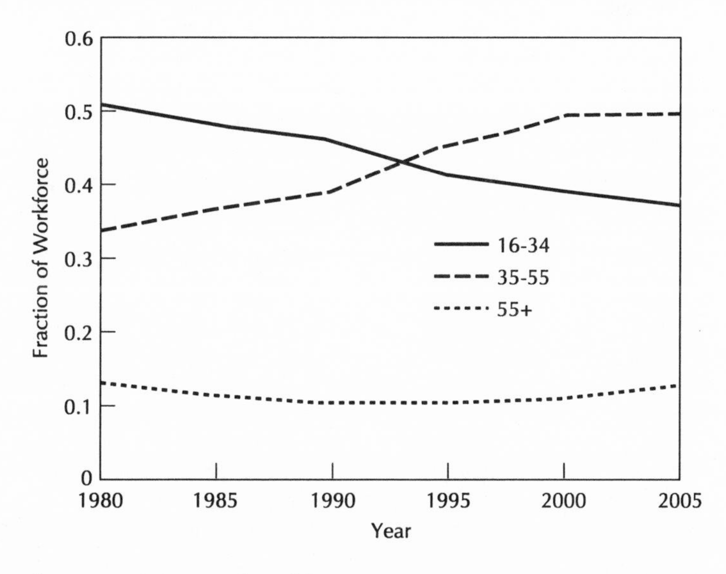

Figure 7.2. Fraction of workforce in various age groups, 1980–2005.

"native intelligence" of these intellectual giants, but they are better equipped to solve a variety of applied statistics problems. In the terms of cognitive science, we can teach people how to use broadly applicable problem-solving schema. This experience will not "improve their intelligence" outside the areas for which the schema are applicable, but it will make them better problem solvers.

These comments are applicable to both the school and the workplace. It is quite possible that, to some extent, the decline in fluid intelligence associated with aging may be alleviated substantially. The key seems to be keeping intellectually active. Extended longitudinal and cohort studies in gerontology have found that cognitive declines with advancing age (here meaning beyond 40!) are smaller in people who keep cognitively active and who have a general feeling that their lives are secure.[42] Adult work experiences, including lifelong retraining, can encourage a positive rather than a negative attitude toward change.

So, from the viewpoint of a cognitive psychologist, I see the amplified view of intelligence as painting a much rosier picture of what *could* happen than the view painted by believers in a highly heritable, monolithic and inflexible general intelligence. The schools and the workplace could be structured to make much better use of our cognitive human capital than we now do. Whether or not this structuring actually will take place is a question for social analysis and would take us far beyond the topic of the current essay.

References

1. Galton, F. (1869), *Hereditary Genius. An Inquiry Into Its Laws and Consequences*, Appleton, New York.
2. Herrnstein, R.J., and Murray, C. (1994), *The Bell Curve: Intelligence and Class Structure in American Life*, The Free Press, New York.
3. Brody, N.P. (1992), *Intelligence*, Academic Press, San Diego, CA.
4. Hunt, E. (1995), *Will We Be Smart Enough? A Cognitive Analysis of the Coming Workforce*, Russell Sage Foundation, New York.
5. Sternberg, R.J. (1990), *Metaphors of Mind: Conceptions of the Nature of Intelligence*, Cambridge University Press, Cambridge.
6. Kaplan, R.M. (1985) The Controversy Related to the Use of Psychological Tests. in Wolman, B.B. (Ed.) *Handbook of Intelligence: Theory, Measurement, and Applications*. NewYork: Wiley-Interscience, 465–504.
7. Wechsler, D. (1975), "Intelligence Defined and Redefined: A Relativistic Approach," *American Psychologist*, 30, 135–139.
8. Seligman, D. (1994), *A Question of Intelligence*, Citadel Press, New York.
9. Hunter, J.E., and Schmidt, F.L. (1990), *Methods of Meta-Analysis*, Sage, Newbury Park, CA.
10. Sternberg, R.J., Wagner, R.K., Williams, W.M., and Horvath, J.A. (1995), "Testing Common Sense," *American Psychologist*, 50, 912–927.
11. Earles, J.A., and Ree, M.J. (1992), "The Predictive Validity of the ASVAB for Training Grades," *Educational and Psychological Measurement*, 52, 721–725.

12. Hunter, J.E. (1986), "Cognitive Ability, Cognitive Aptitudes, Job Knowledge, and Job Performance," *Journal of Vocational Behavior*, 29, 340–362.
13. McHenry, J.J., Hough, L.M., Toquam, J.L., Hanson, M.A., and Ashworth, S. (1990), "Project A Validity Results: The Relationship Between Predictor and Criterion Domains," *Personnel Psychology*, 43, 335–354.
14. Howard, A., and Bray, D.W. (1988), *Managerial Lives in Transition: Advancing Age and Changing Times*, Guilford Press, New York.
15. Cascio, W.F. (1995), "Whither Industrial and Organizational Psychology in a Changing World of Work?," *American Psychologist*, 50, 928–939.
16. Goleman, D. (1995), *Emotional Intelligence: Why It Matters More Than IQ*, Bantam Books, New York.
17. Gardner, H. (1983), *Frames of Mind: The Theory of Multiple Intelligences*, Basic Books, New York.
18. Spearman, C. (1904), "General Intelligence, Objectively Determined and Measured," *American Journal of Psychology*, 15, 201–293.
19. Spearman, C. (1927), *The Abilities of Man*, MacMillan, London.
20. Jensen, A.R. (1980), *Bias in Mental Testing*, Free Press, New York.
21. Thurstone, L.L. (1938), *Primary Mental Abilities*, University of Chicago Press, Chicago.
22. Cattell, R.B. (1971), *Abilities: Their Structure, Growth, and Action*, Houghton Mifflin, Boston, MA.
23. Horn, J.L. (1985), "Remodeling Old Models of Intelligence," in Wolman, B.B. (Ed.), *Handbook of Intelligence. Theories, Measurements, and Applications*, Wiley, New York, pp. 267–300.
24. Horn, J.L. (1986), "Intellectual Ability Concepts," in Sternberg, R.J. (Ed.), *Advances in the Psychology of Human Intelligence. Vol. 3*. Erlbaum, Hillsdale, NJ, pp. 35–77.
25. Horn, J.L., and Noll, J. (1994), "A System for Understanding Cognitive Capabilities: A Theory and the Evidence on Which it is Based," in Detterman, D.K. (Ed.), *Current Topics in Human Intelligence, Vol. 4*, Ablex Publishing, Norwood, NJ, pp. 151–204.
26. Carroll, J.B. (1993), *Human Cognitive Abilities*. Cambridge University Press, Cambridge.
27. Newell, A. (1990), *Unified Theories of Cognition*, Harvard Press, Cambridge, MA.
28. Carpenter, P.A., Just, M.A., and Shell, P. (1990), "What One Intelligence Test Measures. A Theoretical Account of Processing in the Raven Progressive Matrices Test," *Psychological Review*, 97, 404–431.
29. Wigdor, A.K., and Green, B.F., Jr. (1991), *Performance Assessment in the Workplace*, National Academy Press, Washington, D.C.
30. Hartigan, J.A., and Wigdor, A.K. (Eds.), (1989), *Fairness in Employment Testing: Validity, Generalization, Minority Issues, and the General Aptitude Test Battery*, National Academy Press, Washington, D.C.
31. Ackerman, P. (1987), "Individual Differences in Skill Learning: An Integration of Psychometric and Information Processing Perspectives," *Psychological Bulletin*, 102, 3–27.
32. Levidow, B.B, (1994), *The Effect of High School Physics Instruction on Measures of General Knowledge and Reasoning Ability*, Unpublished Ph.D. Dissertation, University of Washington.

33. Joslyn, S. (1995), *Individual Differences in Time Pressured Decision Making*, Unpublished Ph.D. Dissertation, University of Washington.
34. Ericsson, K.A., Krampe, R.Th., and Tesch-Romer, C. (1993), "The Role of Deliberate Practice in the Acquisition of Expert Performance," *Psychological Review*, 100, 363–406.
35. Gardner, H. (1993), *Creating Minds: an Anatomy of Creativity Seen Through the Lives of Freud, Einstein, Picasso, Stravinsky, Eliot, Graham, and Gandhi*, Basic Books, New York.
36. Simonton, D.K. (1984), *Genius, Creativity, and Leadership: Historiometric Inquiries*, Harvard University Press, Cambridge, MA.
37. Detterman, D.K., and Daniel, M.H. (1989), "Correlations of Mental Tests with Each Other and with Cognitive Variables Are Highest in Low IQ Groups," *Intelligence*, 13, 349–360.
38. Reich, R. (1991), *The Work of Nations: Preparing Ourselves for 21st Century Capitalism*, Knopf, New York.
39. Frank, R.H., and Cook, P.J. (1995), *The Winner-Take-All Society*, Free Press, New York.
40. Nickerson, R.S., Perkins, D.N., and Smith, E.E. (1985), *The Teaching of Thinking*, Lawrence Erlbaum Associates, Hillsdale, NJ.
41. Bruer, J.T. (1993), *Schools for Thought: A Science of Learning in the Classroom*, MIT Press, Cambridge, MA.
42. Schaie, K.W. (1994), "The Course of Adult Intellectual Development," *American Psychologist*, 49, 304–313.

PART IV

INTELLIGENCE AND SUCCESS: REANALYSES OF DATA FROM THE NLSY

What accounts for success in life? Most explanations stress the importance of family, education, social class, the network of "who you know," and simply hard work. In *The Bell Curve*, Murray and Herrnstein argue otherwise. They maintain that success is tightly linked to IQ, that education can't do much to alter IQ, and that IQ explains more about why people succeed in getting good jobs, earning money, and staying out of jail than social class or education. To support this argument, they turn to their own analyses of the National Longitudinal Survey of Youth (NLSY).

They draw upon the NLSY data in nine separate chapters, and the regression and logistic regression analyses that they report on undergird their conclusions regarding the dominant role of IQ in determining various life outcomes. The dictates of doing good scientific work have stimulated many others to see if they can replicate these statistical findings.

In Part IV, five sets of contributors re-examine Herrnstein and Murray's statistical analyses of the data from the NLSY. In each of the first four chapters, the authors specify slightly different regression models for the NLSY data and draw from these models quite different insights into what drives success in life. Specifically, Cawley, Conneely, Heckman, and Vytacil re-examine the relationship between cognitive ability and wages, and Cavallo, El-Abbadi, and Heeb look at the role of gender in the relationship between ability, race, and earnings. Winship and Korenman consider the relationship between years of schooling and IQ. Then Manolakes explores the assumed ties between low IQ and crime.

From these chapters we learn about serious flaws and major deficiencies in Herrnstein and Murray's analyses. On the basis of their reanalyses of the NLSY data, and the analyses of others, the authors are unpersuaded by Herrnstein and Murray's argument about the dominant role of IQ in determining life outcomes.

Glymour concludes by offering a critical overview of causal thinking within quantitative social science, including a review of the assumptions implicit in statistical models. These assumptions play an important role in the arguments of *The Bell Curve* and in the surrounding discussions of regression and factor analysis.

chapter eight

Cognitive Ability, Wages, and Meritocracy

JOHN CAWLEY, KAREN CONNEELY,
JAMES HECKMAN, AND EDWARD VYTLACIL

In their controversial book *The Bell Curve*, Richard Herrnstein and Charles Murray summarize an impressive body of research on the correlations between social outcomes and scores on tests of cognitive ability.[1] A remarkable finding of the research they survey is that one linear combination of tests—called *general intelligence*, or *g*—predicts performance almost as well as the full battery of tests.

Charles Spearman was the first to propose that *g* is a common ability that explains performance on all tests of intelligence. General intelligence was also thought to be heritable, although that is a completely separate matter.

Both assumptions have been questioned in the scholarly literature. Theories of multiple abilities go back to Thurstone.[2] Carroll provides a comprehensive discussion of the evidence.[3] The theory of the heritability of intelligence is simplified by, but does not require, unidimensional ability. *The Bell Curve* embraces both *g* and heritability. Moreover, it extends Spearman's work and attempts to demonstrate that differences in *g* explain discrepancies in social outcomes across race.

This chapter tests and rejects two of Herrnstein and Murray's claims about general intelligence. First, we examine their argument that cognitive ability is unidimensional. We show that not much should be made of the fact that *g* explains a majority of the covariance across test scores; this is an artifact of linear correlation analysis, not intelligence. We also find that several other components of measured ability besides "*g*" are statistically significant in predicting log wages.

Second, we test some of the claims about ability and wages made by Herrnstein and Murray. We show that they overestimate the percentage of wage variance that is explained by cognitive ability. Measured cognitive ability, education, and experience combined account for at most one third of the total variance in wages. We also present evidence that ability is unequally rewarded among demographic groups, which is inconsistent with Herrnstein and Murray's claim that the labor market is meritocratic.

We use the same data as Herrnstein and Murray: The National Longitudinal Survey of Youth (NLSY). The NLSY is designed to represent the entire population of American youth and consists of a randomly chosen sample of 6,111 U.S. civilian youths, a supplemental sample of 5,295 randomly chosen minority and economically disadvantaged civilian youths, and a sample of 1,280 youths on active duty in the military. All youths were between 13 and 20 years of age in 1978 and were interviewed annually starting in 1979. The data include equal numbers of males and females. Roughly 16% of respondents are Hispanic and 25% are black. For our analysis, we restrict the sample to those not currently enrolled in school and those persons receiving an hourly wage between $.50 and $1000 in 1990 dollars (all results of this chapter are reported in 1990 dollars). This chapter uses the NLSY weights for each year to produce a nationally representative sample. However, our sample is not nationally representative in age; we only observe an 8-year range of ages in any given year, and the oldest person in our 1993 sample is only 36.

In 1980, NLSY respondents were administered a battery of ten intelligence tests referred to as the Armed Services Vocational Aptitude Battery. We describe the ASVAB subtests in Table 8.1.

Is Ability Unidimensional?

Herrnstein and Murray (henceforth H&M) argue that there is only one significant intelligence factor, called general intelligence or *g*. They fail to mention, however, that many psychometricians who endorse the theory of general intelligence also maintain that there exist other factors of intelligence that have less explanatory power than *g* but are nonetheless both statistically and substantively significant in describing outcomes. For example, Spearman incorporates specific factors *s*, which complement general intelligence *g*.[4] Cattell describes two forms of general intelligence: "fluid," which is applied to all tasks, and "crystallized," which is a combination of fluid intelligence and practice or study of a specific task.[5] Carroll posits a three-stratum theory of

Table 8.1. The Armed Services Vocational Aptitude Battery

Subtest	Minutes	Description (A subtest of ASVAB measuring...)
General Science	11	Knowledge of the physical and biological sciences
Arithmetic Reasoning	36	Ability to solve arithmetic word problems
Word Knowledge	11	Ability to select the correct meaning of words presented in context and to identify the best synonym for a given word
Paragraph Comprehension	13	Ability to obtain information from written passages
Numerical Operations	3	Ability to perform arithmetic computations (speeded)
Coding Speed	7	Ability to use a key in assigning code numbers to words (speeded)
Auto and Shop Information	11	Knowledge of automobiles, tools, and shop terminology and practices
Mathematics Knowledge	24	Knowledge of high school mathematics principles
Mechanical Comprehension	19	Knowledge of mechanical and physical principles and ability to visualize how illustrated objects work
Electronics Information	9	Knowledge of electricity and electronics
ASVAB Testing Time	144	

intelligence in which cognitive abilities range from the narrow to the highly general.

H&M argue that measured cognitive ability is stable throughout one's lifetime. They write, "The AFQT test scores for the NLSY sample were obtained when the subjects were 15 to 23 years of age, and their IQ scores were already as deeply rooted a fact about them as their height" (p. 130). They raise the immutability of cognitive ability when arguing against the effectiveness of social interventions.

Measured cognitive ability is, in fact, far from immutable. For example, cognitive ability test scores rise with schooling. Among NLSY respondents who were 16 years old when they took the ASVAB, those who had completed ninth grade scored over 0.7 standard deviations higher on AFQT and *g* than those who had only completed eighth grade. This difference is statistically significant at the 1% level. Because school attendance is generally compulsory until age 16, little sorting by grade level could occur among these respondents (although in extreme cases those with low ability may have been held back in school). The fact that an extra year of schooling can have a large impact on measured cognitive ability belies the pessimism of H&M about the efficacy of interventions.

There are three commonly accepted ways to estimate *g*. We use principal components analysis, but principal factor analysis and hierarchical factor analysis produce essentially the same results. The principal components method is the least affected by sampling error,[6] but Ree and Earles find that the correlation between each pair of the three estimates of *g* is 0.996.[7] However, no matter which method is used, *g* is only as good a measure of cognitive ability as its constituent tests. Many features of personality and motivation are not captured by the ASVAB.

Because age at the time of test influences performance on the test, we first residualize each of the ASVAB tests on age at the time of the test, separately by race and gender. The residuals were standardized to mean zero and variance one. Principal components were estimated from the standardized residuals. We calculated *g* by multiplying the test score vector by the eigenvector associated with the largest eigenvalue of the matrix of correlations among standardized ASVAB scores.

Ironically, while H&M embrace the theory of *g*, they use a different (though highly correlated) measure of ability in their analysis: the Armed Forces Qualification Test (AFQT) score, which is the sum of the ASVAB subtests Word Knowledge, Paragraph Comprehension, Arithmetic Reasoning, and Mathematics Knowledge. If AFQT is the best measure of general intelligence, then the first principal component should weight each of the four subtests that constitute AFQT by an equal amount and assign zero weights to all other subtests. We do not find such a pattern. Table 8.2 lists the ASVAB weights for the first principal component; these weights suggest that while AFQT is highly correlated with *g* ($\rho = 0.829$), it is a suboptimal measure of general intelligence, which suggests that H&M may underestimate the effect of intelligence on social outcomes.

Table 8.2 also indicates that the first principal component is strikingly similar across race and gender. This has generally been found to be true for different racial populations that share the same language and culture (see Ref. 6). These loadings are similar to those produced if principal components are computed for the sample as a whole rather than separately for each race and gender group. Speeded tests (Numerical Operations and Coding Speed) receive little weight, while the achievement tests that constitute AFQT are heavily weighted.

Each of the principal components assigns nonzero weight to each of the ten tests, but the second principal component heavily weights the speeded subtests for all groups except black females. Carroll describes this commonly found speeded intelligence factor as "numerical facility," reflecting the fact that the speeded tests usually require rapid arithmetic operations. It should be stressed, however, that principal components are mathematical constructs, and it can be misleading to describe principal components in terms of observed human skills.

Table 8.2. Construction of *g* by race and gender

ASVAB subtest	Black females	Black males	Hispanic females	Hispanic males	White females	White males
General Science	0.351	0.338	0.340	0.336	0.343	0.344
Arithmetic Reasoning	0.325	0.319	0.331	0.325	0.356	0.341
Word Knowledge	0.375	0.352	0.346	0.342	0.354	0.347
Paragraph Comprehension	0.360	0.332	0.339	0.329	0.331	0.331
Numerical Operations	0.311	0.292	0.287	0.287	0.277	0.285
Coding Speed	0.281	0.278	0.274	0.286	0.248	0.270
Auto and Shop Information	0.257	0.302	0.304	0.301	0.272	0.264
Math Knowledge	0.343	0.314	0.319	0.309	0.338	0.324
Mechanical Comprehension	0.243	0.304	0.302	0.316	0.311	0.315
Electronic Information	0.289	0.324	0.312	0.327	0.311	0.328

The specification of *g* is robust to the removal of subtests from the matrix; six subtests must be removed before the numerical facility factor becomes the first principal component. Beyond the second factor there are few similarities in the principal components across race and gender groups.

Table 8.3 contains the proportion of variance in ASVAB tests scores attributable to the principal components. *g*, the first principal component, is dominant in the ASVAB test score matrix—it explains between 55.2% and 70.6% of the variation in the test scores of each race-gender group. The amount of variance explained by *g* depends on the similarity of the tests and the range of ability of the persons constituting the sample. Jensen reports that across 20 independent correlation matrices comprising a total of more than 70 tests, the average percentage of variance accounted for by *g* is 42.7% (with a range of 33.4% to 61.4%).

Only for white men and women does the numerical facility factor explain more than 10% of the variance in test scores (11.4% and 10.8%, respectively). In each racial group *g* has more explanatory power for men than for women.

The dominance of the first factor in explaining variance in the test correlation matrix should not be interpreted as convincing evidence in favor of a single factor called general intelligence. For example, Suppes and Zanotti have shown that a scalar measure can always be constructed so that any vector

Table 8.3. Proportion of variance in test scores attributable to principal components

Principal component	Black females	Black males	Hispanic females	Hispanic males	White females	White males
First (*g*)	0.552	0.637	0.650	0.706	0.579	0.639
Second	0.096	0.085	0.079	0.081	0.108	0.114
Third	0.070	0.060	0.054	0.052	0.068	0.059
Fourth	0.063	0.050	0.043	0.037	0.058	0.046
Fifth	0.060	0.035	0.039	0.028	0.043	0.031
Sixth	0.047	0.032	0.036	0.023	0.039	0.030
Seventh	0.033	0.030	0.031	0.021	0.033	0.025
Eighth	0.031	0.028	0.026	0.020	0.031	0.023
Ninth	0.028	0.026	0.024	0.017	0.022	0.017
Tenth	0.019	0.016	0.017	0.014	0.018	0.016

of finite-valued variables are conditionally independent given the constructed measure.[8] This means that there exists one factor that can explain all the dependence across test scores. Thus the existence of a single factor *g* in the present context is not so much a result derived from the nature of intelligence; rather, it is a consequence of mathematics. Further, the key test for a theory of single intelligence is not how well *g* explains performance on the intelligence tests from which it is derived, but how well it predicts social outcomes. This is the subject of the next section.

The Wage Premium for Ability

Central to the theory of meritocracy is the notion that ability is the basis for achievement. One definition of meritocracy is that ability is rewarded uniformly in the labor market, irrespective of the race or gender of the person in whom the ability is embodied. H&M argue that the U.S. labor market satisfies this definition. They claim that "the racial difference [in 1989 wages] disappears altogether" when they control for age, IQ, and gender (p. 326). We do not confirm this result; we reject that *g* and education earn the same wage return across race and gender.

H&M also overestimate the fraction of wage variance explained by cognitive ability. They note that large residuals are common in wage regressions, and speculate

> What then is this [wage] residual, this *X* factor, that increasingly commands a wage premium over and above education? It could be a variety of factors ... but readers will not be surprised to learn that we believe that it includes cognitive ability. (p. 97)

They cite a study of the NLSY by Blackburn and Neumark, which concludes that the rise in the return to education is concentrated among the smartest workers.[9] Elsewhere we dispute the conclusions of Blackburn and Neumark.[10]

If this and the assumption of general intelligence are correct, then the coefficient for g in wage regressions should be numerically important and statistically significant. Previous research has concluded that g is "dominant" in explaining job performance.[7,11] Dominance in this context means that the contribution to R^2 of additional test score components is "small" relative to that of g. Close examination of this work reveals that the additional components are statistically significant and that g explains much less than half of the variance in the outcomes studied (supervisor ratings and success in military occupational training schools).

To examine the relationship between the ability and wages in our sample we estimate the following model of wages:

$$W_{it} = \beta a_i + \gamma X_{it} + \tau_t + \epsilon_{it}$$

$$E(\tau_t \mid a_i, X_{it}) = 0$$

$$E(\epsilon_{it} \mid a_i, X_{it}) = 0,$$

where W_{it} is the log of hourly wages for person i in year t; a_i is measured ability, which may be a scalar or a vector; and τ_t is an intercept term for year t. X_{it} is a set of "human capital" measures, which we specify to include schooling (measured as grades completed), weeks of tenure in the current job, tenure squared, labor market experience (defined by Mincer as age minus schooling minus 6),[12] and experience squared. ϵ_{it} is the error term for individual i in year t, and ϵ_{it} and $\epsilon_{jt'}$ are statistically independent for all $i \neq j$.

Our regression model is motivated by the failure to reject the null hypothesis that the coefficients are equal across years. Because of the panel nature of the data, the error term is correlated across time for individuals. We correct for this by using Eicker–White standard errors generalized for panel data. Because we restrict analysis to individuals who are out of school and employed, each individual is not necessarily in our sample for all 15 years; the panel is unbalanced. The analysis of this chapter focuses on out-of-school workers, because even persons of high cognitive ability are often forced to take low-paying jobs while enrolled. To include such persons in our sample would cause downward bias in ability coefficients. Unemployed workers are also excluded from the sample because their wage is not observed. Also, 0.8% of all person-year observations are excluded due to unemployment, and 24.7% are excluded because of school enrollment. This does not affect our estimates as long as the population of interest is employed, out-of-school workers. However, if the population of interest includes the unemployed and students, then it is necessary to correct for self-selection into the sample. We use a multinomial probit selection model to correct for this bias using Lee's[13] generalization of the two-step method in Heckman[14] and find that these corrected results are similar to our reported results.

Table 8.4 and 8.5 contain the coefficient estimates of our wage model. Table 8.4 uses only the ten principal components as regressors, and Table 8.5 also includes as regressors the human capital variables, controls for the national and local unemployment rates, and a linear time trend. All principal components are normalized to have a mean of zero and a standard deviation of one.

We fit separate regressions for each race-gender group. The F statistics reported at the bottom of each table indicate that we decisively reject the null hypothesis that the wage returns to ability are equal across race and gender groups. An equal gain in cognitive ability is rewarded in significantly different ways across race and gender in the labor market. This result is robust to alternative specifications of ability (see Ref. 15). These results are inconsistent with the definition of meritocracy advanced by H&M.

The results in Tables 8.4 and 8.5 support the theory of multiple strata of intelligence, with "g" dominant in explaining social outcomes. The first principal component, "g", is statistically significant at the 1% level, and positive for all race-gender groups. The sign of the coefficient of the first principal component is meaningful because the first principal component has positive weights on all ASVAB subtests; a negative coefficient unequivocally means that less intelligent workers earn more. The signs of the coefficients of the second through tenth principal components are irrelevant because each principal component can be reconstructed using the negative of its ASVAB weights to explain an equal amount of ASVAB variance. This reconstructed principal component would have a coefficient of equal magnitude but opposite sign.

The coefficient of g is almost always larger than that of any other principal component. On the whole, these results are similar to those found by Ree et al. for job training and job performance; secondary factors are statistically significant but contribute little to the predictive power (R^2) of the model. Because principal components are mutually orthogonal and their variances equal, their marginal contribution to R^2 is proportional to their coefficients in the models with only test scores as regressors.[16]

H&M argue that wage differentials are largely due to differences in ability, but we find that measured cognitive ability, education, experience, and job tenure together account for less than a third of the variance in wages. The highest R^2 from the wage regressions in Table 8.5 is .2851.

The contribution of measured ability to the overall fit of the model is dwarfed by that of other observed characteristics. Table 8.6 provides upper and lower bounds on the contribution of g and AFQT to R^2 in log wage regressions. If ability is the only regressor included, ability contributes between .118 and

Table 8.4. Cognitive ability as a determinant of wages

Variable	Black		Hispanic		White	
(Principal component)	Females	Males	Females	Males	Females	Males
First	0.1952 (0.0088)	0.1647 (0.086)	0.1823 (0.0117)	0.1531 (0.0120)	0.1965 (0.0062)	0.1535 (0.0058)
	$p = 0.0000$	$p = 0.0000$	$p = 0.0000$	$p = 0.0000$	$p = 0.0000$	$p = 0.0000$
Second	−0.0403 (0.0083)	0.0225 (0.0085)	0.0285 (0.0110)	0.0360 (0.0108)	0.0660 (0.0054)	0.0595 (0.0052)
	$p = 0.0000$	$p = 0.0081$	$p = 0.0095$	$p = 0.0008$	$p = 0.0000$	$p = 0.0000$
Third	0.0102 (0.0086)	−0.0198 (0.0086)	−0.0451 (0.0107)	0.0481 (0.0113)	−0.0389 (0.0052)	−0.0010 (0.0051)
	$p = 0.2350$	$p = 0.0221$	$p = 0.0000$	$p = 0.0000$	$p = 0.0000$	$p = 0.8447$
Fourth	−0.0308 (0.0082)	−0.0008 (0.0080)	−0.0098 (0.0104)	0.0082 (0.0113)	0.0072 (0.0057)	0.0279 (0.0050)
	$p = 0.0002$	$p = 0.9249$	$p = 0.3444$	$p = 0.4710$	$p = 0.2041$	$p = 0.0000$
Fifth	−0.0057 (0.0078)	0.0144 (0.0075)	−0.0023 (0.0111)	0.0181 (0.0112)	−0.0058 (0.0051)	0.0328 (0.0051)
	$p = 0.4712$	$p = 0.0541$	$p = 0.8341$	$p = 0.1053$	$p = 0.2598$	$p = 0.0000$
Sixth	−0.0163 (0.0083)	0.0135 (0.0082)	−0.0323 (0.0116)	0.0088 (0.0114)	−0.0329 (0.0052)	−0.0036 (0.0050)
	$p = 0.0484$	$p = 0.0990$	$p = 0.0053$	$p = 0.4430$	$p = 0.0000$	$p = 0.4648$
Seventh	−0.0109 (0.0084)	−0.0080 (0.0079)	0.0003 (0.0102)	−0.0009 (0.0117)	−0.0053 (0.0051)	−0.0043 (0.0050)
	$p = 0.1918$	$p = 0.3131$	$p = 0.9728$	$p = 0.9370$	$p = 0.2973$	$p = 0.3841$
Eighth	−0.0013 (0.0081)	−0.0125 (0.0076)	0.0104 (0.0101)	0.0082 (0.0115)	0.0087 (0.0052)	0.0089 (0.0052)
	$p = 0.8718$	$p = 0.1013$	$p = 0.3045$	$p = 0.4732$	$p = 0.0937$	$p = 0.0840$
Ninth	0.0096 (0.0076)	0.0163 (0.0084)	0.0155 (0.0108)	0.0055 (0.0113)	−0.0116 (0.0053)	0.0199 (0.0052)
	$p = 0.2066$	$p = 0.0508$	$p = 0.1507$	$p = 0.6256$	$p = 0.0296$	$p = 0.0001$
Tenth	0.0040 (0.0086)	0.0016 (0.0079)	−0.0159 (0.0111)	0.0246 (0.0122)	0.0266 (0.0054)	0.0045 (0.0052)
	$p = 0.6403$	$p = 0.8366$	$p = 0.1524$	$p = 0.0438$	$p = 0.0000$	$p = 0.3802$
R^2	$R^2 = 0.1416$	$R^2 = 0.1022$	$R^2 = 0.1157$	$R^2 = 0.0934$	$R^2 = 0.1230$	$R^2 = 0.0947$
No. of observations	10,979	12,477	7,072	8,338	26,783	27,958
$F[50, 93591] = 19.32$						

Sample includes all valid employed out-of-school person-year observations. OLS regression used with stacked person-year observations. Dependent variable is the log of the hourly wage reported for each year in 1990 dollars. Regressions run separately for race-sex groups based on rejection of the hypothesis that coefficients are equal across groups. Reported standard errors in parentheses are Eicker–White robust standard errors generalized for panel data. NLSY sample weights are used.

Table 8.5. Cognitive ability as a determinant of wages

Variable	Black		Hispanic		White	
(Principal component)	Females	Males	Females	Males	Females	Males
First	0.1235 (0.0093)	0.1045 (0.0084)	0.0904 (0.0140)	0.1084 (0.0124)	0.0903 (0.0066)	0.0828 (0.0066)
	$p = 0.0000$	$p = 0.0000$	$p = 0.0000$	$p = 0.0000$	$p = 0.0000$	$p = 0.0000$
Second	−0.0190 (0.0073)	0.0005 (0.0076)	0.0071 (0.0095)	0.0212 (0.0098)	0.0403 (0.0048)	0.0237 (0.0050)
	$p = 0.0092$	$p = 0.9435$	$p = 0.4542$	$p = 0.0313$	$p = 0.0000$	$p = 0.0000$
Third	0.0068 (0.0075)	−0.0062 (0.0078)	−0.0358 (0.0093)	0.0447 (0.0102)	−0.0095 (0.0047)	0.0247 (0.0050)
	$p = 0.3592$	$p = 0.4210$	$p = 0.0001$	$p = 0.0000$	$p = 0.0423$	$p = 0.0000$
Fourth	−0.0130 (0.0075)	0.0025 (0.0072)	−0.0066 (0.0093)	0.0119 (0.0107)	0.0183 (0.0051)	0.0160 (0.0047)
	$p = 0.0835$	$p = 0.7344$	$p = 0.4749$	$p = 0.2678$	$p = 0.0003$	$p = 0.0006$
Fifth	−0.0064 (0.0070)	0.0120 (0.0070)	−0.0039 (0.0094)	0.0204 (0.0102)	−0.0036 (0.0044)	0.0369 (0.0047)
	$p = 0.3569$	$p = 0.0878$	$p = 0.6771$	$p = 0.0451$	$p = 0.4222$	$p = 0.0000$
Sixth	−0.0116 (0.0073)	0.0105 (0.0073)	−0.0194 (0.0102)	0.0091 (0.0103)	−0.0199 (0.0045)	0.0030 (0.0046)
	$p = 0.1096$	$p = 0.1502$	$p = 0.0582$	$p = 0.3817$	$p = 0.0000$	$p = 0.5069$
Seventh	−0.0107 (0.0070)	−0.0081 (0.0072)	0.0053 (0.0091)	0.0069 (0.0105)	0.0067 (0.0045)	−0.0028 (0.0045)
	$p = 0.1233$	$p = 0.2631$	$p = 0.5599$	$p = 0.5134$	$p = 0.1334$	$p = 0.5338$
Eighth	−0.0023 (0.0071)	0.0006 (0.0070)	0.0110 (0.0090)	0.0068 (0.0102)	0.0055 (0.0046)	0.0092 (0.0048)
	$p = 0.7399$	$p = 0.9322$	$p = 0.2208$	$p = 0.5091$	$p = 0.2313$	$p = 0.0543$
Ninth	0.0010 (0.0068)	0.0105 (0.0071)	0.0065 (0.0097)	0.0062 (0.0104)	−0.0122 (0.0047)	0.0053 (0.0048)
	$p = 0.8769$	$p = 0.1411$	$p = 0.5023$	$p = 0.5508$	$p = 0.0090$	$p = 0.2683$
Tenth	−0.0032 (0.0073)	0.0047 (0.0072)	−0.0087 (0.0101)	0.0229 (0.0112)	0.0064 (0.0047)	0.0017 (0.0047)
	$p = 0.6558$	$p = 0.5177$	$p = 0.3863$	$p = 0.0407$	$p = 0.1749$	$p = 0.7196$

Variable	Black		Hispanic		White	
	Females	Males	Females	Males	Females	Males
Grades	0.0721 (0.0058)	0.0625 (0.0048)	0.0463 (0.0066)	0.0561 (0.0062)	0.0772 (0.0033)	0.0716 (0.0032)
completed	$p = 0.0000$	$p = 0.0000$	$p = 0.0000$	$p = 0.0000$	$p = 0.0000$	$p = 0.0000$
Potential	0.0370 (0.0047)	0.0450 (0.0048)	0.0219 (0.0054)	0.0754 (0.0081)	0.0312 (0.0030)	0.0678 (0.0028)
experience	$p = 0.0000$	$p = 0.0000$	$p = 0.0000$	$p = 0.0000$	$p = 0.0000$	$p = 0.0000$
(Potential	−0.0010 (0.0002)	−0.0015 (0.0002)	−0.0008 (0.0003)	−0.0019 (0.0004)	−0.0012 (0.0002)	−0.0020 (0.0001)
experience)^2	$p = 0.0001$	$p = 0.0000$	$p = 0.0009$	$p = 0.0000$	$p = 0.0000$	$p = 0.0000$
Job tenure	0.0019 (0.0001)	0.0015 (0.0001)	0.0017 (0.0001)	0.0014 (0.0001)	0.0017 (0.0001)	0.0013 (0.0001)
	$p = 0.0000$	$p = 0.0000$	$p = 0.0000$	$p = 0.0000$	$p = 0.0000$	$p = 0.0000$
(Job tenure)^2	−0.0000 (0.0000)	−0.0000 (0.0000)	−0.0000 (0.0000)	−0.0000 (0.0000)	−0.0000 (0.0000)	−0.0000 (0.0000)
	$p = 0.0000$	$p = 0.0000$	$p = 0.0000$	$p = 0.0000$	$p = 0.0000$	$p = 0.0000$
National						
unemployment	−0.0011 (0.0016)	−0.0006 (0.0016)	−0.0044 (0.0022)	−0.0014 (0.0020)	−0.0042 (0.0010)	−0.0030 (0.0009)
rate	$p = 0.4815$	$p = 0.7207$	$p = 0.0443$	$p = 0.4823$	$p = 0.0000$	$p = 0.0008$
Local						
unemployment	0.0605 (0.0102)	0.0643 (0.0093)	0.0570 (0.0167)	0.0849 (0.0158)	0.0917 (0.0070)	0.0674 (0.0063)
rate < 6%	$p = 0.0000$	$p = 0.0000$	$p = 0.0006$	$p = 0.0000$	$p = 0.0000$	$p = 0.0000$
Local						
unemployment	−0.0454 (0.0135)	−0.0313 (0.0130)	−0.0906 (0.0183)	−0.1123 (0.0176)	−0.0609 (0.0077)	−0.0903 (0.0081)
rate ≥ 9%	$p = 0.0008$	$p = 0.0160$	$p = 0.0000$	$p = 0.0000$	$p = 0.0000$	$p = 0.0000$
Linear time	−0.0125 (0.0010)	−0.0117 (0.0008)	0.0038 (0.0010)	−0.0215 (0.0010)	−0.0004 (0.0006)	−0.0150 (0.0005)
	$p = 0.0000$	$p = 0.0000$	$p = 0.0002$	$p = 0.0000$	$p = 0.4689$	$p = 0.0000$
R^2	$R^2 = 0.2851$	$R^2 = 0.2210$	$R^2 = 0.2355$	$R^2 = 0.2304$	$R^2 = 0.2667$	$R^2 = 0.2409$
No. of observations	10,802	12,298	6,923	8,216	26,462	27,552
F[95, 92228]=9.28						

Sample includes all valid employed out-of-school person-year observations. OLS regression used with stacked person-year observations. Dependent variable is the log of the hourly wage reported for each year in 1990 dollars. Regressions run separately for race-sex groups based on rejection of the hypothesis that coefficients are equal across groups. Reported standard errors in parentheses are Eicker–White robust standard errors generalized for panel data. NLSY sample weights are used.

.174 to R^2. When we control for human capital measures (education, job tenure, job tenure squared, work experience, and work experience squared), the marginal increase in R^2 due to ability falls to between .034 and .011. H&M dramatically overstate the degree to which differences in wages among individuals can be attributed to differences in their cognitive ability.

Table 8.6. Contribution of ability to wage determination modeled with and without human capital

Group	Modeled with background variables only		Modeled with human capital		Number of observations
	AFQT	g	AFQT	g	
Black females	0.011	0.191	0.007	0.119	10,802
	(0.000)	(−0.000)	(−0.001)	(−0.001)	
	$p = 0.028$	$p = 0.027$	$p = 0.046$	$p = 0.045$	
Change in R^2 =	0.174	0.162	0.033	0.034	
Black males	0.008	0.157	0.005	0.103	12,298
	(0.002)	(0.001)	(−0.000)	(−0.001)	
	$p = 0.027$	$p = 0.031$	$p = 0.030$	$p = 0.031$	
Change in R^2 =	0.126	0.118	0.024	0.027	
Hispanic females	0.009	0.173	0.005	0.084	6,923
	(−0.001)	(−0.002)	(−0.004)	(−0.004)	
	$p = 0.083$	$p = 0.081$	$p = 0.090$	$p = 0.091$	
Change in R^2 =	0.155	0.143	0.017	0.013	
Hispanic males	0.006	0.147	0.004	0.111	8,216
	(0.002)	(0.001)	(−0.001)	(−0.001)	
	$p = 0.110$	$p = 0.114$	$p = 0.106$	$p = 0.108$	
Change in R^2 =	0.131	0.135	0.014	0.024	
White females	0.010	0.185	0.004	0.084	26,462
	(−0.002)	(−0.002)	(−0.004)	(−0.004)	
	$p = 0.063$	$p = 0.070$	$p = 0.061$	$p = 0.063$	
Change in R^2 =	0.165	0.150	0.012	0.012	
White males	0.007	0.141	0.004	0.083	27,552
	(0.000)	(−0.000)	(−0.003)	(−0.003)	
	$p = 0.084$	$p = 0.093$	$p = 0.084$	$p = 0.086$	
Change in R^2 =	0.157	0.138	0.011	0.014	

Sample includes all valid employed out-of-school observations. OLS regression used with Eicker–White robust standard errors generalized for panel data in parentheses. Dependent variable is the log of the hourly wage reported for each year in 1990 dollars. Background variables include local and national unemployment rates and a linear time variable. Human capital includes education, experience, and job tenure with quadratic terms. Regressions run separately for race-sex groups based on rejection of the hypothesis that coefficients are equal across groups.

Table 8.6 also indicates that it makes little difference in terms of predictive power whether *g* or AFQT is used as a cognitive ability measure. The difference in R^2 between them (controlling for education, experience, and job tenure) is less than .02 for each race-gender group.

Conclusions

Our results are consistent with the theory of general intelligence: our version of *g* explains a high proportion of the variance in test scores and *g* is remarkably similar across race and gender. However, our results conflict with the predictions of H&M. Ability factors other than *g* are economically useful. Compared with education, family background, and region of residence, *g* explains little of the variance in wages; if there exists some "X factor" that can explain the large residuals common in wage regressions, it is not measured cognitive ability.

We also find that the returns to *g* differ significantly across race and gender; payment is not made for "ability" alone, which violates the definition of meritocracy advanced by H&M. In summary, our reanalyses of the NLSY data originally analyzed by H&M show measured cognitive ability is correlated with wages but explains little of the variance in wages across individuals and time. This finding is mirrored in Ecclesiastes 9:11:

> ... [T]he race is not to the swift, nor the battle to the strong, neither yet bread to the wise, nor yet riches to men of understanding, nor yet favor to men of skill; but time and chance happeneth to them all.

References

1. Herrnstein, R.J. and Murray, C. (1994), *The Bell Curve: Intelligence and Class Structure in American Life*, The Free Press, New York.
2. Thurstone, L. (1947). *Multiple Factor Analysis: A Development and Expansion of The Vectors of The Mind*, University of Chicago Press, Chicago.
3. Carroll, J.B. (1993), *Human Cognitive Abilities: A Survey of Factor-Analytic Studies*, Cambridge University Press, Cambridge.
4. Spearman, C. (1927), *The Abilities of Man: Their Nature and their Measurement*, Macmillan, New York. Reprinted: (1981), AMS Publishers, New York.
5. Cattell, R.B. (1987), *Intelligence: Its Structure, Growth, and Action*, North-Holland, New York.
6. Jensen, A.R. (1987), "The *g* Beyond Factor Analysis," in R.R. Ronning, J.A. Glover, J.C. Conoley, and J.C. Dewitt (Eds.) *The Influence of Cognitive Psychology on Testing and Measurement*, Lawrence Erlbaum, Hillsdale, NJ.
7. Ree, M.J., and Earles, J.A. (1991), "Predicting Training Success: Not Much More Than *g*," *Personnel Psychology*, 44, 321–332.
8. Suppes, P., and Zanotti, M. (1981), "When are Probabilistic Explanations Possible?," *Synthese*, 48, 191–199.
9. Blackburn, M.L., and Neumark, D. (1993), "Omitted-Ability Bias and the Increase in the Return to Schooling," *Journal of Labor Economics*, 11, 521–544.

10. Cawley, J., Heckman, J.J., Lochner, L., and Vytlacil, E. (1996), *Ability, Human Capital, and Wages*, unpublished manuscript, University of Chicago.
11. Ree, M.J., Earles, J.A., and Teachout, M.S. (1994), "Predicting Job Performance: Not Much More Than *g*," *Journal of Applied Psychology*, 79, 518–524.
12. Mincer, J. (1974), *Schooling, Experience, and Earnings*, Columbia University Press, New York.
13. Lee, L-F. (1983), "Generalized Econometric Models with Selectivity," *Econometrica*, 51, 507–512.
14. Heckman, J.J. (1979), "Sample Selection Bias as a Specification Error," *Econometrica*, 47,153–161.
15. Cawley, J., Conneely, K., Heckman, J., and Vytlacil, E. (1996), *Measuring the Effects of Cognitive Ability*, National Bureau of Economic Research, Working Paper #5645.
16. Goldberger, A. (1968), *Topics in Regression Analysis*, McMillan, New York.

chapter nine

The Hidden Gender Restriction: The Need for Proper Controls When Testing for Racial Discrimination

ALEXANDER CAVALLO, HAZEM EL-ABBADI, AND RANDAL HEEB

In Chapter 14 of *The Bell Curve*, Herrnstein and Murray (H&M) examine the earnings gap between blacks and whites.[1] In one of the most striking and provocative findings in their book, they conclude that this gap results not from racial discrimination, as is often assumed, but rather it stems from an inherent difference in intellectual ability, as measured by AFQT.* H&M find no difference in earnings between blacks and whites after taking into account AFQT, age, and a measure of parental socioeconomic status (SES). They report, "[After controlling for age, education, and parental SES,] ... black wages are still only 84 percent of white wages, again suggesting continuing

*H&M generally refer to their measure of cognitive ability as "IQ." This is not strictly accurate. *IQ* usually refers to a predictor of Spearman's *g*. H&M devote a great deal of effort to determining the weighting factors necessary to combine the ten Armed Services Vocational Aptitude Battery (ASVAB) subtest scores to get the best predicted score for Spearman's *g*. The results of this effort are reported in their Appendix 3 (pp. 579–584). They then disregard this work and instead use a different, arbitrary weighting of these scores, the Armed Forces Qualifying Test (AFQT). Although we have no great objections to using AFQT as a measure of cognitive ability, to label AFQT as IQ is simply wrong. Accordingly, we will refer to the cognitive ability test scores as AFQT, or simply *ability*. Throughout, we are referring to the same test scores that H&M call "IQ."

racial discrimination. And yet, controlling just for [AFQT], ignoring both education and socio-economic background, raises average black wage to 98 percent of the white wage ..." (p. 324).

H&M refer to this finding as part of their indictment of affirmative action policies. In their argument against the need for affirmative action, they claim that "after controlling for [AFQT], it is hard to demonstrate that the United States suffers from a major problem of racial discrimination in occupations and pay" (p. 480). H&M come to their conclusion that no racial earnings differential exists after controlling for cognitive ability by looking at regressions of annual earnings, controlling for age, AFQT, and parental SES. This analysis is replicated in the section on earnings analysis. However, their results do not stand up to closer inspection. On the contrary, their analysis fails to properly control for gender effects.

Repeating H&M's analysis with an appropriate model of the effects of gender reveals substantial differences between earnings of whites and blacks. Simply correcting for gender effects, our analysis shows that black men with average characteristics are predicted to earn about 6% less than comparable white men, after controlling for age, ability, and parental SES. Black women with average characteristics earn about 15% more than comparable white women in the same formulation. In addition, the racial earnings difference increases substantially with age in this version of the model. The earnings gap increases by about 4% per year of age for men and about 2% per year of age for women. However, when men and women are grouped together, these differences largely offset each other and the racial earnings gap is obscured. As a result, H&M obfuscate substantial and intriguing differences in the predicted earnings of the four race-gender groups. In the section on controlling for gender, we correct H&M's analysis by properly controlling for gender effects and show that substantial racial earnings differentials remain even after controlling for AFQT.

Another problem with the H&M analysis is that they propose an unusual and particularly unsatisfactory measure of racial discrimination. H&M compare the mean expected earnings of whites and blacks with "average" characteristics and declare that no racial discrimination is present if these means are similar. Their test is problematic on several grounds. First, there is no measure of statistical precision by which to judge the results of this discrimination test. In addition, by testing for discrimination only at the average of all their control variables, H&M overlook important differences in earnings between whites and blacks with other combinations of age, AFQT and parental SES. In the section on discrimination, we apply a rigorous statistical test for discrimination to H&M's model. We find evidence of racial discrimination even in H&M's own formulation.

Many of the deficiencies of H&M's analysis result from their unorthodox approach, which causes them to overlook gender effects and to fail to apply standard statistical tests to their conclusions. In the section on wage function, we present a more standard labor economics analysis of wages and apply the usual statistical tests for racial discrimination in this model. We find evidence of substantial racial differentials that persist after controlling for cognitive ability. Part of the wage differential is directly attributable to racial discrimination, but most of the wage gap is caused by factors that may or may not be influenced by discrimination.

H&M do not explicitly distinguish between earnings differentials caused by market discrimination and earnings differentials caused by differing endowments of characteristics, such as cognitive ability, education, work experience, and parental SES. Market discrimination occurs when otherwise identical individuals of different races receive different salaries. Racial earnings differences attributable to differing endowments may be only in part associated with racial discrimination. For example, discrimination may cause part of racial difference in accumulation of work experience, education, or SES.

In the section on the racial wage gap, we explicitly decompose the racial wage gap for men into differences caused by market discrimination and differences caused by premarket factors, such as accumulated education and work experience. The portion of the wage gap that results from blacks being treated differently than whites is defined as market discrimination. The remainder of the wage gap, which would exist even if blacks and whites were treated identically, is caused by racial differences in the accumulation or endowments of ability, education, experience, and parental SES.

We find that market discrimination plays a relatively small role in the racial wage gap, amounting to only 8% of the predicted difference in wages between black and white men. We also find, in agreement with H&M, that cognitive ability is an extremely important determinant of wages. However, we refute H&M's more controversial claim that blacks and whites have the same predicted earnings after controlling for ability. On the contrary, substantial earnings differentials remain after controlling for AFQT. These differences are largely explained by variables that H&M exclude from their analysis, such as work experience and education. Our results indicate that premarket factors in attainment of education, experience, and ability play a larger role in racial wage differentials than market discrimination. The question of the degree to which racial discrimination impacts these premarket factors remains unanswered.

Earnings Analysis in *The Bell Curve*

To answer the question of how much of the racial earnings gap remains after controlling for differences in cognitive ability, H&M employ data from the National Longitudinal Study of Youth (NLSY). The NLSY contains detailed earnings surveys as well as a test of cognitive ability, AFQT, for a nationally representative sample of over 12,000 youths who were aged 14 to 22 in 1979. This group of young people has been surveyed every year since 1979. H&M conduct their study by looking at a sample of year-round workers from the 1990 NLSY survey.* The general H&M model is a regression of earnings of year-round workers on a progression of independent variables:

$$\begin{aligned} \text{Earnings} &= X\beta_w + \text{error} \quad \text{(for whites)} \\ \text{Earnings} &= X\beta_b + \text{error} \quad \text{(for blacks)}, \end{aligned} \tag{9.1}$$

where X is a vector of independent variables that includes an intercept, age, and one or more of the following variables: AFQT, parental SES, and education. β is a vector of regression coefficients. For example, for a white worker, the model could be

$$\text{Earnings} = \beta_{0w} + \beta_{1w}age + \beta_{2w}\text{AFQT} + \beta_{3w}\text{SES} + \text{error}. \tag{9.2}$$

For a black worker, the βs may differ. The particular variables in X are varied to show the impact of AFQT on the racial earnings gap, relative to the other factors.

Table 9.1 reproduces the regressions upon which H&M base their discussion.† For this analysis, we constructed the same independent variables used by H&M from the NLSY, including normalized and skew-corrected AFQT scores as a measure of cognitive ability and an index of parental SES.‡ Models I–VI in Table 9.1 are variations on equation 9.1, with different

---

*See the appendix to this chapter for details about data construction. The independent variables, age, AFQT, parental SES, and education (highest grade completed) used in our analysis are constructed to mimic H&M's variables. The dependent variable, annual earnings of year-round workers, is also the same as used by H&M. Our sample differs slightly in that we are more restrictive regarding extremely high or low reported earnings figures. This treatment causes no qualitative or interpretive differences.

†H&M discuss their earnings analysis on pp. 322–326. Their regression results are detailed in Appendix 6, Table 4, p. 651. The relative earnings of whites and blacks reported in this section are the same as H&M report in their Table 4. Our replications of their regressions are reported in our Table 9.1. These replicated regressions are consistent with H&M's interpretations and qualitative results. The relative earnings differences between races are the same. Our sample yields intercepts that are somewhat lower than H&M for all races. The slope coefficients are comparable.

‡H&M standardized all their variables to have mean zero, standard deviation one. In the case of AFQT and SES, which have no natural units, we have followed this practice. For age, education, and, in the later sections, work experience, we have centered the data to have mean zero, without dividing by the standard deviation, so that years remains the unit for these variables. Our coefficients on education, for example, can be interpreted as returns to one additional year of education. To obtain coefficients comparable with H&M, multiply our coefficients by the standard deviation of the variable in question: for age 2.5 and for education 0.96 (p. 651).

Table 9.1. Replication of H&M's earnings analysis

Independent variable	H&M Model I		Model II		Model III		Model IV		Model V (Model III + Edu)		Model VI	
	White	Black	White	Black	White	Black	White	Black	White	Black	White	Black
Intercept	24,667**	19,201**	23,043**	22,548**	22,857**	22,771**	23,696**	19,064**	23,475**	19,715**	23,015**	21,831**
	(347)	(334)	(315)	(510)	(318)	(523)	(310)	(307)	(325)	(368)	(323)	(472)
Age	1,120**	488**	814**	312*	863**	345**	1,043**	484**	1,057**	502**	912**	377**
	(170)	(154)	(160)	(134)	(163)	(132)	(162)	(141)	(162)	(138)	(164)	(132)
AFQT			4,928**	4,759**	4,232**	4,300**					2,647**	3,269**
			(392)	(419)	(436)	(445)					(484)	(476)
SES					1,548**	1,126**			1,257**	1,306**	746	812*
					(456)	(428)			(428)	(381)	(448)	(399)
Education							1,958**	1,728**	1,723**	1,507**	1,169**	794**
							(167)	(196)	(178)	(191)	(198)	(202)
R^2	0.0371		0.1123		0.1178		0.1154		0.1192		0.1325	
Adj. R^2	0.0364		0.1112		0.1163		0.1143		0.1176		0.1305	
n	3,916		3,916		3,916		3,916		3,916		3,916	

Standard errors are in parentheses.
* coeficient significant to .05 probability level.
** coefficient significant to .01 probability level.
See text footnote * on p. 196 for discussion of samples differences from H&M. Edu, education; SES, socioeconomic status. Adj., adjusted.

Table 9.2. Sample means for earnings functions

Variable	Full sample	By race		By race and gender			
				Males		Females	
		White	Black	White	Black	White	Black
Earnings	24,017	19,253	19,253	27,926	20,510	20,094	17,717
Age	28.7	28.7	28.7	28.7	28.6	28.4	28.7
AFQT	0.197	0.327	−0.701	0.341	−0.800	0.307	−0.580
SES	0.173	0.268	−0.470	0.258	−0.483	0.282	−0.491
Education	13.53	13.58	13.16	13.47	12.85	13.74	13.54
n	3,917	2,830	1,087	1,614	588	1,216	499

SES, socioeconomic status.

independent variables.* In Model I, H&M regress 1989 annual earnings on age and an intercept. This model is the baseline, showing the earnings gap before controlling for factors other than age. H&M observe that the predicted earnings for a black worker of average age (about 28.7 years old) is about 23% lower than for a white worker the same age. Model II demonstrates the reduction in the expected earnings gap that results from controlling for age and AFQT. A white of average age earns only about 2% more than a comparable black, after controlling for AFQT.

In Model III, H&M regress earnings on age, AFQT, and parental SES. We call Model III "H&M's preferred model," because it is the model they report in detail in the text (p. 323). In this model they find that the racial earnings gap virtually disappears. Models IV and V, by comparison, show the gap that remains after controlling for education and parental SES, variables that are conventionally thought to be related to racial earnings differentials. After controlling for these two factors, a typical 28-year-old black worker earns about 16% less than a comparable white.

A racial earnings gap exists between whites and blacks. H&M claim that differences in AFQT account for this earnings difference (p. 324). Model VI, which includes education with AFQT, parental SES, and age, is not reported in *The Bell Curve*. This regression shows that the earnings gap reappears in the model that includes both education and AFQT together. Moreover, this

*H&M include Latinos in their analysis. Because the focus of their discussion is on the black/white earnings gap, we will exclude Latinos from our discussion entirely. No qualitative effects arise from this exclusion. Note that the wage and earnings equations for both races (and, where appropriate, for all race-gender combinations) were estimated simultaneously for all regressions reported in this chapter.

regression explains a larger portion of the overall variance in earnings than any other regression reported by H&M.

Controlling for Gender: Testing Implied Restrictions

H&M state that they controlled for gender and found that blacks earn even more than whites after controlling for both gender and AFQT (p. 326). Unfortunately, they do not present these results. As we shall see, this omission proves critical. Consider a gender-enhanced variant of the H&M earnings model:

$$
\begin{aligned}
\text{Earnings} &= X\beta_{wf} + \text{error} \quad \text{(for white females)} \\
\text{Earnings} &= X\beta_{bf} + \text{error} \quad \text{(for black females)} \\
\text{Earnings} &= X\beta_{wm} + \text{error} \quad \text{(for white males)} \\
\text{Earnings} &= X\beta_{bm} + \text{error} \quad \text{(for black males)}
\end{aligned}
\tag{9.3}
$$

where X is a vector that includes an intercept, age, AFQT, and parental SES, and β is a vector of race- and gender-specific coefficients. Recall equation 9.1, the H&M preferred model. Notice that equation 9.1 is a special case of equation 9.3, with the model parameters assumed to be identical for males and females of the same race. In other words, H&M have imposed the restriction that $\beta_{wf} = \beta_{wm}$ and $\beta_{bf} = \beta_{bm}$. The plausibility of this restriction can be easily tested using a Wald test. On implementing this test we find that H&M's assumptions are rejected ($p < .01$).* This result suggests that males and females of different races are apt to be treated differently.

Graphically, the difference between the racial earnings profiles of males and females is quite striking. Figure 9.1 shows the expected earnings by age for blacks and whites, using H&M's earnings equation 9.1, holding AFQT and parental SES at their respective population averages. The lines intersect near the mean, which H&M interpret as evidence that there is no expected difference in earnings between a black worker and a white worker of mean age, AFQT and parental SES. Table 9.2 shows the sample means by race and gender for the variables in the earnings analysis.

Figures 9.2 and 9.3 show the same plot of expected earnings by age for males and females, respectively. Obviously these graphs are very different from the combined data shown in Figure 9.1. For men (Figure 9.2) there is a substantial predicted earnings gap at most ages. This gap grows larger for older men in the sample. In contrast, the predicted earnings gap for women

---

*Because there is heteroskedasticity present in this model (the variance in earnings is not constant across observations), we construct the test statistic using White's heteroskedasticity-consistent covariance matrix.

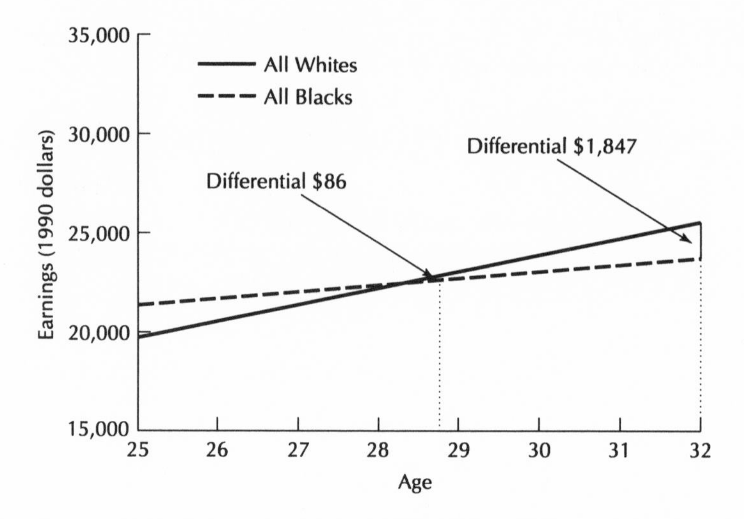

Figure 9.1. Predicted 1989 earnings by age. Using the coefficient estimates of Model III in Table 9.1, the predicted earnings functions are plotted against age, holding AFQT and parental SES constant at their population means of 0. The vertical line at age 28.7 marks the average age in the sample. This point on the chart represents the predicted earnings of the "average" person (i.e., a person of average age, average AFQT score, and average parental SES). At this age, whites earn only $86 more than comparable blacks. The vertical line at age 32 marks the age of the oldest individuals in the NLSY sample. At this age, whites earn $1,847 more than blacks with identical characteristics.

is negative. Black women are predicted to earn more than white women after controlling for age, AFQT, and parental SES. This gap also grows for older women in the sample. Figure 9.4 shows a plot of the predicted racial earnings gap for three samples: men, women, and H&M's gender-pooled sample. The racial earnings differential is of opposite sign for men than for women, after controlling for AFQT, SES, and age. When the two genders are pooled, the racial differentials are largely offset. Because they combined genders, H&M were led to the incorrect conclusion that there is no racial earnings differential at all, after controlling for AFQT and parental SES.

H&M state that they controlled for gender in regressions that were not reported in detail, and found that the black intercept is 101% of the white intercept (p. 326). This result obtains if one runs the regression of earnings from equation 9.1, with the intercept adjusted for each gender* (see Table 9.3, Panel 1). It is important to note that this procedure does not meaningfully

*Specifically, this result obtains in a regression of equation 9.1 with an intercept, a gender dummy variable, and a racial dummy variable.

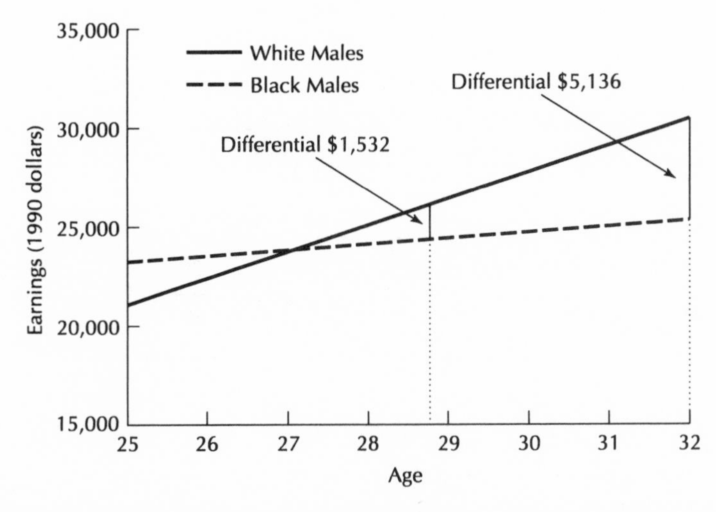

Figure 9.2. Predicted 1989 earnings by age, males only. Using the coefficient estimates shown in Panel 3 of Table 9.3, the predicted earnings functions are plotted against age for males only. AFQT and parental SES are again held constant at their population means of 0. The vertical line at age 28.7 marks the average age in the sample. This point on the chart represents the predicted earnings of the "average" male (i.e., a male of average age, average AFQT score, and average parental SES). At this age, white males earn $1,532 more than comparable black males. At age 32, white males earn $5,136 more than black males with identical characteristics. This figure shows the dramatic effect of proper gender controls. The earnings gap for "average" males is a $1,532 advantage for white males. This is a nearly 20-fold increase when compared with the results in Figure 9.1. Notice, once again, that the earnings differential grows with age to a $5,136 earnings advantage for average white males at age 32. The finding of no racial discrimination is clearly inconsistent with this evidence on the earnings of males.

control for gender in the context of racial differences because it assumes that the effects of gender are the same for both races. This assumption is invalid, as noted earlier. To make matters worse, the race-gender intercepts (e.g., white men, black women, etc.) are unrecoverable from such a regression.

Table 9.3 shows the H&M preferred model, with various controls for gender. Table 9.3, Panel 1, shows that the inclusion of an intercept adjusted for females obtains the result H&M report regarding gender. The Panel 1 model imposes the assumption that the gender effect is the same for both races and that all coefficients are the same for both genders within each race. In this formulation racial differences are obscured. Table 9.3, Panel 2, shows the effect of including separate intercepts for each race-gender combination. This model differs from Panel 1 by relaxing the assumption that the gender effect is the same for both races. The racial earnings differential reappears in

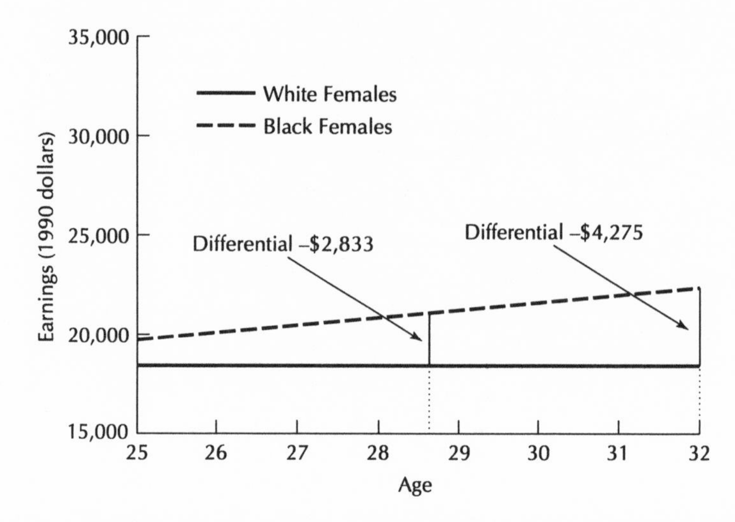

Figure 9.3. Predicted 1989 earnings by age, females only. Using the coefficient estimates shown in Panel 3 of Table 9.3, the predicted earnings functions are plotted against age for females only. AFQT and parental SES are held constant at their population means of 0. The vertical line at age 28.7 marks the average age in the sample. This point on the chart represents the predicted earnings of the "average" female (i.e., a female of average age, average AFQT score, and average parental SES). At this age, white females earn $2,833 less than identically endowed black females. At age 32, white females earn $4,275 less than black females with identical characteristics. A notable feature in this figure is that the white/black earnings gap has reversed direction, becoming an earnings advantage for black females relative to white females. For the "average" female blacks earn $2,833 more than whites, whereas at age 32 black females earn $4,275 more than white females. This dramatic evidence of racial discrimination is obscured when males and females are pooled in the analysis (see Fig. 9.1).

this formulation. Panel 2 still imposes the assumption that the coefficients on variables other than the intercept are the same for both genders within each race. Table 9.3, Panel 3, shows a still better model, which allows the coefficients for each of the independent variables to vary for each race-gender group. The Panel 3 model does not impose any assumed restrictions on the coefficients in that model. The restricted models, Panels 1 and 2, are rejected in favor of the unrestricted model, Panel 3, by a Wald test ($p < .01$ for both tests.)

Detecting Discrimination

H&M propose a simple method for detecting racial discrimination. Their method examines the intercept coefficient for each race. In their analysis, all of the independent variables are centered or standardized to have population mean zero. They make this unusual adjustment so they can interpret the

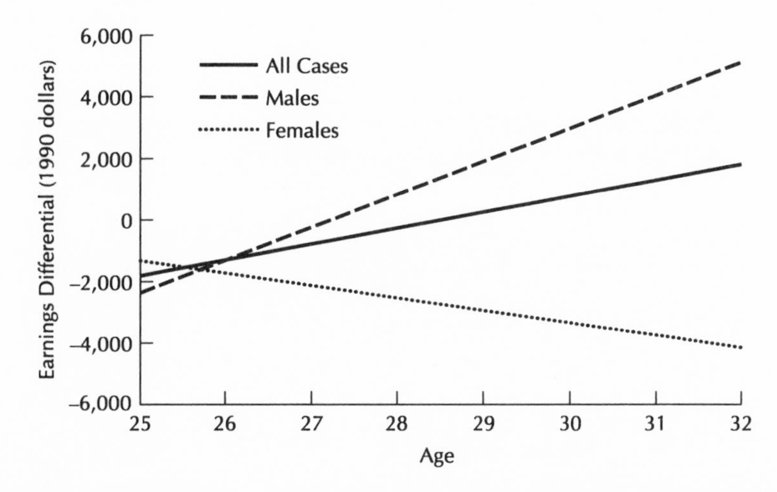

Figure 9.4. White/black earnings differential by age. This figure plots the white/black difference in predicted earnings by age. AFQT and parental SES are held constant at their population means of 0. The line marked "All Cases" refers to the earnings differential implied in Figure 9.1. The lines labeled "Males" and "Females" refer to the earnings differentials implied in Figures 9.2 and 9.3, respectively. The earnings gap for "All Cases" lies in between the earnings gaps for males and females. This shows that white males earn substantially more than black males of identical AFQT and parental SES over a large range of ages, and also that black females earn substantially more the white females of identical AFQT and parental SES over a large range of ages. However, when both genders are combined in estimation, the black/white earnings difference is much smaller in magnitude. Herrnstein and Murray's assumption that the coefficients for males and females are identical for both races conceals substantial racial differences in earnings that appear when proper gender controls are employed.

intercepts as the predicted earnings of a person with population average characteristics for all of the independent variables. If their model is correct, then identical intercepts for whites and blacks imply that whites and blacks with the same (population average) characteristics for the independent variables have the same predicted earnings. If the intercepts are similar, then H&M conclude that no discrimination is present. H&M do not offer any particular degree of similarity between intercepts as the threshold beyond which discrimination can be said to be present, nor do they suggest any level of significance at which the null hypothesis of no racial discrimination should be rejected.

There are two problems with the H&M approach. The first problem is that their method only identifies racial discrimination at the population mean for all the independent variables. This does not imply no racial earnings differential at other values of the independent variables. For example, the coefficients on age in the H&M models are very different for blacks and whites. This implies that the expected earnings differential changes with age, after controlling for

Table 9.3. The effect of gender in H&M–type earnings functions

Indep. variable	H&M–type model w/ female intercept		H&M–type model w/ race-gender intercepts		H&M–type model w/ all coefficients allowed to vary by race and gender			
					Males		Females	
	White	Black	White	Black	White	Black	White	Black
Earnings	25,772**	26,295**	25,963**	24,703**	25,906**	24,374**	18,403**	21,236**
	(432)	(587)	(455)	(565)	(446)	(636)	(352)	(841)
Female	−7,072**	(c)	−7,536**	−3,876**				
	(528)		(598)	(593)				
Age	786**	375**	781**	361**	1376**	316	−18	406*
	(155)	(135)	(155)	(131)	(241)	(179)	(160)	(189)
AFQT	4,147**	4,947**	4,142**	4,655**	3,809**	4,115**	4,488**	5,606**
	(426)	(444)	(426)	(427)	(568)	(497)	(562)	(781)
SES	1,641**	901*	1,647**	1,003*	2,001**	1,213*	1,031	738
	(450)	(424)	(451)	(419)	(633)	(579)	(599)	(601)
R^2	0.1672		0.1687		0.1775			
Adj. R^2	0.1655		0.1668		0.1744			
n	3,916		3,916		3,916			

Standard errors are in parentheses
(c) Female coefficient same for both race groups.
**Coefficient significant to .01 probability level.
* Coefficient significant to .05 probability level.
Adj., adjusted; SES, socioeconomic status.

AFQT and SES. Refer again to Figure 9.1. Although predicted earnings are nearly the same for whites and blacks at the mean age for the gender-pooled regression, the fact that the coefficients for age differ for whites and blacks results in an annual earnings differential of $3,939 by age 32, the age of the oldest NLSY respondents. Moreover, the NLSY is a study of youth. The increasing earnings gap is particularly significant among such a young sample. H&M's test for discrimination is at best incomplete. They consider only the earnings gap among people at the mean of all their independent variables. A substantial differential exists at ages older and younger than the mean.

The second problem with the H&M discrimination test is that it lacks any measure of statistical precision. We cannot determine the confidence one should place in H&M's assertion that there is no racial earnings difference after controlling for AFQT.

Fortunately, both problems can be partially remedied using a statistical test for discrimination. If earnings are predicted by measures of productivity, such as AFQT, education, and work experience, and if the returns to such measures

Table 9.4. Sample means for log-linear wage functions

	Means by race and gender							
	Males		Females		Means by race		Means by gender	
Variable	White	Black	White	Black	White	Black	Males	Females
log(*w*)	2.46	2.19	2.23	2.11	2.36	2.15	2.43	2.22
Age	30.6	30.6	30.5	30.5	30.6	30.6	30.6	30.5
Age^2	944.4	941.3	937.9	937.3	941.7	939.4	944.1	937.8
AFQT	0.367	−0.771	0.284	−0.576	0.332	−0.678	0.241	0.165
SES	0.318	−0.556	0.250	−0.448	0.290	−0.504	0.221	0.154
Education	13.70	12.98	13.76	13.69	13.72	13.27	13.62	13.74
Experience	9.70	8.66	8.25	7.98	8.34	9.09	9.58	8.21
$Exp.^2$	102.57	83.24	78.29	73.67	78.64	92.39	100.42	77.65
P.T. exp.	1.76	1.43	2.81	1.90	1.66	2.20	1.73	2.68
$(P.T.\ exp.)^2$	6.41	4.25	13.74	6.68	5.42	9.48	6.17	12.77
n	1,295	546	994	520	2,289	1,066	1,841	1,514

Means for age, education, and work experience were centered around zero in the regressions.
Exp., experience; P.T., part-time; SES, socioeconomic status.

are the same for both races, then one could say that earnings are "color-blind," in the sense that two otherwise identical people of different races would have the same predicted earnings. Note that this is not the same as saying that there is no racial discrimination in earnings.

Racial discrimination might result in lower earnings, even with a color-blind earnings function. For example, accumulated work experience may be impacted directly by discrimination in hiring, and indirectly by racial discrimination in employer location and investment decisions. Work experience is also affected other factors not associated with discrimination, including individual preferences, local economic prosperity, and differing attitudes regarding the importance of work experience. Educational attainment is affected by factors such as the quality of schools and the accessibility of higher education. It may be difficult to determine the degree to which the root causes of educational differences are discriminatory. Similarly, parental SES may be correlated with other discriminatory factors, such as parents' housing and job opportunities. On the other hand, if the earnings function itself differs by race, this may be interpreted as *de facto* evidence of racial discrimination. It is the color-blindness of the earnings function that we will now test.

Recall equation 9.3, the gender-enhanced version of H&M's preferred model, and assume for a moment that this is the true model. We have shown earlier that the assumption that gender effects are the same for both races is strongly rejected. To test for racial discrimination, we now consider the null hypothesis that the earnings function is "color-blind" in the sense that the coefficients in equation 9.3 are the same for both races. Just as when we tested the assumption that the coefficients were the same for both genders, we now test the assumption that the coefficients are the same for both races, allowing gender differences. Consider the color-blind model:

$$\begin{aligned} \text{Earnings} &= X\beta_f + \text{error} \quad \text{(for females)} \\ \text{Earnings} &= X\beta_m + \text{error} \quad \text{(for males)} \end{aligned} \tag{9.4}$$

The plausibility of the color-blind model can be tested using a Wald test. On implementing this test we reject the hypothesis that the earnings function is color-blind ($p < .01$).* We find *de facto* evidence of the presence of racial discrimination in H&M's preferred model, after correcting for gender. This model predicts that earnings for black men with population average characteristics are about 6% lower than comparable white men at the average age of 28.7, and that this earnings gap grows larger for older men in the sample. For women, the picture is reversed, with black women with population average characteristics earning about 15% more than comparable white women. However, serious deficiencies in the H&M model limit its usefulness for making assessments of racial discrimination. These deficiencies are addressed in the following sections.

A Standard Wage Function

H&M depart from the model most economists would consider a standard wage function. This departure led directly to two significant failures of their analysis: the failure to separate males and females, which causes them to miss the racial earning differences, and the failure to include education and experience variables, which leads to further distortions in the coefficients they estimate. For illustration, we present here a more conventional wage model estimated using the NLSY data.

Log wages are typically chosen as the dependent variable in analyses of individual productivity. The number of hours worked in a year varies significantly, even among full-time workers. Because a worker's annual earnings are the product of the worker's wage and the total number of hours worked, differences in either the wage or the number of hours worked will be reflected in annual earnings. To reduce the impact of the number of hours worked on our calculation of racial differences, we use log hourly wage as our dependent vari-

*Because there is heteroskedasticity present in this model (the variance in earnings is not constant across observations), we construct the test statistic using White's heteroskedasticity-consistent covariance matrix.

able.* The log wage formulation also has the desirable property that extremely high values are "discounted" somewhat, reducing the distortion introduced by outliers.

Independent variables in log wage functions typically include age, work experience, and education. These variables all have important theoretical reasons for inclusion in the model. Earnings profiles are distinctly nonlinear, so quadratic terms in age and work experience are generally included. Work by Murphy and Welch suggests that still higher order terms are necessary to capture the lifetime pattern of returns to these variables, so economists' standard practice may soon change to include quartic formulations.[2] However, because our NLSY sample contains a range of ages (and experience) of only 10 years, only the conventional quadratic terms are included. Quadratic terms should adequately approximate the earnings profiles over such a short time period.

There are sound theoretical reasons for including work experience. We expect workers to gain skill and productivity from practice, as well as from formal and informal training that occurs on the job.[3] We have developed a detailed weekly experience variable from the NLSY, for both full-time and part-time experience, that is used in this analysis. Detailed experience variables like these have been used successfully by Wolpin to predict labor market outcomes.[4]

Education is extremely important for predicting wages. Including education dramatically increases the R^2 of any regression of wages, including H&M's models with AFQT, and so should be included in the regression. Refer again to Table 9.1, Model VI. The R^2 statistic improves noticeably from Model III to Model VI, when education is added to the equation. Moreover, there are strong theoretical reasons for believing that education is important for determining productivity, wages, and earnings.[5] In this analysis we have included the highest grade completed. A better formulation might include variables for degree attained, rather than years of schooling, if one expects that a diploma carries more weight with employers than do years of school experience.

The single most important factor to control in earnings regressions is gender. This factor is so critical that economists almost always estimate separate equations for men and women. There are many reasons for this distinction. Women often leave and then reenter the workforce as a result of childbear-

---

*This is not an innocuous decision. The number of hours worked varies substantially across the race-gender groups. White males work more hours than any other group, even among full-time workers. This difference accounts for a sizable portion of the racial differences in annual earnings. Whether this difference in hours worked is the result of discrimination or other factors is beyond the scope of this chapter.

ing and child-rearing decisions. This indirectly influences their human capital accumulation decisions, their work experience, and their employment opportunities. Women are much more likely than men to specialize in household production rather than market work. Other factors, among them discrimination, social norms for gender roles, and physiological differences, all exert differentiating influences on the labor markets for men and women.[6]

The issue of gender brings us to the theoretically and empirically important issue of self-selection bias. A notable fraction of the population of working-age women chooses not to work. The observed wages of women who do choose to work is affected by the characteristics that determine that choice in the first place. A discussion of the selection issue and the methods of estimating models involving selection is beyond the scope of this chapter.* For more information on selection see Heckman.[7] Also note that selection bias may be an issue for male labor-market entry decisions as well.

Geographical and industry variables are sometimes included in wage functions, although neither are included here. Other variables not included in this example that may also be important are marital status, outside income, family structure, and family background.

Here is a typical wage function, estimated using "standard" methodology and assumptions. The sample used here is from the 1992 NLSY. Because the respondents are 2 years older than in H&M's sample, we have additional work experience data with which to estimate our model. The sample selection criteria are comparable with the 1990 sample used in the regressions reported earlier. Consider the model

$$\begin{aligned} \text{Log wage} &= X\beta_{wf} + \text{error} \quad \text{(for white females)} \\ \text{Log wage} &= X\beta_{bf} + \text{error} \quad \text{(for black females)} \\ \text{Log wage} &= X\beta_{wm} + \text{error} \quad \text{(for white males)} \\ \text{Log wage} &= X\beta_{bm} + \text{error} \quad \text{(for black males)} \end{aligned} \tag{9.5}$$

where X is a vector that includes race-gender specific intercepts, AFQT, education, SES, and both linear and quadratic terms for age, full-time and part-time work experience. Table 9.4 shows the sample means of these variables. The βs are vectors of coefficients that are allowed to vary by race and gender. The results of this regression are presented in Table 9.5.

This model is used to test the restrictions that the coefficients are the same for both genders. Just as they were in the simple earnings model, these restrictions are rejected ($p < .01$). As anticipated, H&M's assump-

*It seems plausible that selection issues are at the heart of the unintuitive result, reported above, that black women earn more than white women after controlling for age, AFQT, and parental SES.

Table 9.5. Log-linear wage functions

Independent variable	Coefficients allowed to vary by race and gender: Males White	Males Black	Females White	Females Black	Coefficients assumed same for both genders: White	Black
Intercept	2.2666**	2.2749**	2.1127**	2.2063	2.1990**	2.2334**
	(0.0292)	(0.0400)	(0.0177)	(0.0310)	(0.0153)	(0.0246)
Age	−0.0514	0.0919	−0.3632	−0.4942	−0.1879	−0.2621
	(0.1785)	(0.2919)	(0.1953)	(0.22864)	(0.1343)	(0.2450)
Age^2	0.0006	−0.0018	0.0057	0.0080	0.0028	0.0041
	(0.0029)	(0.0048)	(0.0032)	(0.0047)	(0.0022)	(0.0034)
AFQT	0.0901**	0.1211**	0.0854**	0.2208**	0.0943**	0.1550**
	(0.0221)	(0.0316)	(0.0225)	(0.0297)	(0.0163)	(0.0233)
SES	0.0570**	0.0425	0.0326	−0.0131	0.0519**	0.0151
	(0.0193)	(0.0284)	(0.0202)	(0.0240)	(0.0142)	(0.0194)
Education	0.0660**	0.0636**	0.0681**	0.0547**	0.0670**	0.0555**
	(0.0392)	(0.0131)	(0.0086)	(0.0130)	(0.0064)	(0.0095)
Experience	C.0496	0.0406	0.0203	0.0019	0.0384	0.0281
	(0.0392)	(0.0352)	(0.0234)	(0.0276)	(0.0209)	(0.0215)
Exp.^{2}	−0.0001	0.0008	0.0017	0.0017	0.0008	0.0010
	(0.0021)	(0.0020)	(0.0014)	(0.0018)	(0.0012)	(0.0013)
P.T. exp.	−0.0173	−0.0150	0.0316	−0.0024	−0.0065	−0.0112
	(0.0038)	(0.0070)	(0.0022)	(0.0034)	(0.0019)	(0.0030)
(P.T. exp.)^2	0.0033	0.0063	−0.0012	0.0036	0.0018	0.0049
	(0.0038)	(0.0070)	(0.0022)	(0.0034)	(0.0019)	(0.0030)
R^2	0.2680				0.2450	
Adj. R^2	0.2594				0.2407	
n	3,354				3,354	

Standard errors are in parentheses.
** Coefficient significant to .01 probability level.
Exp., experience; P.T., part-time; SES, socioeconomic status.
Adj., adjusted;

tion that the genders can be pooled is overwhelmingly rejected in the standard wage model. Testing the null hypothesis of color-blind earnings with the standard model yields more ambiguous results. The joint restriction that all twenty coefficients are equal for both races is not rejected at conventional confidence levels ($p < .18$), although the color-blind earnings hypothesis is rejected when the coefficients are tested individually.

The adjusted R^2 for the regression using the unrestricted standard model is .259, which compares favorably with .167 for a regression of log wage using H&M's preferred model 9.1 on the 1992 sample.* The standard model explains a much larger proportion of the variance in wages than the H&M model. R^2 statistics in the vicinity of .25 to .30 are typical of wage regressions. Clearly, although this a dramatic improvement over the simple AFQT model, a great deal remains unexplained, and so conclusions should be drawn with care and adequate caveats.

The Racial Wage Gap in the Standard Wage Function

To determine the source of the racial wage differential for men, we now decompose the wage gap using the methodology of Oaxaca[8] and Blinder.[9†] This methodology separates the average wage differential into two portions. The first portion is attributed to differences in endowments (differences in group means), and the second portion is attributed to differing returns to characteristics (differences in coefficients). H&M's methodology obscures this distinction. Instead, they interpret their regressions to imply that virtually all of the wage gap is attributable to differing endowments of cognitive ability and not to discrimination.

Table 9.6 shows the detailed results of our decomposition of the racial wage differential for men. Column 5 shows the percentage of the wage gap due to differences in endowments. These values are calculated by multiplying the differences in the group means for each variable (column 1) by the coefficients for the high wage group (column 2) and dividing by the difference in mean log wages of the two groups. These differences are attributable to premarket factors, such as ability (as measured by AFQT) and accumulated work experience. Column 6 shows the percentage of the wage gap attributable to wage discrimination. These values are calculated by multiplying the differences in coefficients by the black mean for each variable. Column 7 is the sum of columns 5 and 6, and shows the percentage of the wage gap attributable to each variable.

Note in Table 9.6 that differences in average AFQT between white men and black men account for 38% of the difference in average log wages. As H&M correctly point out, this difference is the single most important factor in racial wage inequality for men. However, contrary to H&M's conclusion,

---

*This regression is not reported. The coefficients are comparable to Model III in Table 9.1.

†The Oaxaca–Blinder decomposition is described in more detail in the appendix of this chapter.

Table 9.6. Decomposition of wage gap (males only)

Variable	Diffs. in means (white-black) (1)	White male coeffs. (2)	Diffs. in coeffs. (white-black) (3)	Black male means (4)	% Wage gap due to endowments (5)	% Wage gap due to coefficients (6)	% Wage gap due to variable (7)[a]
log(*w*)	0.2711	2.2666	−0.0083	2.1866		−3%	−3%
Age	0.0567	−0.0514	−0.1433	0.0430	−1%	−2%	−3%
Age^2	3.0868	0.0006	0.0024	2.9632	1%	3%	3%
AFQT	1.1385	0.0901	−0.0310	−0.7712	38%	9%	47%
SES	0.8743	0.0570	0.0145	−0.5563	18%	−3%	15%
Education	0.7149	0.0660	0.0024	−0.1601	17%	0%	17%
Experience	1.0312	0.0496	0.0090	1.1938	19%	4%	23%
Exp.^{2}	19.3325	−0.0001	−0.0009	12.6995	−1%	−4%	−5%
P.T. exp.	0.3316	−0.0173	−0.0023	−0.6738	−2%	1%	−2%
(P.T. exp.)^2	2.1541	0.0033	−0.0030	−4.5690	3%	5%	8%
				Total:	92%	8%	100%

[a] Column (7) = (5) + (6). Columns may not sum exactly beause of rounding.
Exp., experience; P.T., part-time; SES, socioeconomic status.
Coeffs., coefficients; Diffs., differences;

AFQT does not constitute the whole story. In total, 62% of the original wage gap remains after controlling for AFQT endowments.

The most important conclusion to draw from Table 9.6 is that only 8% of the racial wage gap for men is attributable directly to racial wage discrimination. The remaining 92% of the gap is attributable to premarket factors. AFQT is the most important of these premarket factors, but it is not the only one. Work experience is also highly important in determining the expected wage gap. Differences in accumulated experience account for 19% of the racial wage difference in our model, with an additional 6% attributable to discrimination against blacks in the form of lower returns to work experience. Hotz et al. have documented the difficulty experienced by young black men in obtaining full-time work experience.[10] Attainment of education explains 17% of the wage gap. Parental SES contributes an additional 15% to the explained wage differential in our model. As noted by Heckman, H&M's parental SES variable is a crude measure that may itself be an indicator of other premarket factors.[11]

H&M present a picture of cognitive ability as a characteristic essentially fixed at birth, with which educators and policy makers must cope, but over which they have no influence. However, this point of view is not universal. The nature-versus-nurture controversy is longstanding and is far from being resolved. For a survey of this debate, see Bouchard[12] and Moore.[13] Neal and Johnson have found evidence in the NLSY itself that AFQT is not entirely immutable, as H&M portray it, but rather is the result of a combination of hereditary, schooling, home environment, and early educational factors.[14] If cognitive ability itself is affected by environmental factors, then even the 38% of the wage gap which is attributable to endowments of cognitive ability may be amenable to policy interventions which would in turn influence the racial wage gap.

Conclusions

In agreement with H&M, we find that relatively little discrimination occurs in the form of lower wage offers to blacks compared with wage offers to otherwise identical whites. However, we refute H&M's assertion that AFQT explains virtually all of the wage gap. Their result that racial earnings differences disappear after controlling for AFQT and parental SES is due to bias caused by choosing a model that does not fit the data. In the corrected model, while the difference in mean AFQT between black and white men is the largest single contributor to the wage gap, it leaves 62% of that gap unexplained.

In contrast to H&M, who argue that the racial wage gap is primarily caused by factors that are immutable, we find that a large fraction of the differential is contributed by factors that may themselves be affected by racial discrimination, or that may be influenced by policy interventions. These premarket factors include accumulated work experience, education, environmental influences associated with SES, and perhaps AFQT itself.

Appendix

Sample Specification

The data source for this study is the National Longitudinal Survey of Youth, survey years 1990 and 1992. The sample we use in the earnings analysis includes all blacks and whites who reported working 52 weeks in 1989, were not students in 1989 or 1990, and were not in the military in 1989. We exclude those cases with 1989 annual earnings below $5,000 or above $500,000 (as measured in 1990 dollars), and also those cases without valid observations for all variables and sampling weights. A total of 3,916 observations remain in this sample for earnings analysis. For the log wage analysis, comparable sample selection criteria are applied to the 1992 survey. The sample size is 3,354.

Description of Variables

We follow Murray and Herrnstein's construction of AFQT and SES measures as outlined in Appendix 3 and Appendix 2, respectively, of *The Bell Curve*. Independent variables used in this study are annual earnings and log average hourly wage. Annual earnings and annual hours worked are survey items in the NLSY. Log wages are calculated by dividing annual earnings by annual hours worked and taking natural logarithms. The measure of education we use in this study is highest grade completed at the survey date. A detailed estimate of actual work experience is constructed by assembling a retrospective weekly work history of all cases, beginning January 1, 1978. Weeks of part-time and full-time work experience are accumulated from the work history. A respondent is classified as a full-time worker if total hours worked at all jobs in a given week are 35 or more. The experience variables used in the regression analyses are expressed as years of work experience from January 1, 1978 to December 31, 1991. The racial variable employed is the racial identification made by the interviewer in the first interview. All independent variables in the study are transformed to deviation from the mean. This allows the intercept coefficients to be interpreted as the expected earnings (or log wage) of a group with independent variables held constant at the overall sample means.

Oaxaca–Blinder Wage Decomposition

Blinder and Oaxaca show that with separate regression models for the high-wage group and the low-wage group, the average wage differential can be decomposed into a portion attributable to differences in endowments of human capital characteristics across groups and a portion attributable to differences in the returns to these characteristics across groups. The regression equations are as follows:

$$Y_i^H = \beta_0^H + \Sigma_{j=1}^n \beta_j^H X_{ji}^H + u_i^H. \tag{A1}$$

$$Y_i^L = \beta_0^L + \Sigma_{j=1}^n \beta_j^L X_{ji}^L + u_i^L. \tag{A2}$$

Evaluating these equations at the means and taking differences, the portion of the average wage differential explained by the regressions is simply

$$\Sigma_{j=1}^n \beta_j^H - \Sigma_{j=1}^n \beta^L j \bar{X}_j^L, \tag{A3}$$

and the portion unexplained by the regression (attributable to discrimination) is

$$\beta_0^H - \beta_0^L. \tag{A4}$$

With some simple algebra it can be shown that

$$\Sigma_{j=1}^n \beta_j^H \bar{X}_j^H - \Sigma_{j=1}^n \beta_j^L \bar{X}_j^L = \Sigma_{j=1}^n \beta_j^H (\bar{X}_j^H - \bar{X}_j^L) - \Sigma_{j=1}^n \bar{X}_j^L (\beta_j^H - \beta_j^L), \tag{A5}$$

and thus the explained wage differential is partitioned into the first term on the right-hand side, which represents the effect of differing endowments of human capital on wages, and the second term on the right-hand side, which is a measure of the effect of differences in the coefficients on wages.

Blinder defined the following quantities:

$$E = \Sigma_{j=1}^{n} \beta_j^H (\bar{X}_j^H - \bar{X}_j^L) \tag{A6}$$

$$C = \Sigma_{j=1}^{n} \bar{X}_j^L (\beta_j^H - \beta_j^L) \tag{A7}$$

$$U = \beta_0^H - \beta_0^L \tag{A8}$$

$$D = C + U, \tag{A9}$$

where E is the portion of the differential due to differing endowments, C is the portion of the differential due to differing coefficients, U is the unexplained portion of the differential, and D is the portion of the differential attributable to discrimination (see Ref. 9).

References

1. Herrnstein, R.J., and Murray C. (1994), *The Bell Curve: Intelligence and Class Structure in American Life*, The Free Press, New York..
2. Murphy, K.M. and Welch, F. (1990), "Empirical Age Earnings Profiles," *Journal of Labor Economics*, 8, pp. 202–29.
3. Mincer, J. (1974), *Schooling, Experience and Earnings*, National Bureau of Economic Research, Cambridge, MA.
4. Wolpin, K.I. (1992), "The Determinants of Black–White Differences in Early Employment Careers: Search, Layoffs, Quits and Endogenous Wage Growth," *Journal of Political Economy*, 100.
5. Becker, G.S. (1975), *Human Capital*, 2nd ed., Columbia University Press, New York.
6. Corcoran, M. and Duncan, G. (1979), "Work History, Labor Force Attachment and Earnings: Differences Between the Races and the Sexes," *Journal of Human Resources*, 14, 3–20.
7. Heckman, J.J. (1980) "Sample Selection Bias as a Specification Error," in J.P. Smith (Ed.), *Female Labor Supply*, Princeton University Press, Princeton, NJ, pp. 206–248.
8. Oaxaca, R. (1973), "Male–Female Wage Differentials in Urban Labor Markets," *International Economic Review*, 14, 693–709.
9. Blinder, A.S. (1973), "Wage Discrimination: Reduced Form and Structural Equation Estimates," *Journal of Human Resources*, 8, 436–455.
10. Hotz, J.V., Xu, L., Tienda, M., and Ahituv, A. (1995), The Returns to Early Work Experience in the Transition from School to Work for Young Men in the US: An Analysis of the 1980s, mimeo, University of Chicago, Chicago.
11. Heckman, J.J. (1995), "Lessons from *The Bell Curve*," *Journal of Political Economy*, 103, pp. 1091–1120.
12. Bouchard, T.J., Jr. (1981), "Familial Studies of Intelligence: A Review," *Science*, 212, 1055–1059.
13. Moore, E.G.J. (1986), "Family Socialization and the IQ Test Performance of Traditionally and Transracially Adopted Black Children," *Developmental Psychology*, 22, 317–26.
14. Neal, D., and Johnson, W. (1996), "The Role of Premarket Factors in Black–White Wage Differences," *Journal of Political Economy*, 104, 869–895.

chapter ten

Does Staying in School Make You Smarter? The Effect of Education on IQ in *The Bell Curve*

CHRISTOPHER WINSHIP
AND SANDERS KORENMAN

Can education increase an individual's IQ? This has been one of the most incendiary and controversial questions in the social sciences in the past few decades. The greatest firestorm occurred after the publication of Arthur Jensen's 1969 article in *The Harvard Education Review*, "How Much Can We Boost IQ and Scholastic Achievement?"[1] The controversy was further fueled by Richard Herrnstein's 1971 *Atlantic Monthly* article, "IQ."[2] Then, after smoldering for two decades, the passion and acrimony reignited with publication in 1994 of *The Bell Curve* by Richard Herrnstein and Charles Murray.[3]

At the center of the dispute are two related issues: the degree to which intelligence is genetically determined, and thus presumably fixed; and the extent to which observed racial differences in intelligence are genetically based. Jensen, Herrnstein, and Herrnstein and Murray have all been sharply rebuked for suggesting that intelligence and observed racial differences in intelligence *might* be largely genetically determined.

Education is widely regarded as a key mechanism to elevate the less well off and to narrow racial differences in social and economic status. A recurring theme in the work of Jensen and Herrnstein has been that although intelligence is a major determinant of economic and social success, because intelligence is largely genetically determined and fixed, schools will have little success in improving individual economic and social success.

In *The Bell Curve*, Herrnstein and Murray argue and present evidence that education has little to no effect on IQ (Appendix 3, pp. 590 and 591). However, Claude Fischer et al., in a book-length critique of *The Bell Curve*, have argued that the effects of education on IQ are substantial and that the AFQT, the measure of IQ Herrnstein and Murray use in *The Bell Curve*, is primarily a measure of educational achievement.[4] The critical question, then, is whether in fact the evidence supports Herrnstein and Murray's assertion of no effect, or is consistent with Fischer et al.'s position.

The purpose of the present chapter is to examine Herrnstein and Murray's analysis of the effects of education on IQ and to review the research literature on this topic. The literature on the effects of educational programs on intelligence is massive. There is, however, surprisingly little consensus as to its findings or their implications. For example, *The Bell Curve* and Stephen Ceci each provide recent extensive reviews but reach sharply different conclusions.[5] Herrnstein and Murray are skeptical about the possibility that formal schooling or compensatory programs can increase the intelligence of individuals: "Taken together, the story of attempts to raise intelligence is one of high hopes, flamboyant claims, and disappointing results" (p. 389). In contrast, Ceci concludes "that schooling exerts a substantial influence on IQ formation and maintenance.... [T]here is now considerable evidence for the importance of variations in schooling on IQ" (see Ref. 5, p. 70).

How are we to explain such divergent summaries of the same literature? A perhaps overly simple answer is that the conclusions reflect political or ideological differences between the scholars. Other factors, however, are important. One critical weakness in the literature has been a lack of focus on determining the size of the key parameter: the effect of schooling on measured intelligence. Herrnstein and Murray as well as Ceci presumably agree that educational programs have some effect on intelligence. The issues in dispute are how much? and under what circumstances? In short, nearly missing from the literature on school effects are sustained attempts to estimate the size of the effect of particular programs. (The literature on Head Start has made such an attempt; see, for example, Ref. 6).

The researchers cited earlier agree that general intelligence, or what psychometricians refer to as *g*, is a meaningful concept and that it is accurately measured by IQ tests. This position has been contested. Howard Gardner[7] and Robert Sternberg,[8,9] among others, have argued that intelligence is multidimensional. We do not contribute to this debate (see, however, the chapters in this volume by John Carroll and Earl Hunt). Our interest in IQ begins instead with the considerable predictive power of measured IQ for different dimensions of social and economic success, such as earnings and income,[10,11] job performance,[12–14] and criminal behavior.[15,16]

Herrnstein and Murray's *The Bell Curve* is the most recent and sweeping analysis to show that IQ predicts a variety of behaviors and adult outcomes. Their estimates of the effects of IQ appear to be robust. In an earlier paper we analyzed the sensitivity of Herrnstein and Murray's estimates to different methods for controlling for family background.[17] We found that even fixed-effect methods based on sibling pairs, perhaps the most powerful way to control for family background, produced estimates of the effect of IQ on adult behaviors and outcomes quite similar to those reported in *The Bell Curve*. We did, however, find effects of family background comparable in size with those of IQ, and quite a bit larger than those reported in *The Bell Curve*. Also, the effect of IQ was substantially reduced by the inclusion of education as a control. We concluded, as did Jencks (see Ref. 11), that the evidence suggests that IQ—along with education and family background—is an important contributor to social and economic success, but not the dominant determinant, as Herrnstein and Murray stress in *The Bell Curve*. Because IQ is one of the important predictors of success, a logical question to ask is, What determines IQ? In particular, is it true that education has only minimal effects on IQ scores?

The present chapter contains two core sections. In the first we review the small set of articles that have used a methodology similar to Herrnstein and Murray's to estimate an effect of education on IQ. We find that in many studies the estimated effect of education on IQ is larger than that reported by Herrnstein and Murray.

In the second core section, we present a series of analyses of the Herrnstein and Murray model. In the first set of analyses we correct technical problems in the original research involving the treatment of missing data, the omission of a variable for the age at which the initial IQ test was taken, and the fact that some individuals are represented in the data more than one time. These corrections result in an estimated increase of 2.5 points of IQ per year of education, more than double Herrnstein and Murray's estimate.

We next examine four issues in model specification: (1) whether the age at which the AFQT was taken belongs in the model, (2) whether parent's socioeconomic status should be included, (3) whether controls for type of initial IQ test are needed, and (4) whether the appropriate educational variable is educational attainment or years in school. Adjusting the model specification for the first three factors leads to an estimate of 2.3 points of IQ per year of educational attainment. When we use years of schooling as opposed to educational attainment, we obtain an estimate of 1.8 points of IQ for each year of schooling.

Finally, we examine the sensitivity of the results to correcting for measurement error in educational attainment and early IQ. Under what we believe are reasonable assumptions about the extent of measurement error, the estimated effect of education is 2.7 IQ points per year of education. We conclude that, although it may be impossible to arrive at a single estimate for the effect of education on IQ, a year of education mostly likely increases IQ by somewhere between 2 and 4 points.

The next section discusses our methods and data. The subsequent section reviews the relevant literature. We then present our analyses of the NLSY. We conclude by summarizing what we have learned from the literature review and the data analysis, and we offer some speculative discussion of the implications of *The Bell Curve* for education policy.

Methods and Data

To conduct our literature search we began with the recent comprehensive reviews of Ceci and Herrnstein and Murray. We collected the papers cited therein and then, using the Social Science Citation Index, we did forward and backward searches to identify studies that provide estimates (or the data to derive such estimates) of the effects of formal education on IQ. An appendix, which is available from the authors, provides a summary table, essentially an annotated bibliography of these studies.

Our empirical analyses are based the National Longitudinal Surveys of Youth—the NLSY. For comparability with *The Bell Curve*, we use data files generously provided by Charles Murray. When necessary we supplement the Herrnstein and Murray data with additional information from the NLSY.

The NLSY is an ongoing longitudinal study of a national baseline sample of 12,686 individuals aged 14 to 20 years as of January 1, 1979.[18] It contains extensive information on individuals' labor market, schooling, and family formation histories. The importance of the NLSY for the research reported here is that it contains a high-quality measure of mental ability; in 1980 the Armed Services Vocational Aptitude Battery (ASVAB) was administered to nearly the entire NLSY sample.

Our interest in this chapter is in the possible effects of formal education on general intelligence, or *g*. Within the psychometric tradition *g* is assumed to be measured by IQ tests. There is considerable consensus in this tradition that many different tests measure *g* and as a result should be considered IQ tests.[5,19–21] The studies we review use a variety of mental ability tests, all of which would be considered IQ tests by psychometricians.

Herrnstein and Murray use the Armed Forces Qualifying Test (AFQT), an equally weighted composite of four items of the ASVAB, as their measure of IQ. They show that the AFQT has the internal structure of an IQ test and that it correlates highly with traditional IQ tests. They further argue that it is one of the best measures of general intelligence or *g* currently available. (See, however, the chapter by John Cawley et al. in this volume, in which they argue that the AFQT is a flawed measure of general intelligence.)

Previous Research

Any attempt to estimate the effect of schooling on intelligence must confront the problem that individuals with higher initial intelligence may select, or be selected into, higher levels of schooling. Although many studies have discussed and examined the possible effects of schooling on intelligence, few have dealt with this selection problem in even a minimally adequate manner. We restrict our primary attention in this chapter to studies that handle this selection problem by controlling for an earlier measure of IQ, the strategy employed by Herrnstein and Murray. Specifically, on pp. 590–591 of *The Bell Curve*, Herrnstein and Murray present a simple model for estimating the effects of education on mental ability (Fig. 10.1).

The model in Figure 10.1 is closely related to the analysis of covariance model used to estimate the effect of a treatment with pretest/post-test data when there is a nonequivalent control group.[22] Herrnstein and Murray's model differs from the standard model in that education has multiple levels, there is considerable time between the pretest (early IQ) and post-test (AFQT), and the time between the two tests and the age at which they are taken vary across individuals. Later we discuss how these differences affect the model specification.

Nearly all the variation in education in the NLSY in 1980, the year that the AFQT was given, is accounted for by differences between individuals in the number of years of high school and college they have completed. As a result, Herrnstein and Murray's analyses, and our reanalysis of their data, (as well as the analyses of most of the other studies we now discuss), involve an assessment of the impact of these years of schooling on IQ.

We have identified six studies that have used an analysis of covariance strategy to estimate the effects of schooling on IQ. Three of the studies are based on U.S. data; the others are from Scandinavian sources. Table 10.1 provides a summary description of this research. The estimated effects range

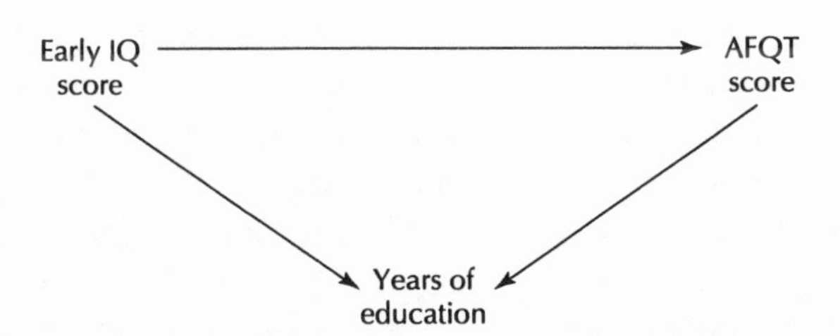

Figure 10.1. Herrnstein and Murray's path model. In this model, the impact of earlier IQ score on later IQ score is magnified by a direct effect and by an indirect efffect mediated through years of education.

from 1 IQ point per year of education in Jencks (Ref. 10) to 4.2 points per year in Husén and Tuijnman (Ref. 23).

With the exceptions of Jencks (see Ref. 10) and Lund and Thrane,[24] none of the studies in Table 10.1 provides a direct estimate of the effect of an additional year of education on IQ. In each case we derived estimates from the results reported. We now discuss each study and explain how we obtained an estimate of the effect of a year of education on IQ.

Lorge's study consists of a sample of 131 boys who were tested in the eighth grade in 1921-22 in New York City, and were retested 20 years later.[25] The IQ score in the 1920s was a weighted composite of the Thorndike–McCall Reading Scale and the I.E.R. Arithmetic Test. In 1941 two tests were administered—the Otis Self-Administering Test of Mental Ability (Higher Examination, Form B), and the Thorndike Intelligence Examination for High School Graduates (Form V).

Lorge does not provide an estimate of the effect of education on IQ. However, from the information presented in the article it is possible to construct a covariance matrix for initial test score, educational attainment, and the two tests given in 1941. We used this information to estimate a regression model to predict the two different test scores, yielding estimates of 2.28 IQ points for the Otis and 2.37 IQ points for the Thorndike per year of education. We made no attempt to correct for measurement error in his variables. Our analysis assumes that the standard deviation of IQ in this sample is 15, something we could not determine from Lorge's analysis. This fact, plus the historical age and small size of the Lorge sample, led us to discount somewhat the value of these estimates.

Harnqvist examines a 10% sample of all Swedes born in 1948 who were given a battery of tests at age 13, just before the beginning of school tracking.[26] About 5,000 young men in the sample were retested 5 years later, in 1966, at the time of military enrollment. Harnqvist finds that education had a substantial effect on IQ. He concludes in a later article that his study indicates that, after controlling for family background and initial IQ, one additional year of school increases IQ by 2 to 3 points.[27]

Jencks constructs a series of path models involving a large number of variables and obtains an effect of 1 IQ point per year of education (see Ref. 10). In his analysis he attempts to account for measurement error in the different variables. Jencks gathers the correlations used to construct the path model from a variety of studies conducted at different times and involving somewhat different samples. He recognizes that pooling of correlations across different samples may be problematic and expresses little confidence in his resulting estimate.

Wolfle[28] uses a similar methodology to Jencks (see Ref. 10). In fact, many of his correlations are derived from the same sources. It is perhaps not surprising,

Table 10.1. Estimates of the effects of education on IQ, controlling for early IQ and possible other variables

Author	Effect	Sample	Conclusion
Lorge (Ref. 25)	2.28 Otis 2.37 Thorndike	131 boys in the 8B classes of a representative sampling of NY public schools in 1921-22.	"The data indicate that increments in adult intelligence test score are related not only to original level of intelligence, but also the extent of education that could be obtained." (p. 492)
Harnqvist (Ref. 26)	2–3 (taken from summary in Ref. 27)	10% sample of the Swedish school population born in 1948, totaling 4,616 men	"When home background, measured with a strong composite of family and residence variables, was controlled, the difference was reduced to 7–8 IQ units. One additional year in compulsory school meant 2–3 IQ units." (Ref. 27)
Jencks (Ref. 10)	1.0 extensive controls	Correlations taken from 10+ different surveys	"Despite our reservations, we tentatively conclude that if students leave school early in adolescence, their verbal and numerical skills do not develop as much as if they remain in school." (p. 88)
Wolfle (Ref. 28)	1.07 controls for father's ed, father SEI, age, sex	Correlations taken from 7 different surveys	"However, education does increase one's general intelligence, and the indirect effect of education on vocabulary through adult IQ was important." (p. 112)
Lund and Thrane (Ref. 24)	3–3.15 2.5–2.8 with correction for measurement error in early IQ	2,485 Norwegian male adolescents who were tested in 1953 at the end of primary school, and 5 years later for compulsory military service	"Though some differences between methods exist with respect to the estimated schooling effects, the results lead to the conclusion that schooling has had a considerable influence on the growth of intelligence." (p. 168)
Husén and Tuijnman (Ref. 23)	4.2 controls for family background	Sample of 671 10-year-old boys in public and private schools in a Swedish city	"The results provide support for the thesis ... that IQ as measured by group intelligence tests is not stable but changes signifcantly between ages 10 and 20 ... the amount and quality of schooling are implicated in the observed changes in measured IQ. . . . " (p. 22)

ed, education; SEI, socioeconomic index

then, that he obtains a similar estimate for the effect of education on IQ, 1.07. The correlations used to estimate his path model are from seven different studies. There are three key correlations for our purposes—that between child and adult IQ, between child IQ and education, and from between and adult IQ. The first two correlations are derived from Duncan,[29] Jencks (Ref. 10), and McCall.[30] The final correlation is from Wechsler.[31] The paper does not report the source of the standard deviation of education, although it is most likely to have come from a series of General Social Surveys. As with Jencks's 1972 estimate, we have little confidence in Wolfle's estimate.

One of the most methodologically advanced studies was conducted by Lund and Thrane. A sample of 7,703 children from a southeastern region of Norway, including Oslo, were given a set of military classification tests at the end of grade seven. Four were conventional IQ tests. The same tests were given to 3,400 of the boys approximately 5 years later, at the time of military induction. The final sample with information on both tests and educational attainment contains 2,485 individuals. Lund and Thrane analyze the data in several ways. One of the methods they use is an analysis of covariance model, done with and without corrections for measurement error in the early IQ score. Without the measurement error corrections they obtain an estimate of from 3 to 3.15 IQ points per year of education.* When they correct for measurement error in early IQ, they obtain estimates of from 2.5 to 2.8 IQ points per year of education. The other methods of analysis they use produce somewhat smaller estimates.

Herrnstein and Murray cite Husén and Tuijnman as supporting their contention that education has a minimal effect on IQ (p. 591, footnote 17). However, Husén and Tuijnman's summary of their results contradicts Herrnstein and Murray's claim. Specifically, they state

> that not only child IQ has an effect on schooling outcomes but also that schooling per se has a substantial effect on IQ tests scores. Hence, schools not only confer knowledge and instrumental qualifications but also train and develop students' intellectual capacity. (Ref. 23, p. 22)

Husén and Tuijnman proceed to discuss how their research and that of others demonstrates that IQ does change from childhood to adulthood and how education appears to be a central factor in this change.

The Husén and Tuijnman study is a reanalysis of data originally analyzed by Husén.[32] The sample is a cohort of Swedish males (n = 671) who were tested at 10 years of age in 1938 and retested 10 years later at the time

*These estimates are based on the post-test distribution of scores. This is comparable with our use of the age-standardized AFQT scores. Estimates scaled relative to the pretest distribution show somewhat larger effects.

of induction into the military. IQ in childhood was measured using four tests: word opposites, sentence completion, perception of identical figures, and restoring disarranged sentences. Adult IQ was measured by four tests: synonyms, concept discrimination, number series, and matrices. Husén and Tuijnman use LISREL to analyze the data. They report an estimate of 0.28 for the effect of education on IQ. Rescaling this estimate (multiplying by 15), we get an effect of 4.2 IQ points per year of education.

A problem with the Husén and Tuijnman study is that education is scaled so that it is only approximately the case that 1 unit is equivalent to 1 year of education. In particular, individuals who have less than a seventh grade education and individuals who have some university education are collapsed, respectively, into single categories at the bottom and at the top of their scale. The effects of doing this are ambiguous. On the one hand, collapsing a continuous or interval scale variable typically reduces the association between it and other variables. This would result in Husén and Tuijnman underestimating the effect of education. On the other hand, individuals in the top and bottom categories of the scale differ by 5 points on Husén and Tuijnman's scale, but probably on average differ by more than 5 years in their actual educational attainments. The effect of this, then, is to overstate the effect of education.

A similar problem is that in the case of some of the intermediate categories it is possible that a single category of education can represent different amounts of educational attainment. For example, their category "incomplete secondary education" can represent 1 or 2 years beyond the 7 years of compulsory education. As with the two extreme categories, the effects of collapsing are ambiguous. Given that the Husén and Tuijnman's estimate is the highest reported in Table 10.1, it is likely it overstates the effect of education on IQ in their sample, perhaps considerably so.

Four other studies worth mentioning have provided estimates of the effect of education on IQ using other methods. DeGroot[33] and Green et al.[34] use "experiments in nature" to estimate the effects of school on IQ. DeGroot exploits the fact that World War II interrupted the schooling of many children in the Netherlands. He compares the IQs and education of successive cohorts of applicants to a training school. He finds an approximate 5-point drop in IQ among individuals most affected by the war, and a deficit of 1.5 years in the amount of time they have spent in school translating into a 3.33-point loss in IQ per year of school. DeGroot, however, makes no attempt to determine whether the war might affect applicants' IQ through other mechanisms in addition to the suspension of schooling.

Green et al. analyze the impact of the decision by the school board of Prince Edwards County, Virginia to shut down all schools from 1959 to 1964 to avoid racial integration. As a result, many black children received no education, whereas white children attended newly opened private schools.

Ceci, in summarizing the results of this study, concludes that a 1 year loss of education resulted in the 6-point drop in IQ. Unfortunately, the schools in Prince Edwards County were unwilling to make available earlier IQ scores for children in the study. As a result, estimates are based on a comparison of the IQs of children in Prince Edwards County with similar children in adjacent counties. Although Green et al. go to great lengths to achieve comparability between these samples, it is unclear that they succeed, casting doubt on their unusually high estimate of the effect of schooling.

Cahan and Cohen[35] and Neal and Johnson[36] capitalize on the fact that rules about when children are allowed to enter school create a discontinuity in the relationship between age and years of education. For instance, if school policy states that a child can start first grade only if he or she is 6 prior to January 1st, then a child born in December will have one additional year of education compared with a child who is born only one month later in January. Cahan and Cohen use this effect of school policy in a regression discontinuity design. They estimate effects of attending fifth or sixth grade (much earlier years of education then we have been considering) on twelve tests. The range of estimates across these tests is between 1.65 and 7.5 points per year of education, with a mean estimate of 4.1 points per year of education.

Neal and Johnson use the same school policy variation to create an instrument for education based on an individual's quarter of birth. They use the NLSY data restricted to individuals born 1961 or later, a sample approximately double in size of that used by Herrnstein and Murray. Using their instrumental variable estimator, Neal and Johnson obtain an effect of 3.3 IQ points for men and 3.75 IQ points for women per year of education.

Reanalysis of the Herrnstein–Murray Model

Herrnstein and Murray use the fact that for a small portion ($n = 1408$) of the NLSY sample, IQ scores at earlier ages are available from school records, along with a score on the AFQT test and a measure of educational attainment in 1980, the time at which the AFQT was taken. They are thus able to estimate a model of the form in Figure 10.1 for a subsample of the NLSY. This approach is very simple. It might be desirable to have a more detailed model that could capture the dynamic evolution with age of IQ and educational attainment. For the purpose of this chapter, however, we will work with variants of the model in Figure 10.1.

Herrnstein and Murray report estimates from two models. In the first, they estimate the effect of early IQ on later IQ without controlling for education. In the second model they add education. On page 590 they state that their analyses control for years elapsed between when the initial test and the AFQT were taken, the age at the first test, and the particular type of initial test. In both cases their dependent variable is IQ (AFQT score) measured in percentiles. For the sake of comparability, we report results using IQ score in percentiles as the dependent variable. In addition, we present analyses in which IQ is measured in the more traditional form of standard IQ units, where IQ is scaled to have a standard deviation of 15 within each age group. Later we focus on this latter set of estimates, reported in column (3) of Table 10.2.

Herrnstein and Murray's estimates for the effect of education on AFQT, net of the effects of early IQ, are small. They find that an additional year of education increases an individual's percentile ranking on the AFQT by 2.3 percentage points. They report that when the dependent variable is measured in standardized units, the effect of education on IQ is equal to 1 IQ point per year of education. Model 1 of Table 10.2 reports the effects of education and early IQ taken from our replication of their results.

Technical Corrections

We found it difficult to replicate Herrnstein and Murray's results. There were two problems. First, Herrnstein and Murray did not include age at which the first test was taken as a control variable in their analysis, contrary to the statement on page 590. Murray (personal communication) stated that it was unnecessary to control for age at first test because both the initial and AFQT test scores were age standardized. We return to this issue later and explain in detail why we disagree with Murray. Here we simply investigate the effects of including age at first test in the model.

The second problem we identified is that Herrnstein and Murray had improperly handled missing data in their original analysis. This was independently discovered by Murray. Their original analysis included seven individuals whose education was coded as −5 instead of as missing. In three cases it is possible to impute values for education in 1980 because these individuals' education was the same in 1979 and 1981.

A third issue results from the inclusion of individuals who had more than one measure of early IQ. Herrnstein and Murray entered these individuals into the data one time for each separate observation of early IQ, treating each record for an individual as an independent observation (Murray, personal communication). Some individuals contribute two or three observations. The problem in doing ordinary least squares (OLS) in this situation is twofold. First, the OLS formula for standard errors of the slope coefficients is wrong because it fails to take account of correlation across errors for different observations on the same individual. Second, OLS gives too much weight to individuals with several early IQ scores (though all individuals only have one observation on AFQT, the dependent variable) decreasing the efficiency of the estimate.

Table 10.2. Effects of education of early IQ and AFQT for different model specifications: coefficients and standard errors

	Dependent variable				
	AFQT percentiles		AFQT measured in IQ units		
Model	**Education**	**Early IQ**	**Education**	**Early IQ**	**Description**
Original Model					
(1)	2.280	0.752	1.110	0.381	Years elapsed between tests
	0.221	0.015	0.117	0.008	Dummies for type of test as controls
Technical Corrections					
(2)	3.185	0.736	1.600	0.372	Missing data correction
	0.273	0.016	0.145	0.008	Model 1 controls
(3)	4.892	0.711	2.506	0.360	Model 1 controls + age at first test
	0.371	0.016	0.197	0.008	
(4)	4.746	0.716	2.461	0.360	Corrects for multiple observations on same individual in data set
	0.401	0.018	0.216	0.010	All subsequent models make this correction
Model Respecification					
(5)	4.746	0.716	2.461	0.360	Model 1 controls + age at AFQT
	0.401	0.018	0.216	0.010	
(6)	4.240	0.664	2.192	0.332	Model 4 controls + SES
	0.396	0.020	0.207	0.011	
(7)	4.407	0.655	2.294	0.329	Model 5 controls without dummies for test type
	0.400	0.020	0.208	0.011	
(8)	3.666*	0.684	1.800*	0.344	*Effect of years in school
	0.412	0.020	0.218	0.011	Model 6 controls
Measurement Error Corrections (data collapsed so there is only one observation per individual)					
(9)	6.680	0.892	3.545	0.446	Model 6 controls, reliabilities
	0.642	0.021	0.354	0.012	Education = .8; early IQ = .8
(10)	5.182	0.767	2.716	0.385	Model 6 controls, reliabilities
	0.478	0.020	0.247	0.011	Education = .9; early IQ = .9
(11)	8.527	0.722	4.469	0.361	Model 6 controls, reliabilities
	0.752	0.020	0.406	0.011	Education = .8; early IQ = .9
(12)	4.005	0.936	2.126	0.469	Model 6 controls, reliabilities
	0.401	0.019	0.223	0.011	Education = .9; early IQ = .8
(13)	2.860	0.954	1.518	0.479	Model 6 controls, reliabilities
	0.291	0.019	0.162	0.011	Education = 1; early IQ = .8
(14)	9.783	0.607	5.096	0.303	Model 6 controls, reliabilities
	0.810	0.019	0.432	0.010	Education = .8; early IQ = 1

Early IQ is measured in percentile units. SES, socioeconomic status.

Model 2 in Table 10.2 shows the effect of correcting for missing data. The most recent edition of *The Bell Curve* (see Ref. 3) provides—without comment—estimates using the corrected data. Examining column 3 we see that the effect of education is increased substantially from 1.1 to 1.6 points of IQ.

It may seem surprising that correcting the missing data on seven individuals in a sample of 1,408 should have such a large effect. The influence of a point, however, is proportional to its distance from the centroid (mean values) of the data distribution.[37,38] In this sample, the mean education is 11.4. The education of these seven individuals is coded as -5, more than 16 years from the mean. As a result, they have a substantial influence on the estimate of the effect of education.

Model 3 of Table 10.2 reports the effect of education when age at first test is controlled. Note that this is what Herrnstein and Murray state they did in the first edition of their book, but, in fact, did not do. The effect of education is increased further—to 2.5 points of IQ per year of education—more than double the estimate in Herrnstein and Murray's original model.

In order to correct for the fact that approximately 10% of the sample is represented more than once in the data set, we weight each record with a weight equal to the reciprocal of the number of times the given individual appears in the data set. The weights for each individual sum to one. To account for correlated errors we use STATA's Huber procedure for estimating standard errors in the presence of clustered data.[39] This procedure is a generalization of Huber's[40] method (see also Ref. 41) of calculating standard errors with heteroskedastic data. Model 4 provides these estimates. The effect of education on the AFQT is only minimally smaller and still more than twice Herrnstein and Murray's original estimate. Its associated standard error is only slightly larger.

Model Respecification

Models 1, 2, and 3 in Table 10.2 include only a minimal set of controls. In addition, some of the variables are age standardized and some are not. We now consider the effects of respecifying the model in several ways.

As noted, Murray has argued that because the IQ variables in his model are age standardized, there is no reason to include in the model controls for age at either the early test or at the time of the ASVAB (in 1980). The issues here are complicated and are analogous to the problem of detrending time-series data in econometric models. If we have a model of the form:

$$\text{AFQT} = b_0 + (b_1 \times \text{ early mental ability}) + (b_2 \times \text{ education}) + (b_3 \times \text{ age}) + e,$$

where all the variables are in nonstandardized forms, two issues arise. First, early mental ability should be measured at the same age for all individuals in the study. In the NLSY, early IQ is measured across a range of ages. Murray is correct that standardizing early mental ability by the age at which the test taken is an appropriate way to adjust for these differences. Second, it may be desirable to measure AFQT in age-standardized units for reasons of interpretation or because we are used to dealing with IQ measures in age-standardized form. If so, and if we want to estimate consistently the coefficients in the above equation, then either all the variables in this equation need to be age standardized with respect to the age at which the AFQT was taken or, equivalently, the age at which the AFQT was taken needs to be included in the model. This is similar to time series analysis, in which one must detrend both the dependent and independent variables in order to obtain consistent estimates.[42]

Model 5 in Table 10.2 shows the effect of including age at which the AFQT was taken as one of the control variables. As in models 3 and 4, the estimated effect of education on AFQT is more than twice as large as in Herrnstein and Murray's original analysis. In fact, the estimates in model 5 are identical to those in model 4. The reason is that, because age at AFQT = age at first test + years elapsed between tests, controlling for any two of these three variables is equivalent.

In Section III of *The Bell Curve* Herrnstein and Murray present extended analyses of the effects of the AFQT and parents' socioeconomic status (SES) on a variety of measures of social and economic success. Their measure of SES is an equally weighted combination of mother's and father's education, parental income, and occupational status of the head of household (see Ref. 17 for a discussion of their parental SES measure). Herrnstein and Murray, however, do not control for parents' SES in estimating the effects of education on the AFQT score. We would expect that SES would affect both individuals' educational attainment and their later mental ability. Model 6 in Table 10.2 adds parents' SES as a control. As we would anticipate, the effect of education falls, in this case to 2.2 IQ points per year of education.

Models 1 through 6 use a set of dummy variables to control for the type of IQ test initially taken (as did Herrnstein and Murray's analyses). Students took any of ten tests. In this analysis, as in Herrnstein and Murray's, the IQ tests have been standardized to percentile scores.* Given this standardization, it is not clear why one should add dummies for the type of test taken. In fact, there is an argument against controlling for type of test. There may be a spurious correlation between the type of test taken and measured IQ. Spurious correlation could arise if, for example, students in one region of the country

*In our analysis we used both the percentile and standard deviation forms of the early IQ score. When we used the standardized version, tests for nonlinearity indicated the need for transformation. This was not true when we used the percentile version.

where IQs are lower were more likely than students in other regions to take one type of test. Model 7 in Table 10.2 estimates the effect of education on IQ, omitting the controls for the type of test. The effect of education is now modestly higher, at 2.3 IQ points per year of education.

All the models we have discussed as well as those in the literature have estimated the effect of educational attainment, not years in school. Because very bright students may skip a grade and less able students may repeat grades, educational attainment is a function of both the number of years one has been in school and the rate of progress through school, which presumably is a function of how bright one is. The question is then whether using educational attainment in 1980 as opposed to years in school in 1980 gives a biased estimate of the effect of schooling on IQ scores. The bias could be positive or negative. On the one hand, educational attainment measures years of schooling with error, and thus the effect of schooling should be downward biased. On the other hand, educational attainment partly reflects early IQ, and because our measure of early IQ is imperfect, early IQ is imperfectly controlled in our model; as a result, the effect of schooling should be upwardly biased.

We were able to construct a measure of years in school from the NLSY. Because full information on school attendance is not available prior to 1980, and the measure must be constructed from retrospective reports on school attendance collected in different years, our measure is likely to contain substantial measurement error. Model 8 reports the results of using years in school as opposed to educational attainment to estimate the effect of school. The estimate is 1.8 points of IQ per year of schooling, much smaller than our estimate in model 7 of 2.3. We have been unable to determine whether this difference is due to greater measurement error in our years-in-school variable (as compared with the educational attainment variable), or because the effect of educational attainment is biased upward because it is partly a measure of mental ability. We intend to investigate this in future research.

Measurement Error Corrections

Measurement error raises difficult problems. There is strong reason to believe that our measures of both educational attainment and early IQ contain measurement error and that this affects our estimates. Unfortunately, we do not know the reliability (the proportion of variance that is "true" variance relative to the total variance, which consists of true variance and measurement error) of either of these two variables in the NLSY. Work by other researchers is of some help. Examination by Jencks (Ref. 11, Table A2.14) of a variety of different sources produces estimates of the reliability of education that range from .854 to .933. Work by Orley Ashenfelter and Alan Krueger on twins suggests a reliability of 0.9 for educational attainment. IQ measures typically have reliabilities above 0.9.[43]

In models 9 through 14 we carry out analyses to gauge the sensitivity of the estimates of the effects of education to different assumptions about measurement error in education and early IQ. For individuals with multiple values on early IQ, we use the mean value of their early IQ and the mean length of time between their tests and the AFQT. This procedure reduces the number of observations in our data set ($n = 1253$), though it allows us to compute correct standard errors.*

The estimated effect of education ranges from 5.1 IQ points per year of education when we assume that education has a reliability of 0.8 and early IQ is perfectly measured, to 1.5 IQ points per year of education when we assume that education is perfectly measured and early IQ has a reliability of 0.8. Less extreme assumptions produce intermediate estimates. Our preferred model is 10, where we assume that both education and early IQ have reliabilities of 0.9, consistent with values reported in the literature. Here the estimated effect is 2.7 points of IQ per year of education. Note that both the overall amount of measurement error in the two variables and the relative amounts of error in the two variables affect education's coefficient. If we assume that both education and early IQ have reliabilities of 0.8 instead of 0.9 (model 9), the estimate of the effect of education is 3.5 instead of 2.7 points of IQ per year of education. If, instead, we assume that the reliability of education is 0.8, but the reliability of early IQ is 0.9 (model 11), then the estimated effect is 4.5 IQ points per year of education.

We also investigated the effects of measurement error in parental SES. As noted, Herrnstein and Murray's measure of SES is an equally weighted combination of mother's education, father's education, occupational status of the head of household, and family income. Jencks reports reliabilities for these variables ranging from .72 to .96 (see Ref. 11). Because SES is a composite of these variables, its reliability will be higher. We estimated models 9 through 14 in Table 10.2 assuming that the reliability of SES was .8 and that it was .9.

---

*The regression routine that we used to estimate models with measurement error (STATA's eivreg) does not have an option for computing standard errors with clustered data. This is a problem, given that some individuals appear multiple times in our data set. We have two options. One possibility is to estimate our models using weights, as we have in models (4) through (8), and to report incorrect standard errors. The other option is to collapse the data set so that each individual appears only once in the data set. Specifically, for individuals with multiple values on early IQ, we could use the mean value of their earlier IQ along with the mean length of time between their tests and the AFQT for their values on these independent variables. This produces somewhat less efficient estimates because it reduces the number of observations in our data set ($n = 1,253$), though it allows us to compute correct standard errors. In Table 10.2 we present estimates using the second approach. Estimates for the effect of education using the first approach are somewhat higher than those shown in column 3 of Table 10.2 for models 9 through 14, though never by more than .15.

When we assumed a reliability of .8 for SES, the estimates for the effects of education in column 3 of Table 10.2 changed minimally, always by less than .1, often substantially less. When we assumed a reliability of .9 for SES, probably a more realistic assumption, the estimates in column 3 changed even less, always by less than .05.

Our analyses here, although more extensive than those of Herrnstein and Murray, are still primitive. We have used only one method to correct for selection in the pretest/post-test design: analysis of covariance. Other methods are available and can produce different results (see Ref. 22). An important issue we have addressed only superficially through age standardization is the fact that early IQ is measured at different ages for different individuals. We have yet to attempt to examine whether the effects of schooling differ by race or ethnicity. We have also made no attempt to investigate the importance of differences in the quality of schooling individuals receive or whether other aspects of family background, aside from SES, might affect IQ. In analyses not presented here, we did examine whether the effects of education might be nonlinear and/or whether attending high school and college might have different effects. Although we found large differences, they were statistically insignificant. As a result, we are unable to draw any clear inference about the differential effect of different levels of education on IQ.

Conclusions

Neither our review of the literature nor our analyses of the NLSY data provide a single number for the effect of education on IQ. If we ignore the most extreme estimates in the literature or in our own analysis, our "best guess" estimate would be somewhere between 2 and 4 points of IQ per year of education. From our own analysis our preferred model is model 10, which assumes reliabilities of .9 for both early IQ and education, and has an estimated effect of 2.7 points of IQ per year of education.

Our analysis to this point leaves open the question of which is the more important determinant of one's AFQT score—education or early (potentially "inherited") IQ. As noted in the introduction, Fischer et al. have taken the opposite position from Herrnstein and Murray. Specifically, Fischer et al. have argued that the effects of education on IQ are considerably more important than the effect of early IQ on later IQ. They cite our earlier work, in which we first reanalyzed Herrnstein and Murray's model as supporting this position. We disagree.

All the models in Table 10.2 suggest that both education and early IQ have important effects on later intelligence as measured by the AFQT. It is difficult, if not impossible, however, to compare unstandardized coefficients. We can get around this problem by standardizing both early IQ and educational attainment to have standard deviations equal to one. If we assume that each year of education increases an individual's IQ by 2.7 points, and that the coefficient on IQ is 0.36 (the median estimate in models 5 through 14) then the rescaled effect of education is 5.1 and the rescaled effect of early IQ is 10.6, more than twice as large.

Herrnstein and Murray argued in *The Bell Curve* that education has little or no effect on IQ. Through examining the literature and reanalyzing the data, we have shown that this conclusion is not supported. But is the effect large or small? In particular, does this range of estimates suggest that education's effect on IQ is modest, and that education may be quite limited as a policy mechanism for reducing inequality?

Dickens et al. argue that if IQ is as important a determinant of social and economic success as *The Bell Curve* suggests, then investments that increase IQ *even modestly* will have substantial payoffs.[44] It then becomes critical to know how responsive IQ is to education. Dickens et al. point out that Herrnstein and Murray's most conservative estimates suggest that a single point of IQ increases annual earnings by $232. Thus, if a year of school increases IQ by 2.7 points, our preferred estimate, the annual payoff from a year of schooling—through its effect on IQ alone—is $626. Assuming the cost of a year of high school is between $5,000 and $8,000, these estimates suggest a substantial monetary payoff.

Furthermore, Herrnstein and Murray demonstrate that IQ is an important determinant of a variety of dimensions of social and economic success such as reduced involvement in crime. Therefore, education may have a considerable effect, through IQ alone, on these other outcomes. (See Ref. 45 for estimates of the social benefits of reducing criminal involvement, one of the outcomes Herrnstein and Murray examine.) Moreover, the payoffs from additional education discussed so far all work through increasing IQ, and hence do not include the "direct" benefits that education might confer. In our earlier analyses of sibling differences in IQ and education, we found that an additional year of educational attainment raises annual earnings by about $1,300 (compared with one's sibling), controlling for differences in AFQT score.

Ironically, then, if the effect of education on IQ is within the broad range we have estimated, *The Bell Curve*'s demonstration of the importance of IQ for social and economic success (in combination with other evidence of substantial "direct" effects of education) provides evidence for the importance of educational investment as a policy instrument, quite contrary to the conclusion that one might reach from reading *The Bell Curve*.

References

1. Jensen, A.R. (1969), "How Much Can We Boost IQ and Scholastic Achievement?," *Harvard Educational Review*, 39, 1–123.

2. Herrnstein, R.J. (1971), "IQ," *Atlantic Monthly*, September, 43–64.
3. Herrnstein, R.J., and Murray, C. (1994), *The Bell Curve: Intelligence and Class Structure in American Life*, The Free Press, New York; (1996) paperback edition.
4. Fischer, C.S., Hout, M., Janowski, M.S., Lucas, S.R., Swidler, A., and Voss, K. (1996), *Inequality by Design: Cracking The Bell Curve Myth*, Princeton University Press, Princeton, NJ.
5. Ceci, S. (1991), "How Much Does Schooling Influence General Intelligence and its Cognitive Components? A Reassessment of the Evidence," *Developmental Psychology*, 27, 703–722.
6. Currie, J., and Thomas, D. (1995), "Does Head Start Make A Difference?" *American Economic Review*, June, 85, 341–364. Currie, J., and Thomas, D. (1996), *Head Start and Cognition Among Latino Children: Interactions with Language and Culture*, mimeo, February 1996.
7. Gardner, H. (1983), *Frames of Mind: The Theory of Multiple Intelligences*, Basic Books, New York.
8. Sternberg, R.J. (1985), *Beyond IQ: A Triarchic Theory of Intelligence*, Cambridge University Press, Cambridge, England.
9. Sternberg, R.J. (1990), *Metaphors of the Mind: Conceptions of the Nature of Intelligence*, Cambridge University Press, Cambridge, England.
10. Jencks, C. (1972), *Inequality: A Reassessment of the Effects of Family and Schooling in America*, Basic Books, New York.
11. Jencks, C. (1979), *Who Gets Ahead? The Determinants of Economic Success in America*, Basic Books, New York.
12. Gottfredson, L.S. (1986), "Societal Consequences of the g Factor in Employment," *Journal of Vocational Behavior*, 29, 379–410.
13. Hunter, J. (1986), "Cognitive Ability, Cognitive Aptitudes, Job Knowledge, and Job Performance," *Journal of Vocational Behavior*, 29, 340–363.
14. Hunter, J., and Hunter, R.F. (1984), "Validity and Utility of Alternative Predictors of Job Performance," *Psychological Bulletin*, 96, 72–98.
15. Gordon, R.A. (1976), "Prevalence: The Rate Datum in Delinquency Measurement and its Implications for a Theory of Delinquency," in M. Klein (Ed.), *The Juvenile Justice System*, Sage Publications, Newbury Park, CA, pp. 201–284.
16. Gordon, R.A. (1980), "Labeling Theory, Mental Retardation, and Public Policy: Larry P. and Other Developments Since 1974," in W.R. Grove (Ed.), *The Labeling of Deviance: Evaluating a Perspective*, Sage Publications, Newbury Park, CA, pp. 175–224.
17. Korenman, S., and Winship, C. (1995), *A Reanalysis of* The Bell Curve: *Intelligence, Family Background, and Schooling*, NBER Working Paper 5230, revised August 1996.
18. Center for Human Resources Research (1994), *NLS Handbook 1994*, Columbus, OH, The Ohio State University.
19. Brody, N. (1992), *Intelligence*, 2nd ed., Academic Press, San Diego, CA.
20. Carroll, J.B. (1992), "Cognitive Abilities: The State of the Art," *Psychological Science* 3, 266–270.
21. Gottfredson, L.S. (1994), "Mainstream Science on Intelligence," *The Wall Street Journal*, December 13.
22. Judd, C.M., and Kenny, D.A. (1981), *Estimating the Effects of Social Interventions*, Cambridge University Press, Cambridge, England.
23. Husén, T., and Tuijnman, A. (1991), "The Contribution of Formal Schooling to the Increase in Intellectual Capital," *Educational Research*, 20, 17–25.

24. Lund, T., and Thrane, V.C. (1983), "Schooling and Intelligence: A Methodological and Longitudinal Study," *Scandinavian Journal of Psychology*, 24, 161–173.
25. Lorge, L.L. (1945), "Schooling Makes a Difference," *Teacher's College Record*, 46, 483–492.
26. Harnqvist, K. (1968), "Relative Changes in Intelligence from 13 to 18," *Scandinavian Journal of Psychology*, 9, 50–82.
27. Harnqvist, K. (1977), "Enduring Effects of Schooling: A Neglected Area in Educational Research," *Educational Researcher*, 6, 5–11.
28. Wolfle, L.M. (1980), "The Enduring Effects of Education on Verbal Skills," *Sociology of Education*, 53, 104–114.
29. Duncan, O.D. (1968), "Ability and Achievement," *Eugenics Quarterly*, 15, 1–11.
30. McCall, R.B. (1970),"Childhood IQs as Predictors of Adult Educational and Occupational Status," *Science*, 197, 482–483.
31. Wechsler, D. (1958), *The Measurement and Appraisal of Adult Intelligence*, 2nd ed., Williams and Wilkins, Baltimore, MD.
32. Husén, T. (1951), "The Influence of Schooling upon IQ," *Theoria*, 17, 61–88.
33. DeGroot, A.D. (1948), "The Effects of War upon the Intelligence of Youth," *Journal of Abnormal Social Psychology*, 43, 311–317. DeGroot, A.D. (1951), "War and Intelligence of Youth," *Journal of Abnormal and Social Psychology*, 46, 596–597.
34. Green, R.L., Hoffman, L.T., Morse, R., Hayes, M.E., and Morgan, R.F. (1964), *The Educational Status of Children in a District without Public Schools*, (Co-Operative Research Project No. 2321), Office of Education, U.S. Department of Health, Education, and Welfare, Washington, D.C.
35. Cahan, S., and Cohen, N. (1989), "Age Versus Schooling Effects on Intelligence Development," *Child Development* 60, 1239–1249.
36. Neal, D.A., and Johnson, W.R. (1996), "The Role of Pre-Market Factors in Black-White Wage Differences," *Journal of Political Economy*, 104, 869–895.
37. Belsley, D.A., Kuh, E., and Welsch, R.E. (1980), *Regression Diagnostics*, John Wiley, New York.
38. Bollen, K.A., and Jackman, R.W. (1990), "Regression Diagnostics: An Expository Treatment of Outliers and Influential Cases," in J. Fox and J.S. Long (Eds.), *Modern Methods of Data Analysis*, pp. 257–291.
39. STATA: Statistics, Graphics, Data Management. Version 4.0. (1995), Stata Corporation, College Station, Texas.
40. Huber, P.J. (1967), "The Behavior of Maximum Likelihood Estimates under Non-standard Conditions," Proceedings of the Fifth Berkeley Symposium on Mathematical Statistics and Probability, 1, 221–233.
41. White, H. (1980), "A Heteroskedastic-Consistent Covariance Matrix Estimator and a Direct Test for Heteroskedasticity," *Econometrica*, 48, 817–830.
42. Goldberger, A.S. (1991), *A Course in Econometrics*, Harvard University Press, Cambridge, MA.
43. Ashenfelter, O., and Krueger, A.B. (1994), "Estimates of the Economic Returns to Schooling from a New Sample of Twins," *American Economic Review*, 84, 1157–1173.
44. Dickens, W.J., Kane, T.J., and Schultze, C. (1997), *Does The Bell Curve Ring True? A Reconsideration*, Brookings, Washington, D.C.
45. Freeman, R.B. (1996), "Disadvantaged Young Men and Crime," NBER Conference on Youth Unemployment and Employment in Advanced Countries, Winston-Salem, NC, December 12–14, 1996.

chapter eleven

Cognitive Ability, Environmental Factors, and Crime: Predicting Frequent Criminal Activity

LUCINDA A. MANOLAKES

Richard Herrnstein and Charles Murray's *The Bell Curve: Intelligence and Class Structure in American Life* has revived the ongoing debate over the appropriateness and usefulness of IQ as an explanatory variable in models predicting behavior.[1] The book posits that a core human cognitive ability exists and is "one of the most thoroughly demonstrated entities in the behavioral sciences and one of the most powerful for understanding socially significant human variation" (p. 14). The work's main thesis is that an individual's intelligence—no less than 40% and no more than 80% of which is inherited genetically from his or her parents—has more effect than socioeconomic background on future life experiences, including criminal actions.

The authors claim that "[h]igh cognitive ability is generally associated with socially desirable behaviors, low cognitive ability with socially undesirable ones" (p. 117). Logistic regression analyses are performed on a subsample of white males of the National Longitudinal Survey of Youth (NLSY), and a relationship between IQ and criminality is found. Herrnstein and Murray find that for both of their measures of criminal activity—self-reported by the respondent or via an interview conducted with an incarcerated respondent between 1979 and 1990—socioeconomic status has an insignificant effect after controlling for IQ and IQ itself is a strong predictor of criminal behavior.

In my view, the authors' argument is based on an incomplete set of variables and simplistic statistical analyses. Regarding the former, the authors include only parental education, occupational prestige of parents, and household income in order to measure the environment of an individual. Social life, however, is comprised of many other effects in addition to these three individual-level variables and includes macro-level structural variables. In order to draw conclusions regarding the variables that predict criminality, a theoretical model must be expanded to include the variety of variables necessary to represent the environment in which respondents find themselves. With respect to the latter critique, Herrnstein and Murray have not provided for interactions between the independent effects selected for statistical analysis. I argue that permitting interactions between and among explanatory variables (both individual and structural) in a logit analysis is necessary in order to better the fit between the designed model and the observed data. This methodology provides for theoretical specificity with respect to predicting crime by considering a wider range of explanatory variables.

Note that it is not my intention to provide an overarching theory of crime. Rather, I hope to offer some methodological groundwork for criminological theory construction by critiquing the methods employed in *The Bell Curve* and illustrating the applicability of interaction effects to studying criminality. To this end, the variables considered for inclusion in the logistic regression equation have not been selected idiosyncratically. The following concise review of the literature reveals that it is rich with variables that criminologists find both interesting and important for discussions of crime; choosing from among them will prove most useful to my efforts.

In studies prior to those of Herrnstein and Murray, IQ has been found to be a strong predictor of criminal behavior and/or juvenile delinquency. Hirschi and Hindelang review much of the literature linking IQ and criminality, and conclude that IQ can perform as well in predicting official delinquency as do race or social class variables.[2] (See also Refs. 3–6). Stattin and Klackenberg-Larsson concur, concluding that cognitive ability must be acknowledged as playing a crucial role in criminal activity because the negative relationships between IQ and criminality appear in studies using different designs, different subject samples, different measures of intelligence, and/or different aspects of delinquency."[7] Having conducted an extensive longitudinal study, the authors find that at age 3, boys' cognitive ability is highly correlated with future criminality and remains so through the age of 17 (except between the ages of 5 and 8, when the correlation falls just below the level of significance). Among the respondents, non-offenders were consistently found to have the highest IQs at all ages and the frequent offenders had relatively low IQ scores (see Refs. 5 and 8).

Many sociologists rarely consider IQ for models designed to explain human behavior. Instead, demographic and/or social environment characteristics are of utmost importance in constructing theories, including those attempting to explain why individuals break the law. For example, Messner and Rosenfeld[9] and Cloward and Ohlin[10] stress that the individual inclination to engage in delinquent or criminal behavior cannot adequately explain the phenomenon. The social environment must be considered a contributor to the development of delinquency insofar as it is an enabling factor—a learning environment in which access to delinquent roles and training in delinquency is available.

Similarly, social disorganization theorists expand upon the importance of the social environment in predicting criminal behavior. Delinquency is more frequent in urbanizing areas where neighborhoods deteriorate and living conditions become overcrowded.[11] Faris and Dunham, working in this perspective, claim that urban areas are found to be more prone to crime and vice than are rural areas.[12] This may be due to the fact that "[t]he high-delinquency area maximizes the temptations for illegitimate behavior and minimizes external inhibitions. In situations like this, [everyone is] engaged in what is literally a battle for survival" (Ref. 13, p. 110); therefore, the racial/ethnic composition of the deteriorated and chaotic areas of the city is not to blame,[14] but simply "the nature of social life in the areas themselves" (see Ref. 12, p. 70). In a given disorganized area, delinquency rates remain relatively high as nationalities successively move in and out of the area. As a result, one must consider the individual-level variables and structural-level variables that influence *who* commits crimes.

The demographic composition of crime-prone urban areas makes a discussion of the relationship between race and crime necessary as well. A larger percentage of blacks, both young and unemployed, live in urban areas than do whites.[15] Therefore, with a greater proportion of blacks in poverty and unemployed or underemployed, "the economic well-being argument would seem to have special relevance for blacks" (Ref. 16, p. 161). Russell extends the discussion of race further, criticizing the numerous empirical studies that establish a link between race and crime—arrests, conviction, and incarceration—for consistently failing "to develop a broad-based analytical and theoretical framework for explaining the phenomenon of disproportionality."[17] She suggests that the treatment of minorities in the United States, as well as the internalization of that treatment, must be investigated because both are surely related to criminal participation.[18,19] In the same vein, Walker et al. state that racial inequalities persist and that overt and subtle discrimination continues to plague the criminal justice system.[20] Young black men are often considered suspect by the police until they prove otherwise, a feat not easily achieved. In many communities, to be white is to be given the benefit of the doubt and eligibility "for consideration and for much more deferential treatment [by the police] than that accorded blacks in general" (Ref. 21, p. 194).

Yet African Americans are not the only segment of the population overrepresented in the criminal justice system. Men are more likely than women to exhibit criminal behavior, especially those forms considered most serious.[22] Many have concluded that women commit a small proportion of violent crimes and are more likely to commit crimes considered less serious (e.g. prostitution, shoplifting), which pose less threat to society.[23,24] As a result, much of the research has lagged behind current trends in female criminality.[13,25]

This concise review of the criminological literature cannot begin to indicate its vastness because the few variables selected represent only some of the major concepts deemed theoretically significant by criminologists. To reiterate, the independent variables chosen for consideration in the logistic regression equation predicting the likelihood for criminality include IQ, residence (rural or urban), race, gender, and parental educational attainment. To study these explanatory effects, they must be operationalized via the available dataset.

Method

Herrnstein and Murray use data collected by the NLSY, which at the time of their analyses had been conducted annually between 1979 and 1993, and includes 12,686 participants who were between 14 and 22 years of age in 1979. For their analyses on a subsample of white males, the authors use respondent IQ scores obtained from the Armed Forces Qualification Test (AFQT), a test described as the Spearman *g*-loaded* subtest of the Armed Forces Vocational Aptitude Battery (ASVAB). The variable codes the test score as a percentile and was measured in 1980.†

To formulate their second independent variable—the socioeconomic index—Herrnstein and Murray use variables that refer to the status of the respondents' parents.

> Since the purpose of the index was to measure the socioeconomic environment in which the NLSY youth was raised, the specific variables [used were]

---

*There are many scholars who accept that *g* is the expression of a core human mental ability and is synonymous with what most people mean when they use the term ***intelligence*** (p. 22).

†As Herrnstein and Murray write, "In 1989, the armed forces decided to rescore the AFQT so that it consisted of the word knowledge, paragraph comprehension, arithmetic reasoning, and mathematics knowledge subtests. The reason for the change was to avoid the numerical operations subtest, which was both less highly *g*-loaded than the mathematics knowledge subtest and sensitive to small discrepancies in the time given to subjects when administering the test (numerical operations is a speeded test in which the subject completes as many arithmetic problems as possible within a time limit)" (p. 570).

> total net family income, mother's education, father's education, and an index of occupational status of the adults living with the subject at the age of 14. (p. 574)

All of these variables were measured in 1979, except for family income, which was an average of the total net family income for 1978 and 1979. For the purpose of creating an index, each of the aforementioned variables were standardized, summed, and averaged (p. 574). Age (as measured in 1979) is also included as an independent variable to account for the fact that

> the AFQT was designed by the military for a population of recruits who would be taking the test in their late teens, and younger subjects in the NLSY sample got lower scores for the same reason that high school freshmen get lower SAT scores than high school seniors (p. 571).

The dependent variable used by Herrnstein and Murray that I have chosen to discuss involves an additive index of those variables measured in 1980 pertaining to the self-reported criminal activity of white males in the prior year. With respect to this measure, the respondent is either in the top-decile of frequency of criminal activity or not. It must be noted that the precise method by which this additive index is constructed is not specifically defined in the text.* However, attempts have been made to replicate the statistical analyses presented in *The Bell Curve* with respect to criminal activity. Although I have been unable to produce Herrnstein and Murray's exact results, similar statistical results have been achieved.

For instance, a logistic regression model,† which includes standardized IQ scores, a standardized socioeconomic scale,‡ and standardized ages of the respondents (measured in 1979) as independent variables to predict the aforementioned dependent variable (self-reported criminal activity) somewhat substantiates the conclusions contained in *The Bell Curve*. According to this model, an increase in IQ yields a decrease in the likelihood for frequent criminal activity (parameter estimate $= -0.1463$; probability $= 0.0518$), whereas an increase in socioeconomic status yields an increase in the likelihood

---

*Herrnstein and Murray also construct a dependent variable based on whether the white male NLSY respondents were ever interviewed while in a correctional facility between 1979 and 1990. In order to manage the length of this chapter, I have chosen not to evaluate this second dependent variable at this time (see Ref. 26 for an analysis of Herrnstein and Murray's equation and conclusions regarding this second dependent variable).

†Equation calculated for white male NLSY respondents with less than a college education ($n = 1,707$).

‡Four variables—household income, mother's education, father's education, and maximum occupational prestige of adults living with the respondent at age 14—were standardized, summed, and averaged in an attempt to replicate Herrnstein and Murray's calculation of a socioeconomic scale.

for top-decile criminal behavior (parameter estimate = 0.3220; probability = 0.0251).

Because the purpose of this paper is to evaluate Herrnstein and Murray's logistic regression analyses as well as their conclusions regarding criminality in a more sociological manner, the NLSY data serve as the dataset for this project. Rather than use a social environment index, I shall treat all independent variables individually so that their singular effects on the dependent variable can be analyzed, and so that the effects of interaction terms between independent variables may be investigated.

Variables

The variable* measuring the intelligence levels of the male NLSY respondents is their AFQT percentile scores as revised by the armed forces in 1989.

The second independent variable included in the model predicting criminal activity is parental education level when the respondent was 14 years of age (measured in 1979). This variable equals the maximum educational attainment of the subject's father and/or mother and is coded as the highest grade of school completed. Including parental education is more effective than using parental occupational prestige in that it not only implies employment status and income, but also denotes the learning and cultural environment in which each respondent lived as a child.

The third independent variable expresses the relationship between the type of environment in which the respondent resided at age 14 and his area of residence in 1979 (each coded as 1 = urban and 2 = rural). It is constructed to measure both the recency and longevity of the respondents' urban residence and is treated as a categorical variable rather than a continuous variable in the logit model. Categories report whether the respondent lived in an urban area at the age of 14 and in 1979 (coded as 1), lived in a rural area at the age of 14 but reported an urban residence in 1979 (coded as 2), lived in an urban area at the age of 14 but reported a rural residence in 1979 (coded as 3), or lived in a rural area at the age of 14 and in 1979 (coded as 4). Using this variable I investigate the extent to which the particular environment in which one resides contributes to criminal activity.

Herrnstein and Murray, in order to justify limiting their statistical analyses to *white* males, cite the work of Hindelang, which explains that black juveniles are known to underreport the frequency with which they engage in criminal acts.[27,28] The work of Hindelang and colleagues emphasizes that criminal

*All of the variables discussed in the Introduction are located in or can be created from the NLSY dataset.

self-report scales are predominantly based on less serious crimes of high frequency.[28,29] These are the types of crimes to which whites are more likely to admit. Blacks, on the other hand, are more likely to admit to less frequent but more serious crimes. Ignoring the fact that all of the NLSY variables regarding crime—including those related to breaking the law and being booked, charged and/or convicted of a crime—are based on the self-reporting of the respondents, the authors chose to accept only *some* of the data by including the responses of certain *segments* of the respondents.

I find Herrnstein and Murray's supposition that analyses cannot include various racial categories to be both unconvincing and unjustified. As Genovese notes, the co-authors exclude race from consideration on the basis that it will not withstand scientific analysis yet proceed to discuss outcomes in terms of those excluded racial categories.[30] Rather than include all criminal activity variables and exclude blacks from the analysis as Herrnstein and Murray do, I have heeded the warnings of Hindelang and colleagues by excluding the least serious criminal activity variables that are committed on a more frequent basis—running away from home, skipping school, drinking alcohol while under age, and using marijuana—from the additive index of self-reported frequency of criminal activity. By controlling for the underreporting of blacks in this way, I have enabled race to be included as an independent variable in the logit model (coded as 1 = white and 2 = black; measured in 1979).*

The dependent variable has been constructed in an attempt to reflect what I assume to be the procedures employed in *The Bell Curve*. There are twenty variables in the dataset pertaining to self-reported criminal activity. Although the questions are asked of the respondents in the 1980 survey, they inquire about criminal activity that has taken place in the prior year. The variables measure the frequency with which the individual engaged in the following activities: (1) running away from home, (2) skipping school, (3) drinking alcohol, (4) intentionally damaging property, (5) fighting at work or school, (6) shoplifting, (7) stealing other's belongings worth less than $50, (8) stealing other's belongings worth more than $50, (9) using force to obtain things, (10) seriously threatening to hit and/or actually hitting someone, (11) attacking someone with the intent to injure or kill, (12) smoking marijuana/hashish, (13) using other drugs/chemicals to get high, (14) selling marijuana/hashish,

*Caucasians and African-Americans are the only racial categories included in the model because the third racial category ("Other") combines all nonwhite and nonblack races (i.e., Hispanic, Asian, Indian, etc.). My excluding this last category must not be interpreted as a lack of interest in the criminal behavior of this group (or groups) but as an attempt to avoid the extreme heterogeneity of this broadly defined racial category and to limit the complications involved in discussions of statistical results.

(15) selling hard drugs, (16) attempting to "con" someone, (17) taking an automobile without the owner's permission, (18) breaking into a building, (19) knowingly selling/holding stolen goods, and (20) aiding in a gambling operation. (All are coded as 0 = never, 1 = once, 2 = twice, 3 = 3 to 5 times, 4 = 6 to 10 times, 5 = 11 to 50 times, and 6 = more than 50 times.)

For the purposes of creating an additive index of criminal behavior, the categories for each variable in the above list are recoded according to their central values. In other words, for each of the crime variables, the first three categories (0, 1, and 2) remain unchanged. The fourth category (3) is recoded as 4 times, the fifth category (4) is recoded as 8 times, and the sixth category (5) is recoded as 30.5 times. The final category (6) is recoded as 75 times.*

With the self-report crime variables recoded in this way, they are totaled in order to provide an additive index.† A binary dependent variable is constructed according to the additive index on the basis of whether or not the subject is in the top-decile of self-reported crime (1 = yes, 2 = no).

Logit Model: Interpretation

Because the dependent variable contains only two categories, logistic regression techniques are used to provide the expected natural log (logit) of the ratio of the two dependent variable probabilities.[31] A saturated logit model is calculated using the variables operationalized earlier and interactions are deleted from the model in a hierarchical manner until those remaining are statistically significant. The final model calculated for 4,355 male respondents includes IQ, parental education, residence, and race as well as interactions between IQ and

---

*The upper bound of the seventh variable category (6) is unknown because it designates any frequency greater than 50; therefore, a value of 75 has been arbitrarily selected as the central value.

†The number of times the respondent ran away from home in the past year, skipped school in the past year, drank alcohol in the past year, and/or used marijuana in the past year were excluded from the additive index for two reasons. First, those questions pertaining to running away, skipping school, and drinking alcohol while under age (a) were only asked of those respondents under the age of 18 years at the time of the interview, (b) are exhibited by a large proportion of the sample population, and (c) must not be considered on the same level of criminality as the other variables used to measure criminal activity. Second, although the question pertaining to marijuana use was asked of all the respondents, regardless of age, it too must not be considered on the same level of criminality as the other variables used to measure illegal activity.

To add the crime variables together without recoding them in the way I have described earlier would have been problematic because the original categories of the variables were ordinal and represented wide frequency ranges of criminal activity. Totaling the original ordinal categories across the list of crime variables would have rendered sums without meaningful interpretations.

Table 11.1. Means and standard deviations of variables used in logit model

Variable	Mean	Standard deviation
IQ	41.5273	29.7873
Parental education	11.7747	3.4152
Change in residence	1.6489	1.0294
Race	1.2682	0.0443

parental education, IQ and race, and parental education and race (Tables 11.1 and 11.2).*

Standardized estimates are calculated for the variables in the final model in order to compare the strength of the associations between the dependent variable and each independent variable net the effect of all others. It is clear that IQ has a significant effect on criminal behavior: for men, the standardized estimate is equal to 0.4665.† However, Herrnstein and Murray's hypothesis that IQ is the *only* explanatory effect with the ability to predict criminal behavior is inconsistent with the data. Even if one were to accept the authors' interpretation of IQ, results show that the effects of IQ are not sufficient to explain the likelihood of criminal behavior.

For example, the standardized estimates show the interaction between IQ and parental education (-0.5717) to have a stronger effect on the dependent variable than IQ. Similarly, the interaction between parental education and race also strongly affects the dependent variable probabilities (0.3690), although not to the extent that IQ does. In addition, parental education as a predictor variable is also shown to have a rather strong effect on the dependent variable probabilities (0.1968). Thus, the standardized estimates clearly indicate that IQ is not the only variable affecting the likelihood for criminality,

*The likelihood ratio statistic reported in Table 11.2 (0.9999) illustrates that the logit model fits the observed data well; that is, we are assured that the data predicted by the logit model reflects the observed data. This likelihood ratio, rather than R^2, is the more appropriate method by which to assess the logit model.

†It must be noted that the relationship between IQ and criminality may be due, in part, to the fact that the NLSY data do not include any questions regarding white collar crime, organized crime, corporate crime, consumer fraud, or those acts that serve as precursors to such crimes. Therefore, I contend that selection bias occurs via the survey instrument in that respondents with higher IQs are not asked to report about the crimes they are more likely to commit.

Table 11.2. Maximum likelihood analysis of variance table and standardized and nonstandardized parameter estimates

Source	DF	Chi-square	Prob
Intercept	1	100.41	0.0000
IQ	1	18.44	0.0000
Parental education	1	14.57	0.0001
Change in residence	3	16.24	0.0010
Race	1	2.40	0.1214
IQ * parental education	1	19.80	0.0000
IQ * race	1	5.83	0.0158
Parental education * race	1	8.35	0.0038
Likelihood ratio	2204	1923.12	0.9999

Effect	Estimate	Standard error	Chi-square	Probabilities	Standardized estimate
Intercept	−3.2142	0.3208	100.41	0.0000	—
IQ	0.0279	0.0065	18.44	0.0000	0.4665
Parental education	0.1063	0.0278	14.57	0.0001	0.1968
Change in residence	0.2479	0.0727	11.62	0.0007	0.0926
	0.1513	0.1101	1.89	0.1693	0.0392
	−0.3120	0.1192	6.85	0.0088	−0.0826
Race	−0.3801	0.2454	2.40	0.1214	−0.1838
IQ * parental education	−0.0022	0.0005	19.80	0.0000	−0.5717
IQ * race	−0.0056	0.0023	5.83	0.0158	−0.1307
Parental education * race	0.0620	0.0215	8.35	0.0038	0.3690

nor are its effects simply direct; the effect of IQ is mediated by other factors, such as the education of one's parents.

Such findings necessitate a discussion of the nonstandardized parameter estimates. Changes in residence are shown to have strong effects on the dependent variable probabilities. Living in an urban area at the age of 14 and in 1979 increases the logged odds of being in the top-decile of criminal activity by 0.2479. Having lived in a rural area at the age of 14 but reporting an urban residence in 1979 also increases the logged odds of engaging in frequent illegal acts (0.1513), although this parameter estimate is not statistically significant. Moving from an urban environment (age 14) to reside in a rural environment (in 1979) decreases the logged odds of the dependent variable by 0.3120.

Parental education affects the dependent variable probabilities as well. An increase of 1 year in parental education yields a 0.1063 increase in the logged odds of exhibiting top-decile criminality. The interaction between parental education and race is statistically significant and increases the logged odds of the dependent variable by 0.0620.

An increase of one unit in IQ increases the logged odds of being in the top-decile of self-reported criminal activity by 0.0279. The interactions between IQ and race, and IQ and parental education, decrease the logged odds of placing in the upper decile of illegal activity by 0.0056 and 0.0022, respectively.

Because those coefficients are linearly related to the logged odds of the dependent variable, as well as because variables included in the logit model are effect coded,* interpretation of the parameter estimates in and of themselves is both abstract and complicated. To facilitate the analysis, the calculated logit values are converted into probabilities of the dependent variable—the likelihood that a respondent is in the top-decile of frequency of self-reported criminal activity. The following discussion details the effects that the independent variables and interaction terms have on the dependent variable probabilities.†

Logit Model: Evaluation

Change in Residence

The variable measuring the recency and longevity of urban residence for the NLSY male respondents is significant in predicting criminal behavior (Fig. 11.1). Regardless of race, those men who had lived in an urban area at the age of 14 and in 1979 are the most likely to exhibit top-decile criminal behavior; whites have an 18% chance and blacks have a 16% chance.‡ For men who had lived in a rural area at the age of 14 but reported an urban residence in 1979, white men have a 16% chance and thus were slightly less likely to engage in illegal acts on a frequent basis (whites have a 16% chance and blacks have a 14% chance).§ Those men who had lived in an urban area at the age of 14 but moved to a rural residence by 1979 are least likely to exhibit top-decile criminality (whites have an 11% chance and blacks have a 10% chance). Finally, the likelihood for top-decile criminal behavior increases for white men living

---

*If variable A has k levels or categories, its main effect has $k-1$ degrees of freedom; therefore, the design matrix contains $k-1$ columns, which correspond to the $k-1$ levels or categories of variable A. The ith design matrix column contains a 1 in the ith row, a -1 in the last row, and zeroes in all other rows.

†For ease of presentation, any discussion of IQ in the text includes only probabilities that the respondent is in the highest decile of criminality for respondents at the lower, median, and upper quartile of the IQ distribution.

‡The interaction between change in residence and race is not statistically significant; results for the separate racial categories are given in Figure 11.1 for the purposes of presentation only.

§This parameter estimate is not statistically significant (probability = 0.1693).

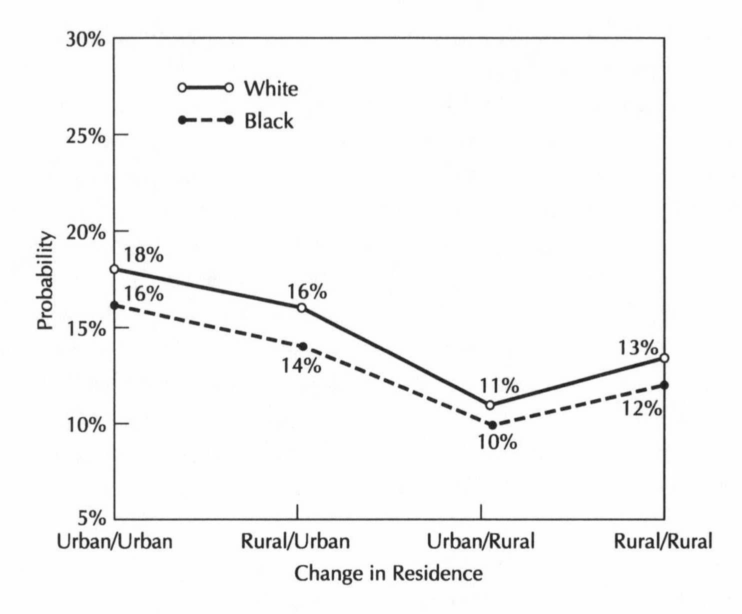

Figure 11.1. Probability of being in the top-decile of self-reported criminal activity, by change in residence for whites and blacks.

in a rural area at the age of 14 and in 1979 (13%) as well as in black men with similar residence patterns (12%).

It is clear that both for African-Americans and Caucasians, residing in an urban area has a great impact on the likelihood for frequent illegal activity. Social disorganization theorists and some urban anthropologists agree that a city is oftentimes an anonymous and disorganized place wherein social networks are either nonexistent or temporary, and individuals are freed from the constraints of social control.[32] According to Merry, "[c]ities are characterized by anomie: by a sense of normlessness, both in the sense of the individual's lack of attachment to a moral code and of a collective loss of moral consensus" (Ref. 33, p. 63). The anonymity of the city is said to provide for limitless opportunities for crime; illegal acts can be committed with little fear of being recognized by victims and/or apprehended by authorities. As Merry describes, individuals in urban areas are more able to victimize others in their own neighborhoods due to the fact that it is unlikely that victims will recognize their assailants in other social contexts.

Similarly, nonexistent, detached, or loose social networks within urban communities detract from the implementation of sanctions against those who choose to break the law for at least two reasons. First, the lack of social networks produces more emphasis on individualism than on community. As a result, those living in urban areas are more likely to protect themselves because, as Balthazar and Cook emphasize, urban living is a battle for survival (see Ref. 13). Therefore, preoccupied with their own well-being, individuals are less concerned with those around them. Secondly, the ability to punish those who break the law is compromised when a community is fragmented and lacks strong social networks; as Merry explains, it is highly unlikely that punishment is wielded across social boundaries.

IQ, Parental Education, and Race

Although there are no statistically significant three-way interactions in the final logit model, IQ, parental education and race interact with one another and have strong effects on the likelihood for frequent criminality. For example, one important interaction effect for predicting top-decile criminal behavior involves the relationship between IQ and race (Fig. 11.2). For white men, as IQ increases from the lower to the upper quartiles, their chances of exhibiting frequent illegal activity decreases (lower quartile = 17%, median IQ = 16%, upper quartile = 15%). It is quite interesting to note, however, that the same relationship does not hold true for African-American NLSY males. For black men, as IQ levels increase, so do the chances for top-decile criminal behavior. Black men with lower IQs have a 12% chance of committing illegal acts on a frequent basis whereas those with a median IQ have a 14% chance. Those black men with an above average IQ are the most likely of the three groups to be placed in the top-decile of frequency of criminal activity (a 17% chance).

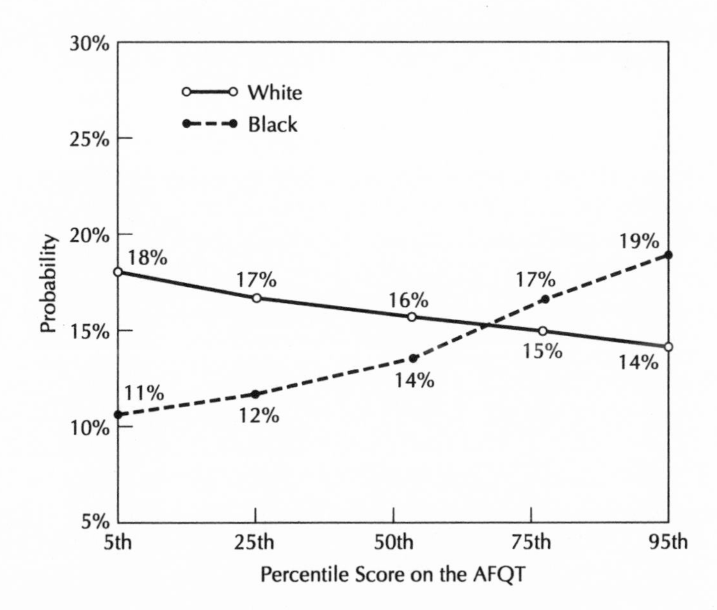

Figure 11.2. Probability of being in the top-decile of self-reported criminal activity, by percentile score on the AFQT for whites and blacks.

This finding appears to directly contradict the findings that Herrnstein and Murray present in *The Bell Curve*. According to the authors, "high cognitive ability protects a person from becoming a criminal even if other precursors are present" (p. 244). They justify such a statement by arguing that "[h]olding socioeconomic status constant does little to explain away the relationship between crime and cognitive ability" (p. 235). Yet we see that holding race constant does provide interesting results regarding the relationship between the likelihood for top-decile criminality and IQ. For the African-American males in the NLSY sample, the *more* "intelligent" the respondent, the *greater* his chance to exhibit frequent illegal activity. This finding points to a crucial flaw in Herrnstein and Murray's theory; the argument that IQ protects an individual from becoming a criminal is true only for those individuals on whom statistical analyses were conducted (white males). The error in generalizing this theory could not have been detected by the authors because they chose to exclude black men from their statistical analyses.

The interaction between parental education and race is also significant in predicting the frequent criminality of the NLSY males (Fig. 11.3). For whites, there is a direct relationship between parental education and criminal behavior. White men born of parents with less than a high school education have a 13% chance of committing illegal acts on a frequent basis. The likelihood for top decile criminal activity increases to 16% for those white men born to a high

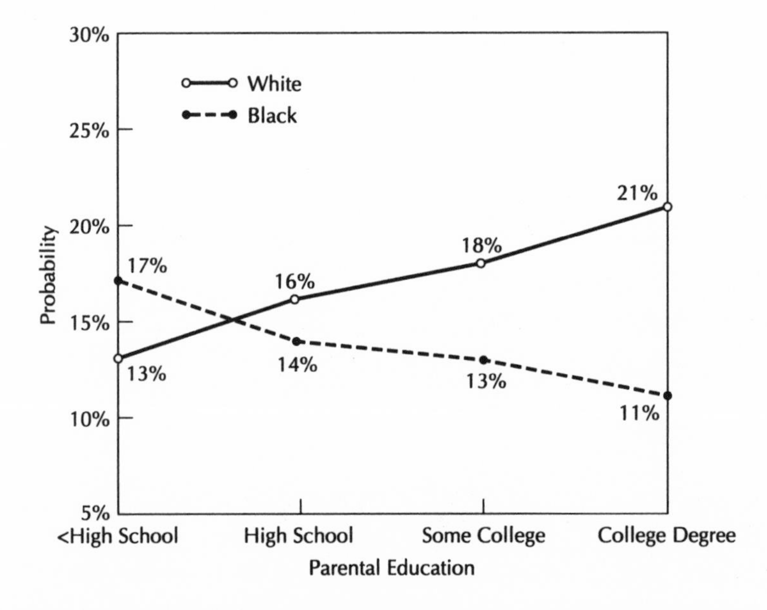

Figure 11.3. Probability of being in the top-decile of self-reported criminal activity, by parental education for whites and blacks.

school graduate and increases to 18% for those individuals with parents with some college training. Those white NLSY males born of parents with a college degree or more are the most likely (21%) to exhibit frequent criminal behavior.

The effect of parental education on criminality is reversed for the black NLSY respondents in that there is an inverse relationship between parental education and top-decile criminal behavior. Those black men with the least educated parents are most likely to commit frequent illegal acts (17%). This likelihood decreases to 14% for those blacks with parents who graduated from high school and decreases to 13% for those with parents with some college education. The black NLSY respondents with the most educated parents are the least likely to be placed in the top-decile of frequency of criminal activity (an 11% chance).

The interaction between IQ and parental education is found to be significant in predicting the criminality of the NLSY males. In order to investigate the effect of this interaction term, it is most instructive to hold the IQ variable constant and to investigate the ways in which the dependent variable probabilities vary as parental education varies (Fig. 11.4). For those men in the lower quartile of IQ, the likelihood for top-decile criminal behavior increases as parental education increases. The likelihood for lower IQ males to exhibit frequent criminal activity if their parents have less than a high school education is 12.3%. Lower IQ sons of individuals who either graduated from high school

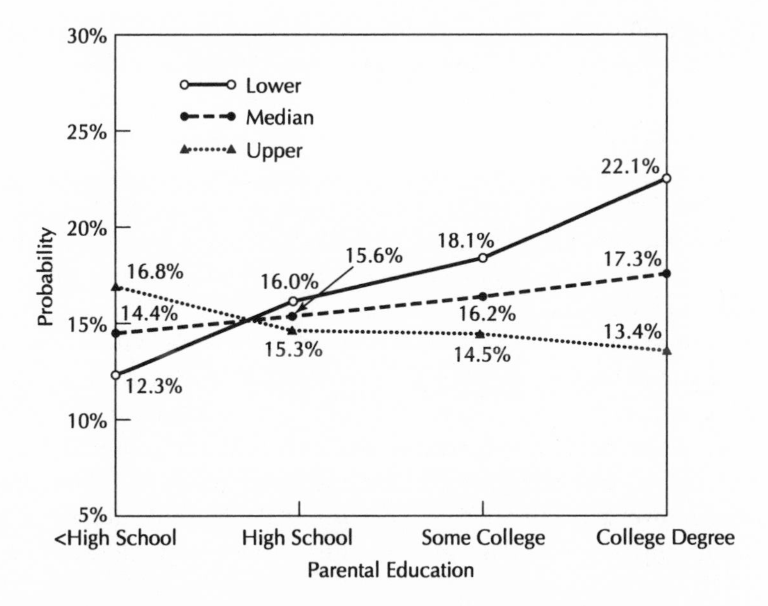

Figure 11.4. Probability of being in the top-decile of self-reported criminal activity, by parental education for quartile scores on the AFQT.

or had some college training have a 16.0% or 18.1% chance of placing in the top-decile of illegal activity, respectively. The lower IQ men who have at least one parent with at least a college degree have a 22.1% chance of committing frequent crimes. For those respondents with a median IQ level, the probabilities increase less dramatically as parental education increases (the probabilities of being in the top-decile of self-reported criminal activity are as follows: a parent with less than a high school diploma, 14.4%; a parent with a high school diploma, 15.6%; parental education consisting of some college instruction, 16.2%; parental education consisting of a college degree or more, 17.3%).

It is most interesting that for those male respondents who had an upper quartile IQ score, parental education has an inverse effect on their chances for displaying top-decile criminal behavior. In other words, "bright" male respondents born of parents who had not graduated from high school have a 16.8% chance of displaying frequent criminality. The likelihood for those higher IQ men with at least one parent who graduated from high school is 15.3%. The probability that a higher IQ male respondent will commit illegal acts on a frequent basis decreases to 14.5% when he is born of at least one parent who had some college education, and to 13.4% if born of at least one parent who had a college degree and/or graduate training. Figure 11.4 illustrates that for lower and higher IQ men, as a respondent's intelligence level (according to his IQ percentile score) approaches that of at least one of his parents (according to maximum parental education), the likelihood for top-decile criminality decreases.

When controlling for race, other interesting results are obtained, although the three-way interaction involving IQ, parental education, and race is not statistically significant. For white men, regardless of their IQ, a direct relationship between parental education and the likelihood for frequent criminal behavior is evident; however, there is little change for higher IQ men (Fig. 11.5). The likelihood that white, lower IQ males display top-decile criminal activity more than doubles as parental education increases from less than high school (probability of 12.3%) to at least a college degree (probability of 27.3%). The likelihood that white, median IQ males exhibit frequent criminal activity increases from 13.4% to 20.5% as parental education increases from less than high school to at least a college degree, respectively. Finally, for those upper quartile IQ white men, the chances for top-decile criminal activity barely increase as parental education increases (probabilities increase from 14.7% for least educated parents to 15.1% for most educated parents). For black men, however, an inverse relationship between parental education and top-decile criminal activity is detected, irrespective of the IQ levels of the respondents (Fig. 11.6)

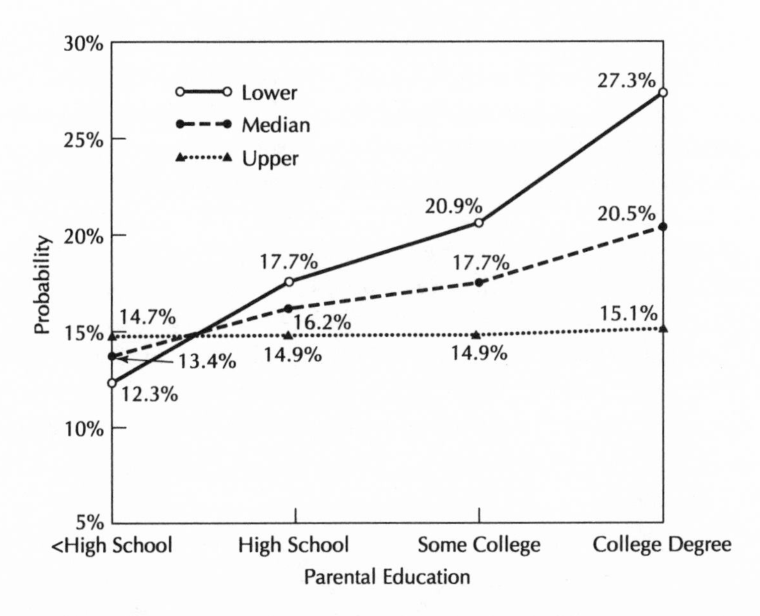

Figure 11.5. Probability of being in the top-decile of self-reported criminal activity, by parental education for quartile scores on the AFQT for whites.

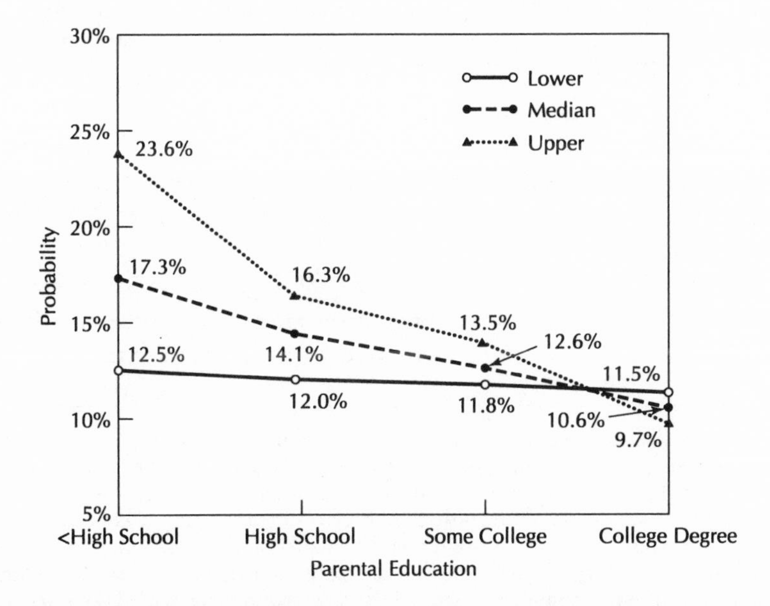

Figure 11.6. Probability of being in the top-decile of self-reported criminal activity, by parental education for quartile scores on the AFQT for blacks.

For black men with lower IQs, as parental education increases from less than high school to at least a college degree, probabilities decrease only slightly (from 12.5% to 11.5%, respectively). For median IQ black men, the likelihood for frequent illegal activity decreases from 17.3% for those born of parents with less than a high school education to 10.6% for those born of parents with at least a college degree. Most striking are the results for the black, upper quartile IQ respondents. For these men, the chance for displaying top-decile criminal behavior is greatest when their parents have less than a high school diploma (23.6%). The probability for frequent criminality decreases to 16.3% and 13.5% as parental education increases to high school diploma and some college instruction, respectively. Finally, for higher IQ black men born of parents with at least a college degree, the likelihood for top-decile criminal activity is 9.7%.

Conclusions

Of course, the effect of interaction terms and independent variables on the likelihood for top-decile criminality may be explained by numerous theoretical perspectives not mentioned here. But I leave that task for the experienced criminologist. Regardless of the ways in which the predictor variables influence the dependent variable, the fact remains that the inclusion of interaction effects and other independent variables is necessary to improve the statistical analyses presented in *The Bell Curve*. By limiting the independent variables introduced in a logistic regression equation to IQ, a socioeconomic status scale, and age, Herrnstein and Murray ignore the explanatory power of numerous other effects that criminological theory has determined to be quite important in predicting or explaining criminal behavior.* Such an oversight leads the authors to conclude that socioeconomic background has a negligible influence on illegal activity once the effects of cognitive ability are taken into account. It must also be noted that by limiting their analyses to *white* men only, Herrnstein and Murray are unable to determine the ways in which the effects of variables differ for separate racial categories.

This chapter is an attempt to replicate the logistic regression analyses presented in *The Bell Curve* while improving the range of independent variables included in the model. However, there were problems in understanding some

*Fischer et al. also argue that an individual's social background involves more than parental education, occupational prestige, and household income. The authors expand the notion of the social environment and include other factors, such as number of siblings, poverty status, school quality, region of residence, years of education, etc., in a logistic regression equation predicting the likelihood that an NLSY male was incarcerated. It is not surprising that statistical results indicate that "at least three other factors were jointly more important [than IQ] in predicting who ended up in jail: the schooling men had completed at the time they took the AFQT, the kind of high schools they last attended, and having been poor just before the 1980s" (see Ref. 26).

of the authors' definitions of variables. Herrnstein and Murray neglect to describe specifically their construction of the dependent variable measuring self-reported criminal activity. As a result, I have sought to replicate it in the most sociological and statistically responsible manner, assuming that the authors did so as well. Herrnstein and Murray also operationalize some variables in ways that I, without the restriction of reflecting their analyses, would not; however, I have not attempted to improve upon their choice.

Although there is an inverse relationship between IQ and the likelihood for criminal behavior when only these two variables are considered, a more sensitive sociological analysis of the data shows IQ to be only one of many variables that affect the chances of the NLSY men to frequently engage in illegal acts. Parental education and change in residence also have statistically significant relationships with a male's chances for criminal activity, as do the interaction terms involving IQ and parental education, IQ and race, as well as parental education and race. Clearly, regardless of one's definition of IQ, it must not be considered the only important variable in predicting crime.

What is most disconcerting is that the social policy recommendations that Herrnstein and Murray make in *The Bell Curve* are based upon the unsubstantiated notion that IQ can effectively predict crime and socioeconomic background cannot do so. The authors claim that in general, for social policy to be effective, its focus must shift from the *social* causes of inequality or negative outcomes, to the *cognitive* causes (p. 527). They propose that the criminal justice system must "be made simpler" because

> a person with comparatively low intelligence, whose time horizon is short and ability to balance many competing and complex incentives is low, has much more difficulty following a moral compass in the 1990s than he would have in the 1950s.... People of limited intelligence can lead moral lives in a society that is run on the basis of "Thou shalt not steal." They find it much harder to lead moral lives in a society that is run on the basis of "Thou shalt not steal unless there is a really good reason to" (p. 544).

Therefore, Herrnstein and Murray argue that the meaning of criminal offenses, as well as the consequences, must be made clearer and more objective—in essence, simpler—so that individuals of all cognitive levels can understand not only what constitutes a wrongful act, but what the punishment will be for engaging in such behavior. However, the authors do not offer specific ways in which the criminal justice system is to be altered as such.

In light of the findings presented in this chapter, such a policy recommendation—to "simplify" the criminal justice system—seems incomplete. Even using the logic of Herrnstein and Murray, it is evident that such a solution would not be effective because the results illustrate that for black men, an *increase* in IQ accompanies an *increase* in the likelihood for frequent criminal activity. Similarly, *net the effect of IQ*, parental education and change in residence each have a significant effect on the likelihood for top-decile illegal activity on the part of the NLSY males. Therefore, it is obvious that simplifying the criminal justice system is not what is needed in order to alleviate the crime problem in the United States. Because the social environment is shown to be as influential as IQ in predicting criminal activity, social policy must continue to battle against social, racial, and economic inequalities that pervade life in America if crime rates are ever to be reduced in an effective manner.

References

1. Herrnstein, R., and Murray, C. (1994), *The Bell Curve: Intelligence and Class Structure in American Life*, The Free Press, New York.
2. Hirschi, T., and Hindelang, M.J. (1977), "Intelligence and Delinquency: A Revisionist View," *American Sociological Review*, 42, 571–587.
3. Hirschi, T. (1969), *Causes of Delinquency*, University of California Press, Berkeley.
4. Reiss, A.J., and Rhodes, A.L. (1961), "The Distribution of Juvenile Delinquency in the Social Class Structure," *American Sociological Review*, 26, 720–732.
5. West, D.J. and Farrington, D.P. (1973) *Who Becomes Delinquent?*, Heinemann, London.
6. Wolfgang, M., Figlio, R.M., and Sellin, T. (1972), *Delinquency as a Birth Cohort*, University of Chicago Press, Chicago.
7. Stattin, H., and Klackenberg-Larsson, I. (1993) "Early Language and Intelligence Development and their Relationship to Future Criminal Behavior," *Journal of Abnormal Psychology*, 102, 369–378.
8. Moffitt, T. E., Gabrielli, W.F., Mednick, S.A., and Schulsinger, F. (1981), "Socioeconomic Status, IQ, and Delinquency," *Journal of Abnormal Psychology*, 90, 152–156.
9. Messner, S.F., and Rosenfeld, R. (1994), "A Society Organized for Crime," in *Crime and the American Dream*, Wadsworth, Belmont, CA, pp. 1–18.
10. Cloward, R.A., and Ohlin, L.E. (1960), *Delinquency and Opportunity: A Theory of Delinquent Gangs*, Free Press, New York.
11. Shaw, C. (1929), *Delinquency Areas*, University of Chicago Press, Chicago.
12. Faris, R.E.L., and Dunham, H.W. (1994), "Natural Areas of the City," in S.H. Traub and C.B. Little (eds.), *Theories of Deviance*, 4th ed., F.E. Peacock Publishers, Itasca, IL, pp. 63–72.
13. Balthazar, M.L., and Cook, R.J. (1984), "An Analysis of the Factors Related to the Rate of Violent Crimes Committed by Incarcerated Female Delinquents," *Journal of Offender Counseling, Services and Rehabilitation*, 9, 103–118.
14. Stark, R. (1987), "Deviant Places: A Theory of the Ecology of Crime," *Criminology*, 24, 893–909.
15. Covington, J. (1995), "Radical Classification in Criminology: The Reproduction of Racialized Crime," *Social Forces*, 10, 547–568.
16. LaFree, G., Drass, K.A., and O'Day, P. (1992), "Race and Crime in Postwar America: Determinants of African-American and White Rates, 1957–1988," *Criminology*, 30, 178–199.
17. Russell, K.K. (1992), "Development of a Black Criminology and the Role of the Black Criminologist," *Justice Quarterly*, 9, 667–683.

18. Messner, S.F., and Golden, R. (1992), "Racial Inequality and Racially Disaggregated Homicide Rates: An Assessment of Alternative Theoretical Explanations," *Criminology*, 30, 421–427.
19. Staples, R. (1975), "White Racism, Black Crime, and American Justice: An Application of the Colonial Model to Explain Crime and Race," *Phylon*, 36, 14–22.
20. Walker, S., Spohn, C., and DeLonc, M. (1996), *The Color of Justice: Race, Ethnicity, and Crime in America*, Wadsworth, Belmont, CA.
21. Anderson, E. (1990), "The Police and the Black Male," in *Streetwise: Race, Class, and Change in an Urban Community*, University of Chicago Press, Chicago, pp. 190–206.
22. Daly, K., and Chesney-Lind, M. (1988), "Feminism and Criminology," *Justice Quarterly*, 4, 497–538.
23. Alexander, R., Jr., and Nickerson, N. (1993), "Predictors and Nonpredictors of Recidivism in Female Offenders," *Free Inquiry in Creative Sociology*, 21, 141–148.
24. Champion, D.J. (1990), *Corrections in the United States: A Contemporary Perspective*, Prentice-Hall, Englewood Cliffs, NJ.
25. Thornburg, H. (1975), *Development in Adolescence*, Brooks/Cole Publishing, Monterey, CA.
26. Fischer, C.S., Hout, M., Jankowski, M.S., Lucas, S.R., Swidler, A., and Voss, K. (1996), *Inequality by Design: Cracking The Bell Curve Myth*, Princeton University Press, Princeton, NJ.
27. Hindelang, M.J. (1978), "Race and Involvement in Common Law Personal Crimes," *American Sociological Review*, 43, 93–109.
28. Hindelang, M.J. (1981), "Variations in Sex-Race-Age–Specific Incidence Rates of Offending," *American Sociological Review*, 46, 461–474.
29. Hindelang, M.J., Hirschi, T., and Weis, J.G. (1979), "Correlates of Delinquency: The Illusion of Discrepancy Between Self-Report and Official Measures," *American Sociological Review*, 44, 955–1014.
30. Genovese, E. (1995), "Living with Inequality," in Russell Jacoby and Naomi Glauberman (eds.), *The Bell Curve Debate: History, Documents, Opinions*, Times Books, New York, pp. 331–334.
31. Knoke, D., and Bohrnstedt, G.W. (1994), "Nonlinear and Logistic Regression," in *Statistics for Social Data Analysis*, 3rd ed., F.E. Peacock Publishers, Itasca, IL, pp. 321–354.
32. Wirth, L. (1938), "Urbanism as a Way of Life," *American Journal of Sociology*, 44, 1–24.
33. Merry, S.E. (1988), "Urban Danger: Life in a Neighborhood of Strangers," in G. Gmelch and W.P. Zenner (eds.), *Urban Life: Readings in Urban Anthropology*, 2nd ed., Waveland Press, Prospect Heights, IL, pp. 63–72.

chapter twelve

Social Statistics and Genuine Inquiry: Reflections on *The Bell Curve*

CLARK GLYMOUR

The Bell Curve by Herrnstein and Murry put American academic social scientists in an uncomfortable place.[1] The conclusions of the book are unwelcome, while the methods of the book appear to be the standbys of everyday social science. The unstated problem for many commentators is how to reject the particular conclusions of *The Bell Curve* without also rejecting the larger enterprises of statistical social science, psychometrics, and social psychology. In some sense, that is the general problem addressed in various ways in this collection of essays. The hard issues lurking behind the discussion are whether large parts of the social sciences and their methods are bogus, phony, pseudo-scientific, and whether, if and insofar as they are, they must be.

Varieties of Pseudo-Science

Pseudo-science comes in a lot of varieties, not equally irremediable. The cold fusion episode[2] represents one sort of pseudo-science, the sort in which competent, serious scientists step outside their range of expertise and make unskillful and incompetent use of techniques that others more expert can and do use reliably to address the same questions. Astrology is quite another sort of pseudo-science, the sort that has a technology that no one—not even the greatest astrologer—can use reliably to gain useful information, because it is premised on radically false claims. There is a third kind of pseudo-science, of which no exact historical examples come to mind (chemistry from 1810 until 1860 or so was something like this), characterized by techniques that work reliably only in rare domains but are used much more widely, where they succeed, if at all, only by chance. The first employment of the young Leibniz, later a discoverer of the calculus, illustrates a fourth kind of pseudo-science. Leibniz was charged with devising a proof that a certain person, and no other, should be selected for a political office, a proof Leibniz completed some years after the selection had been made. In Leibniz's kind of political science, methods are designed and applied with the intention of justifying prefixed conclusions: conclusions drive inquiry rather than inquiry conclusions. Leibniz's example is only the extreme of a range of cases in which data are not permitted to speak.

Besides all these, there is a kind of meta–pseudo-science that, without proof, declares vast terrains of inquiry ever beyond exploration by any scientific method. J.M. Dumas was the most influential French chemist of the middle of the nineteenth century—his textbook is still in print in France—and he ruled that the atomic composition of matter is unknowable. If he were master, he wrote, the word *atom* would be banned from chemistry because it presupposes something beyond all experience. Dumas' view was not that atomism was false, or nonsensical; he *knew* it was unknowable. At the end of the nineteenth century the great German physical chemist, W. Ostwald, held the same opinion. In the middle of the twentieth century, B.F. Skinner ruled that whatever mental phenomena intervene between stimulus and response are unknowable—no doubt something goes on in between, but no scientific method could discover what. Critics rightly pointed out that behaviorism was an elaborate self-deception: as if to defy Descartes' *cogito*, John Watson, the first American behaviorist, went so far as to claim there are no minds. In practice behaviorists everywhere attributed inner states to people and creatures, and used those attributions in the design and assessment of their experiments. They were forsworn, nonetheless, from actually thinking about what they were doing, and so from any chance of doing it better.

Karl Pearson's legacy in statistics is much the same. Pearson wrote perhaps the silliest influential book of philosophy ever published, *The Grammar of Science*, in which he maintained, repeatedly and without any sense of incongruity, that the entire world is nothing but subjective sensation, and sensation is a production of *brains*.[3] He coupled an acute critical sense with a wonderful tolerance for conceptual incoherence, nowhere more damaging than in his strong convictions about particular causal relations (the hereditary causes of virtue—Pearson was a keen eugenicist) and, incongruously, with his equally strong conviction that causation is nothing but correlation, so that in his judgement Yule's and Spearman's searches for structure, for example, were vain efforts deserving the scorn he gave them. *The Grammar of Science* is still the semi-official philosophy of some of professional statistics, and the opinions of several contemporary eminences seem close to Pearson's own.

Almost unanimously, social scientists criticizing *The Bell Curve* have treated the book as a cold fusion episode, in which people who should have known better used the competent methods of the social sciences—factor analysis, regression, and logistic regression—incompetently. I have read any number of perfectly sound criticisms of this sort, and yet I think they do not refute, but only repress, the terrible thought that *The Bell Curve* signals something fundamentally wrong with much of contemporary social science. Part of what troubles me about the cold fusion simile is that equally sound criticisms could be—and have been—made of many celebrated pieces of empirical social science: the arguments that smoking causes lung cancer were soundly ridiculed by statistically sophisticated critics; reanalysis of the influence of lead exposure on children's IQ fails to find any significant effect when reasonable measurement error is allowed; the regression model of the *American Occupational Structure*[4] cited by the National Academy of Science as *primo* social science, fails almost any statistical test, and so on. What troubles me more is that the principal methods of causal analysis used in *The Bell Curve* and throughout the social sciences are either provably unreliable in the circumstances in which they are commonly used or are of unknown reliability. But I'm getting ahead of the story.

Others who are not social scientists, and who have no stake in separating the larger enterprises from *The Bell Curve*, claim that some of the methods of that book and of social statistics more generally—factor analysis, for example—are purely astrological, so much mathematical rigamarole that is not, and could not be, an instrument of scientific inquiry. Stephen Jay Gould claims that factor analysis produces conjectures about the existence of unobserved properties solely because the properties, if they existed, would explain features of data; in his phrasing, factor analysis "reifies" unobserved quantities, and he thinks "reification" is a Big Mistake.[5] I wonder whether he thinks atoms and molecules and their weights are Big Mistakes as well, and if not, why not.

Gould and many others have a further argument that factor analysis is astrological pseudo-science: factor analysis generates an infinity of alternative explanations of the data. At least two things seem wrong with this routine point. First, the infinity of alternatives may all share important and interesting features (e.g., the number of latent common factors), so that if the truth were reliably somewhere among the alternative hypotheses the method yields from a data set, then the method would provide valuable information. And second, for *every* interesting scientific theory in every subject, there are an infinity of alternatives that will save the observations used to argue for the scientific theory. Most of the time we simply ignore most of the alternatives to our theories: paleontologists ignore the hypothesis of Phillip Gosse, a nineteenth century divine, according to which God created the world with fossils in place; physicists ignore a plethora of hidden variable theories that save the phenomena as well as quantum theory. Ignoring alternatives is a powerful, but not necessarily reliable, method of inference, a method we use all the time, most often without notice. In ignoring alternatives throughout the sciences we are either making wanton inferences, or we are discriminating among theories by other criteria than their fit with the data. The extra criteria may be substantive or formal, for example, various forms of simplicity and continuity principles. What objection can it be to factor analysis that it unveils what we usually hide: rather than suppressing alternatives, *that* method reminds us that they exist, and even shows us how to find them, while leaving it to us make discriminations on the basis of further substantive or formal considerations? None, I think. A fair assessment of factor analysis and other analytic techniques may not in the end be kinder, but it must be more subtle than Gould's objections.

Another objection to factor analysis is given in the chapter by Cawley et al. in this volume. They say, according to a well-known theorem due to Suppes and Zannotti: "a scalar measure of ability can always be constructed to fully explain the variance in a battery of test scores."[6] Unfortunately, the theorem says nothing of the kind. It says (for the binary case) there always exists a variable conditional on which a set of measured variables is independent. Even then the theorem does not hold if various simplicity or monotonicity conditions are required of any model.

Because of philosophical *caveats*, or because of ill-considered objections, some social scientists have retreated to incoherence or triviality in defense of their own methods. Thurstone did as much in his first book on factor analysis, claiming it was only a method of simplifying the presentation of data, but then, rather disingenuously, titling it *The Vectors of Mind*.[7] In a later book, he was more direct and candid about the causal role he assigned to factors.[8] It has also been suggested that the aim of factor analysis is only to provide a hierarchical classification with no implications for causation, constitution, or prediction, making the scientific point of the enterprise opaque. In a very recent (and in many other respects sensible) book, Leslie Hayduk claims the unobserved features produced using factor analysis are only "concepts," meaning, apparently, utter fictions.[9] He then worries at length about which estimators are unbiased. I wonder just why one would care whether an estimator of the distribution of an utter fiction is or is not biased, or what that would mean.

The social science literature is full of other ill-considered retreats from scientific seriousness in the face of still other objections. The ideal of social sciences that can tell us what causes what, and how much, and what we can and cannot do about it, is among the most important scientific ambitions of our time; it ought not to be abandoned because of philosophical confusions. The real harm of euphemisms and equivocations about the claims of inquiry is that they deflect attention from the fundamental questions about methods of data analysis. Whatever euphemisms are proposed, factor analysis and other methods of analysis *are* used to construct causal stories, and principal questions ought to be (and what use is a philosopher except for oughts?): under what conditions are the methods reliable tools to that end? and are better tools possible? The rest of this essay is about those questions.

The Aims of Inquiry

In one conception the aim of inquiry is to change opinion, full stop. Another conception adds a vague codicil: the aim of inquiry is to bring opinion closer to more of the interesting truth. The elaborated and vaguer conception is the less cynical, and the only basis upon which social scientists claim the vast majority of their funding. But either conception is consistent with two different attitudes, one disposed to rhetoric and the other to discovery. The rhetorically disposed investigator knows the opinion to which others should be brought, and the acquisition and analysis of data is simply a persuasive device to that end—the enterprise is essentially Leibniz's. In contrast, inquiry that aims at discovery has no outcome prefixed by the investigator.

Everyone must know of examples of the rhetorical attitude. My first was meeting a contract researcher for the U.S. Navy whose job was to prove that advertising improves enlistment rates, which he did by applying every kind of multiple regression to every set of variables for which he had data until he found a method and variable set in which advertising dollars had a statistically significant partial regression coefficient. Perhaps not much social science is deliberately pseudo-scientific in this way, although I think parts of *The Bell Curve* certainly are; but through one or another kind of observer artifact, a larger body of social science may be unwittingly (or perhaps half-wittingly) of this kind.

Physicists learned about observer effects from nineteenth century astronomy, and the point was driven home by any number of episodes—false discoveries of extra interior planets, nonexistent radiations—that resulted from the influences of desire and expectation on perception. Good experimentalists have since employed elaborate strategems to reduce observer effects in obtaining and analyzing data in physical experiments, and equally elaborate controls are standard in good psychological experimentation. Investigator effects in the social sciences are more fundamentally entangled with what passes for philosophy of science. I leave aside familiar problems of data selection in the social sciences, where experiments are difficult and liable to all kinds of sample attrition and noncompliance, and nonexperimental samples are often not selected at random from the population about which generalizations are offered. The larger sources of pseudo–social science are in the analysis and explanation of data.

Reliability and Social Theory

In Leibniz's procedure a single hypothesis is entertained, and procedure makes sure the hypothesis is confronted with no evidence that might refute it. The step is rather small from there to the most common sort of argument in the empirical social sciences, in which a very few hypotheses are confronted with evidence that might decide among them, but the remaining possibilities are ignored, or serious inquiry is confined to a single hypothesis confronted with evidence that might have told against it, but does not, but also does not tell against myriad other, unmentioned, alternative hypotheses. Leibniz's procedure will find the truth only if Leibniz's single hypothesis is true. These social scientific procedures will find the truth only if one of the few hypotheses considered is true. Social scientists call their sort of thing *modeling*, and extol it with no embarrassment, without the least sense of fraud, and even defend it as an ideal. The defense is that social science requires *substantive knowledge* or theory.

Substantive knowledge and theory exclude most hypotheses. *A priori* exclusions reduce the informativeness of data, sometimes almost to Leibniz's vanishing point—the empirical data are permitted only to discriminate among a very few hypotheses, or to provide estimates of parameters within a context of causal relations prefixed by the investigator. Where "substantive knowledge" is true and "theory" is correct, nothing is lost and something may be gained by its use. Where "substantive knowledge" is illusion and "theory" false, the result is an investigator artifact; where artifacts are endemic, the practice is pseudoscience. The most important methodological issues in the empirical social sciences have to do with when, and how much, received opinion can be relied upon, and how inquiry may be conducted when "theory" cannot be trusted.

We know lots of things about people and societies. We know the rudiments of human biology and anthropology—sex, and how humans age, and cluster, and the differences of activities typical of different ages. We know fundamental categories of human relations: agency, trust, deference, candor, deceit, power, promising, obeying, and on and on. We understand the elementary relations of desire, expectation, and action, and we understand typical human desires. We understand that human perception, belief, and attribution are limited, stylized, and prone to various errors. We use the myriad categories of this sort of folk social psychology to label events, to connect descriptions, and to make predictions. This sort of knowledge is not in doubt, and no data analysis technique can substitute for it. Sometimes we know still more that is relevant, such as the order of occurrence of events (but for the variables in social studies there is often no time order that makes sense: of cognitive ability and education, for example, neither precedes the other). In still other contexts there may be reliable expert knowledge that limits the possible hypotheses; with the aim of predicting mortality, I once examined a medical database on hospitalized pneumonia patients that involved hundreds of variables, the majority of which were eliminated by physicians on substantive grounds, and I had no reservation in relying on their judgements when analyzing the data.

When social scientists speak of "theory," however, they seldom mean either common-sense constraints on hypotheses or constraints derived from laboratory sciences or from the very construction of instruments. What they do mean varies from discipline to discipline, and is often at best vaguely connected with the particular hypotheses in statistical form that are applied to data. Suffice it to say "theory" is not the sort of well-established, severely tested, repeatedly confirmed, fundamental generalizations that make up, say, the theory of evolution or the theory of relativity.

That is one of the reasons for the suspicion that the uses of "theory" or its euphemism, "substantive knowledge," are so many fingers on the balance in social chemistry, but there are several other reasons. One is the ease of finding alternative models, consistent with common-sense constraints, that fit nonexperimental data as well or better than do "theory"-based models. (I will pass on illustrations, but in many cases it's *really* easy.) Another is that when one examines practice closely, "theory"-based models are quite often really dredged from the data—investigators let the data speak (perhaps in a muffled voice) and then dissemble about what they have done.

Distinct disciplines emphasize different variables and constraints in explaining behavior—anthropologists one sort of thing, sociologists another, economists still others, with distinct mechanisms. In any concrete case there is no way to know, *a priori*, which mechanisms prevail, or if others championed by no particular discipline are the more important. Still another consideration comes from research on individual human judgement and decision making, both lay and expert. The evidence is nearly overwhelming that, on average, "mindless" regressions do as well or better than experts at predicting complex behavioral outcomes, whether degree completion or recidivism.[10] And individual behavior in experimentally controlled circumstances—notably in experimental economics—frequently contradicts theoretical expectations. Finally, and perhaps most important, the theoretically guided statistical models developed to explain data do not have a stunning history of predictive success. Post-hoc explanation is the social scientific standard.

Algorithmic Social Science

Long before there were so many professors so profligate with theories, other means of realizing the goals of social inquiry were developed. Factor analysis and multiple regression, each with innumerable variations, are *fin de siecle* inventions, but they remain the pre-eminent algorithmic techniques for guiding hypothesis formation in the social sciences, and in particular they are the methods of *The Bell Curve*. Factor analysis and regression are strategems for letting the data say more, and for letting prior human opinion determine less, methods conceived in a spirit that was philosophically objectivist and optimistic, and substantively heavily prejudiced. (Yule and Spearman, for example, began with the expectation that empirical inquiry and objective data analysis methods would reveal the secrets of the human condition. But, likewise, Spearman never doubted that there is a single, innate common factor responsible for the correlations he found among various psychometric tests, and at one or another time in the first third of this century most of those he influenced firmly believed that the innate abilities of Europeans are superior to those of persons from other geographies.) Neither method eliminates human judgment, each reduces its scope: factor analysis requires "interpretation" of the output and decisions that influence how many factors the procedures find; regression requires that the investigator separate the variables into those that are effects and those that are potential causes of the effects, and those that will be ignored altogether.

There are only two kinds of questions of real scientific importance about factor analysis and regression: Under what conditions are these families of methods reliable? Are there methods that require still less of human judgment and are more generally reliable? For regression, the answers to technical versions of the first question seem to be well understood; for factor analysis, if the answer to the first question is known, I can't find it. The answer to the second question is that there are, in fact, methods that require less—or at least different—judgements from the investigator, and that are reliable under more general conditions than either factor analysis or regression. It says something—something sad—about the social sciences today that more reliable methods are almost never used. But I'm ahead of myself once more.

Factor Analysis and *The Bell Curve*

The thesis of *The Bell Curve* is that a variety of social ills are caused by want of cognitive ability and that the distribution of cognitive ability in the population cannot be much altered by social interventions. Murray and Herrnstein embrace Spearman's thesis that IQ tests measure a single common cause, g. Their argument—rather cavalier in *The Bell Curve*, but extensive and detailed elsewhere—for that conclusion is based on factor analyses of correlations among scores on mental tests for many samples of people.

I understand the *serious* claims of factor analytic psychometric studies to be (1) that there are a number of unmeasured features fixed in each person but continuously variable from person to person; (2) that these features have some causal role in the production of responses to questions on psychometric tests, and the function giving the dependence of measured responses on unmeasured features is the same for all persons; (3) that variation of these features within the population causes the variation in response scores members of the population would give were the entire population tested; (4) that some of these unmeasured features cause the production of responses to more than one test item; (5) that the correlations among test scores that would be found were the entire population to be tested is due entirely to those unmeasured features that influence two or more measured features.

Suppose, for the moment, we grant the first five of the psychometric assumptions. The reliability of factor analysis does not follow. For factor analysis to find the truth, a number of other conditions are necessary, including these: (6) the measured variables must be normally distributed linear functions of their causes; (7) measurement of some features must not influence the measures found for other features; neither the values of measured features nor the values of their unmeasured causes should influence whether a person is sampled; (8) two or more latent factors must not perfectly cancel the effects of one another on measured responses. These conditions are necessary, but I doubt they are sufficient for any sort of factor analysis to yield the truth (in sufficiently large samples) about the number of factors, or about what measured variables each latent factor influences, or the strengths of those influences.

So there are really two questions. First, when the eight assumptions just mentioned are granted, how reliable is factor analysis? And, second, what credence should we give to the assumptions?

Computer simulation provides the best way I know to come to some understanding about the reliability of the methods given the assumptions. Specify a number of alternative structures as directed graphs, identifying nodes as latent or measured; specify means and variances for each of the exogenous (graphically, that is, zero indegree) variables, and for each directed edge specify a nonzero real number representing the corresponding linear coefficient. Then, for each such structure, calculate the correlation matrix of the measured variables, give the matrix to factor analysis programs, and count the error rates of the procedures for the various features that factor analysis is supposed to reveal. Do the same again using the structures and a random number generator to generate sample correlation matrices for samples of various sizes.

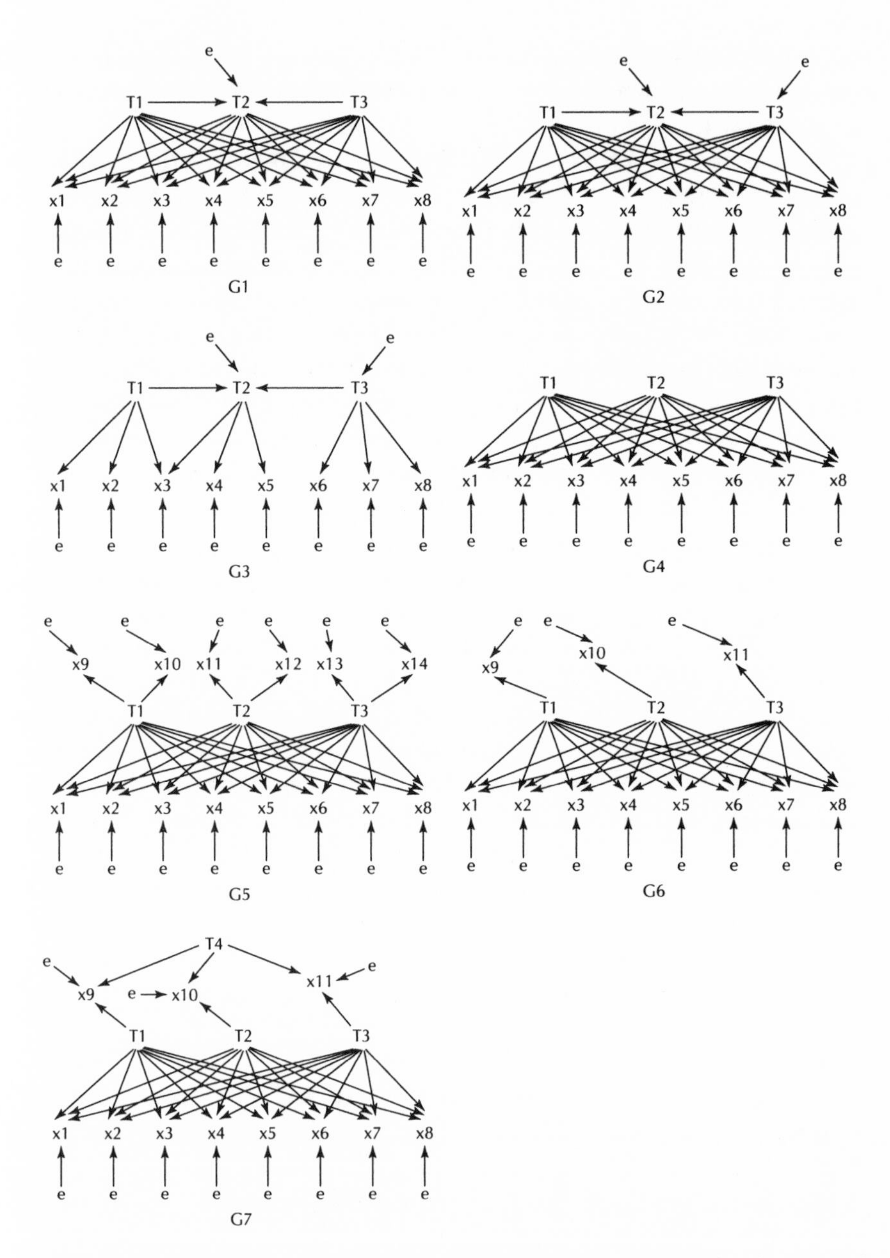

Figure 12.1. Alternative causal structures, depicted as directed graphs, for a set of latent variables (T), observed variables (x) and stochastic components (e). In these structures, "parents" refer to the number of latent variables, of which the set that are independent are "zero indegree".

I will at least illustrate what I have in mind. Figure 12.1 provides some simple examples of such graphs (each "e" is a distinct, independently distributed variable).

I gave each of the exogenous variables 0 mean and unit variance, randomly assigned linear coefficients between 0.5 and 1.5, generated the covariance matrices, and gave them to two programs: the default principal factors (4M) program in BMDP and the principal components program in EQS. EQS has a default calculation of the number of factors; BMDP has a constant used in determining the number of factors, and suggests the constant be set equal to unity divided by the number of measured variables, which was done. So far as I can tell, this rule for setting the constant is about optimal for these cases.

How do these programs do at determining the number of latent common factors in the structures that generated the covariances? The graphical representation (which will seem gratuitous to many statisticians, but is not) brings to notice an ambiguity in the question. How do we count the number of common causes of a set of measured variables. Do we count the actual number of zero indegree ancestors of any two or more members of the set, or do we count the number of parents of any two or more members of the set? By the first count, by zero indegree ancestors, the structure G1 has two common causes of the x variables; by the second count, by parents, it has three. Table 12.1 reports the number of factors reported by the programs, and the number of latents in the two ways of counting.

The only generalization that seems true is that both programs report no more factors than there are parents of two or more measured variables. Only twice was BMDP right about the number of zero indegree ancestors, and only twice was it right about the number of parents of two or more variables. EQS was right four times of seven about the number of zero indegree ancestors, and two times of seven about the number of parents.

Table 12.1. The number of latent factors identified by two statistical packages (BMDP and EQS) from simulated data generated using the structures in Figure 12.1.

	G1	G2	G3	G4	G5	G6	G7
0 ind	2	1	1	3	3	3	4
Parent	3	3	3	3	3	3	4
BMDP	1	1	3	2	*	3	3
EQS	1	1	1	1	3	3	2

The asterisk indicates BMDP would not run on this case.

This example is only an illustration of the sort of thing that would have to be done far more extensively, and come out far better, to afford any evidence that, on the eight assumptions described earlier, factor analysis is reliable in various respects. (The BMDP program I used, for example, assumes the latent variables are uncorrelated, and a procedure that allows "oblique rotation" might do better on the first three structures; but notice the BMDP procedure did no better on the four structures in which the latent factors were uncorrelated than on the three structures in which the latents were correlated.) Although there are fragmentary simulation studies of special cases, so far as I can tell, studies of this sort are rarely done, never described in the documentation for commercial factor analysis programs, and an adequate study of this kind, surveying a reasonable variety of structures, a variety of factor analysis procedures, and the sundry properties they are supposed to discover, has never been done at all.

The only simulation tests I have found of the reliability of programs at finding the number of latent factors fail to make clear what that means, and assume besides that the factors are uncorrelated. The assumption seems unwarranted. If we adopt for the moment the first four basic psychometric assumptions, then on any of several pictures the distribution of unmeasured factors should be correlated. Suppose, for example, the factors have genetic causes that vary from person to person; there is no reason to think the genes for various factors are independently distributed. Suppose, again, that the factors are measures of the functioning or capacities of localized and physically linked modules. Then we should expect that how well one module works may depend on, and in turn influence, how well other modules linked to it work. Even so, a great number, perhaps the majority, of factor analytic studies assume the factors are uncorrelated; I cannot think of any reason for this assumption except, if wishes are sometimes reasons, the wish that it be so.

What credence should we give to the assumptions identified earlier? The eighth—that two or more latent factors must not perfectly cancel the effects of one another on measured responses—seems quite harmless and common to almost all good sciences; one can find its ancestor in Isaac Newton's Rules of Reasoning. The seventh—essentially that there is no sample selection bias—could be warranted by random sampling from the population, although I think that is rarely done. The sixth—normality and linearity—is harder to justify, but at least indirect evidence could be obtained from the marginal distributions of the measured variables and the appearance of constraints on the correlation matrix characteristic of linear dependencies, although tests for such constraints seem rarely to be done. In any case, the other issues could be repeated for nonlinear factor analysis. The fifth assumption—that all correlations are due to unmeasured common causes—is known to be false of various psychometric and sociometric instruments, in which the responses given to earlier questions influence the responses given to later questions. The fourth—that other features of persons influence their scores on psychometric tests—is uncontroversial; the third—that the function giving the dependence of manifest responses on hidden features is the same for all persons, is without any foundation—if the dependencies were actually linear, however, differing coefficients for different persons would not much change the constraints factor models impose on large sample correlation matrices. The best evidence for the second assumption—that the features of persons that produce their responses to psychometric test questions are fixed, constant, within each person—is the high test-retest correlations of IQ scores, but that argument meets a number of contrary considerations, for example, the dependence of scores on teachable fluency in the language in which the test is given.

There is another quite different consideration to which I give considerable weight. I have found very little speculation in the psychometric literature about the mechanisms by which unmeasured features—factors—are thought to bring about measured responses, and none that connects psychometric factors with the decomposition of abilities that cognitive neuropsychology began to reveal at about the same time psychometrics was conceived. Neither Spearman nor later psychometricians, so far as I know, thought of the factors as modular capacities, localized in specific tissues, nor did they connect them with distributed aspects of specific brain functions. (It may be that Spearman thought of his latent *g* more the way we think of virtues of character than the way we think of causes.) One of the early psychometricians, Godfrey Thomson, thought of the brain as a more or less homogeneous neural net, and argued that different cognitive tasks require more or less neural activity according to their difficulty. Thomson thought this picture accounted not only for the correlations of test scores but also for the "hierarchies" of correlations that were the basis of Spearman's argument for general intelligence.[11] The picture, as well as other considerations, led Thomson to reject all the assumptions I have listed. I think a more compelling reason to reject them is the failure of psychometrics to produce predictive (rather than post-hoc) meshes with an ever more elaborate understanding of the components of normal capacities. Psychometrics did nothing to predict the varieties of dyslexias, aphasia, agnosias, and other cognitive ills that can result from brain damage.

Drawing conclusions about factor analysis is a dangerous business because the literature is too large for anyone with any other interest in life to survey. For all I *know*, asymptotic reliability proofs may exist, and excellent and thorough simulation studies may have been done, but I have not found much that addresses the central questions. With a very few exceptions, what I find instead are very modest simulation studies of special cases, statistical studies of the properties of estimators—studies that presuppose exactly what is in doubt, the credibility of factor models—and introductory discussions that raise the question of reliability only to evade it. An enterprise a century old whose necessary assumptions are in doubt, the reliability of whose methods, even given those assumptions, are unproven (and little investigated), and which is without major predictions or explanations that tie it to more secure sciences, deserves, I think, to be judged a kind of pseudo-science.

Regression and *The Bell Curve*

The principal statistical technique of *The Bell Curve* is not factor analysis but regression, both linear and logistic. Regression is a wonderful method for extracting causal information from data, provided very strong assumptions are warranted, for example, that none of the regressors are effects of the outcome variable, and that there are no unrecorded or neglected factors that influence both the regressors and the outcome variable, and that various distribution assumptions are met.

There is an enormous literature on ways of detecting erroneous distribution assumptions in regression models—non-normality, nonlinearity, heteroscadicity, autocorrelation, etc.—and heuristics for fixing some of these flaws. Despite this attention, statistical textbooks routinely preach against using regression as a method for inferring either the existence or strength of causes from nonexperimental data. The reasons have to do with the sensitivity of regression conclusions to causal assumptions that cannot be checked by the usual regression diagnostics. The most frequent worry of this sort is, in statistical jargon, *correlated error*—the error term in the regression model may be correlated with one or more of the regressors. Sampling variation aside, neither I nor most people believe correlations come from nothing, and I understand *correlated error* to mean that the regression model omits variables that influence both the outcome variable and one or more of the regressors, so that the association between the regressors and the outcome may be due, in whole or part, to omitted influences. Mosteller and Tukey, for example, devote an entire chapter to examples of fallacious causal inferences from regression, and when unpacked each of their cases involves an omitted common cause of regressor and outcome variables.[12] Another, less commonly noted but equally serious, concern is that in observational samples the values of the outcome variable may have influenced which units appear in the sample, resulting in a bias in regression estimates of linear dependencies.

Correlated error and sample selection biased by the outcome variable are only particular issues within a more general body of concerns; the estimates obtained using a regression model depend in intricate ways, rarely discussed in the statistical literature, on whether the causal claims of the model are a correct account of how the data were generated. A great deal of social science, including *The Bell Curve*, is in no position to make a case for the causal assumptions necessary to use regression reliably in causal inference.

One issue is a kind of double sample-selection bias, in which both the outcome variable and one or more regressors influence membership in the population. If both regressor and outcome influence sample selection, regression applied to the sample (no matter how large) will produce an (expected) nonzero value for the linear dependence, even when the regressor has no influence at all on the outcome variable. The same goes for nonlinear influence in logistic regression.

Another issue, conditional error correlation, involves a sort of logical defect in regression. In multiple regression, estimates of the linear influence of a variable R_i on outcome Y are based on the association of R_i and Y, *conditional* on the other regressors. If there is an omitted common cause of a regressor R_i and the outcome, then the estimate of the influence on the outcome of *any other regressor*, R_j will be biased either if R_j influences R_i or if there is an omitted common cause of R_j and R_i. The R_j partial regression coefficient will then be an incorrect estimate of the linear influence of R_j on outcome, even though R_j is not correlated with the error variable in the regression, because R_j will be correlated with the error conditional on the other regressors. The same point holds for logistic regression.

Still a third issue concerns cases, which may easily arise in nonexperimental social data, in which the regression model has one of the causal directions wrong—the outcome variable influences one of the regressors. If so, then the backwards regressor acts just like R_i in the previous paragraph: regression will yield the wrong estimates of influence not only for the backwards regressor, but also for any regressor that influences the backwards regressor, or for any regressor for which there is an omitted variable that influences both it and the backwards regressor.[13]

These issues about causation and misspecification all arise in the analyses of *The Bell Curve*, and they are at the heart of what is wrong with the scientific argument of the book. The central claims of *The Bell Curve* are that a variety of social ills are caused by cognitive disability. The analyses all presuppose a common causal schema (Fig. 12.2):

Here X is some social ill—poverty, illegitimacy, whatever—Z is some covariate, often age. The variables in circles or ovals are unmeasured. The directed edges indicate hypothetical causal relations, while the undirected edges represent unexplained correlations. The SES Index is a linear function of the variables that are its parents in the graph. In addition, Herrnstein and Murray assume that the variance of V is small, so that IQ score is a good proxy for cognitive ability. The arguments of the book are almost all based on regressions of various X variables on IQ, SES index, and, for each choice of X, some choice of Z.

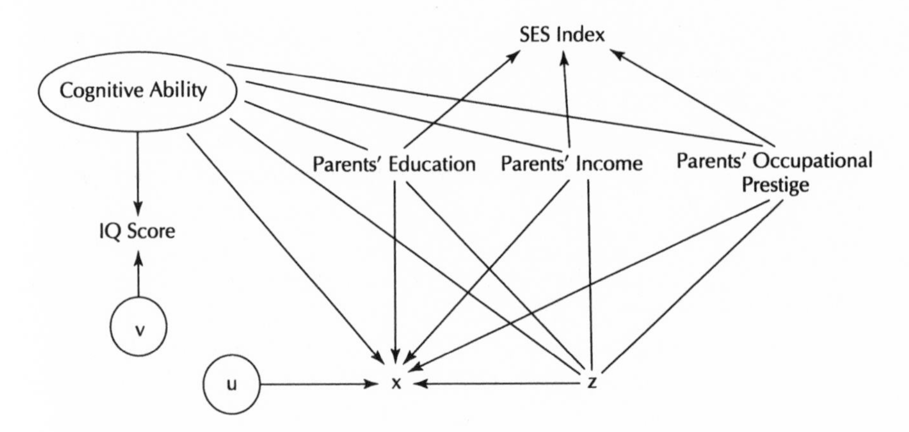

Figure 12.2. *The Bell Curve* causal model.

The obvious objection is that the regression model leaves out variables that influence both IQ score and X. The list of possibilities is endless: besides education, mother's character, attention to small children, the number of siblings, the place in birth order, the presence of two parents, a scholarly tradition, a strong parental positive attitude towards learning, where rather than how long parents went to school, might influence both cognitive ability and X, and so might a lot else. Herrnstein and Murray write: "A major part of our analysis accordingly has been to anticipate what other variables might be invoked and seeing if they do in fact attenuate the relationship of IQ to any given social behavior" (p. 123). In fact, however, the only variable they consider from among those just listed is education.

Adding education or other variables to the set of regressors in *The Bell Curve* would, I am sure, have changed the estimates of the influence of cognitive ability on social ills, but it would have done little to improve the scientific case. Suppose, for example, education is added to the set of regressors. Then, if education is a cause of IQ scores and X, a "spurious correlation" between IQ scores and X, will have been removed. But what if, in addition, unmeasured cognitive ability influences IQ scores and also influences education, and furthermore, there is some other omitted common cause of education and X?

In that case, illustrated in Figure 12.3, multiple regression of X on IQ, Education, SES Index, and Z will again yield a biased estimate of the (direct) influence of cognitive ability on X, even assuming with Murray and Herrnstein that IQ is a good proxy for cognitive ability. (The example is easy to work out in the linear case, although it holds as well for logistic regression: assume standardized variables, with linear coefficients, a, b, c, d corresponding to the edges in Figure 12.3—e.g., c is the coefficient of Cognitive Ability in the equation for Education, etc.—and e representing the correlation between Ed-

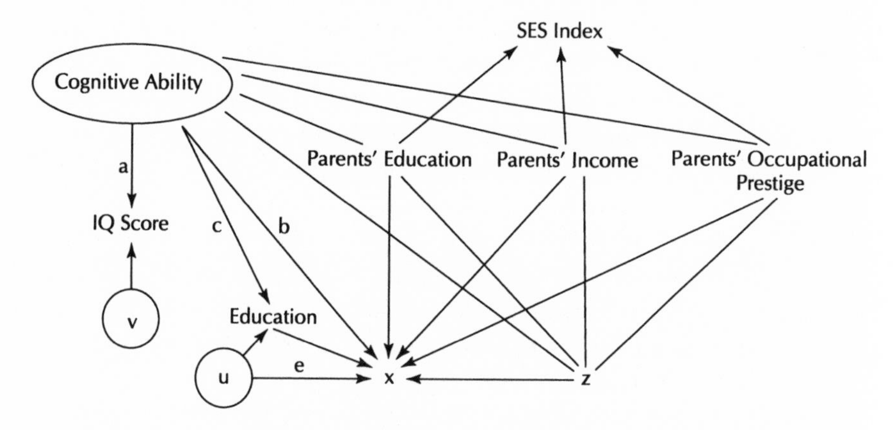

Figure 12.3. An alternative to *The Bell Curve* causal model (see Fig. 12.2).

ucation and X produced by the unrecorded common cause, U; for simplicity, assume with Herrnstein and Murray that $a = 1$; then the partial correlation of IQ with X, controlling for Education, SES, and Z, is not b but $b + ce$). Or, what if, as in Figure 12.4, unmeasured cognitive ability influences IQ scores and also influences education, and furthermore, X and education influence one another? In that case, again, adding education to the set of regressors will also *add* a spurious correlation between IQ scores and X.

In general, adding a variable to a regression can introduce spurious correlations as well as remove them, and one needs to know a lot about the actual causal structure to know the effect of adding a variable on estimates of the influences of other regressors.

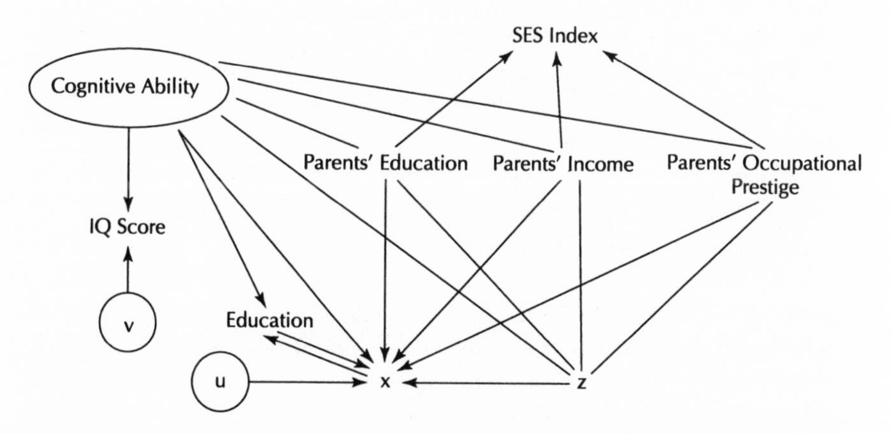

Figure 12.4. Another alternative to *The Bell Curve* causal model (see Figs. 12.2 and 12.3).

The difficulty is quite general, and not particular to *The Bell Curve*: regression provides no reliable means to determine when an unconfounded set of regressors has been found. The results of alternative regressions provided by Cawley et al., in this volume, provide illustrative reasons not to believe the regressions of *The Bell Curve*, but they do not provide any good reason to believe that the alternative regressions have captured the influence of cognitive ability. It is sometimes recommended that correlated error be checked by seeing if an original set of regression coefficients are stable under the addition for further regressors. If you imagine an infinite process of adding variables (and ignore sampling variation), that is a reliable procedure only if you do not condition on a variable influenced by both a regressor and the outcome variable. In practice only a handful of variables are ever considered, and the recommendation is unreliable: a regression coefficient can be correct but unstable, or incorrect but stable.

There is a straightforward nonregression test for what economists call exogeneity—for whether a regressor R is unconfounded, and not a backwards regressor: Find two variables, say U and V, such that U and V are independent but U, V are *dependent* conditional on R. If $\{U, V\}$ are not independent of Y but are independent of Y conditional on R, then there is no common cause of R and Y, and R is not an effect of Y (see Ref. 13). I have never seen the test applied by social scientists.

Scientific Search

Regression is a method that reliably turns data into causal information, provided certain conditions are met; the trouble is that the conditions are very strong, and may well not hold even approximately in many empirical cases. One pseudo-scientific response is to claim that social statistics, therefore, can give no evidence about causation; absent controlled experiments, social causes go the way Dumas would have done with atoms and Skinner would have done with minds. Pearsonian pseudo-science finds in the unreliability of regression—even fitted with the standard diagnostics—grounds for concluding that no reliable method exists. Critics of "causal modeling" claim that methodologies for causal inference from observational data are impossible: there may be good cases of causal inference, where people get the right answer, but there is no systematic method to guide social inquiry. Many social scientists seem to have bought this claim (not the argument—there seems to be no argument, except that if there were such methods regression would surely be it); hence, the repeated litany about the need for "substantive assumptions" in empirical social science, and hence, also, the Leibnizian quality of much of social inquiry.

In contrast, assuming the goal of inquiry is to understand and predict, the rational and scientific response is to investigate what information can be extracted from data under assumptions weaker (or at least very different) than those needed for regression, assumptions that may more often be warranted in empirical applications. There are several enterprises of this kind, with interesting results, both positive and negative.

One line of investigation has examined the case in which measured variables are influenced by latent variables. Assuming the values of both latent and measured variables are ordered, it is assumed that (1) measured variables do not influence one another; and (2) the ratio of the probabilities of any higher to any lower value of a measured variable increases monotonically with increasing values of any latent, other latents constant; and (3) measured variables are independent conditional on any set of values of all of their latent common factors; and (4) the latent variables are consistently estimatable as the number of measured variables increases. The problem is, then, to find additional conditions on the joint marginal distribution of the measured variables necessary and sufficient for the existence of a model, consistent with the assumptions and the marginal, and having a single latent factor. Such conditions have been found.[14] Still to be solved is the problem of finding a comparable characterization for n latent variables, for arbitrary positive n.

The conditional independence assumption (3) is a special case of a more general Markov assumption used in almost every form of causal model; the special case says that variables that do not influence one another are independent conditional on all of their common causes. Another special case of the Markov assumption—variables that neither influence one another nor have a common cause are independently distributed—is the most fundamental in experimental design.

These necessary and sufficient conditions for the existence of a model with a single latent variable do not guarantee that *only* such a model will do, and some additional simplicity assumption would be required to exclude models with multiple latent common causes of measured variables. Another line of investigation imposes a general version of the Markov condition, formulates an explicit simplicity criterion, and then characterizes features of all models consistent with these and other assumptions and with various joint distributions of measured variables. The simplicity assumption most thoroughly investigated—sometimes called *faithfulness*—is that causal mechanisms do not perfectly cancel one another, so that, for example, two variables will not be independent in virtue of the cancellation of multiple common causes. The general form of this simplicity assumption is that all conditional independencies in the joint distribution result from the Markov condition, or equivalently, that the conditional independencies in the distribution must be unchanged by small perturbations of the parameters of the model. These assumptions entail that the causal structure of the model implies conditional independence constraints on the joint probability distribution of all variables, and vice-versa.

Either alone or in combination with further distribution assumptions (e.g., linearity), the conditional independencies among variables, measured or latent, in turn may imply conditional independencies or other forms of constraints on the joint marginal distribution of measured variables alone. The result is that, under the assumptions, features of data constrain the posterior probabilities of alternative causal models, and asymptotically determine a family of causal hypotheses, that is, determine unequivocally those causal relations shared by all consistent causal models.

The faithfulness assumption can be understood as a condition on the prior probability distribution of models and parameters. To it one might naturally add others, for example, the assumption that the parameters of a model are independent in the prior probability distribution, and the assumption that if two models assign the same parents (direct causes, in other words) to a variable, x, then the parameters giving the conditional probabilities of any value of x on any set of values of the parents of x have the same prior probability distribution for the two models.

These mathematical features have been used in at least three different sorts of search procedures. One kind of Bayesian exploratory procedure estimates the posterior distribution of a large number of alternative models.[15] Predictions can then be based on the weighted average of the sample of models. A second Bayesian procedure finds the class of alternative models estimated to have the highest posterior probability.[16] (For combinatorial reasons it is impossible to compute the posterior distribution over all models, and so the search procedures must be selective in one way or another.) These procedures typically use large sample approximations to the posterior probability, and have so far only been implemented for multinormal and multinomial distributions under the assumption that there are no latent common causes, no sample selection bias, no feedback, and that the data are not from a mixture of distinct causal structures. So far as I know, these procedures have not been applied to empirical problems where the causal structure was independently known, but they have been applied to simulated data meeting the assumptions; they do splendidly. Work extending the procedures to include the possibility of unmeasured common causes of measured variables is well under way.[17]

A third kind of search uses sample data to decide whether various constraints hold in the marginal distribution of measured variables and then computes the class of causal models consistent with the constraints.[18–21] For multinormal and multinomial distributions, these investigations have led to asymptotically correct searches (1) when it is assumed there are no unmeasured common causes of measured variables, (2) when no such assumption is made, (3) when sample selection bias is allowed,[22] and (4) when the data may be from a mixture of distinct causal structures. Assuming multinormality, an asymptotically correct constraint search has also been found when linear "nonrecursive" models with feedback may obtain.[23] Other work has shown the search procedure correct as well for a class of models with discrete variables.[24] With some additional assumptions, there is a correct search procedure available even for causal relations among latent variables themselves.[21,22,25]

These investigations are scarcely exhausted. The constraint-based strategy can be implemented over any family of distributions, closed under marginalization and conditionalization, for which adequate tests of conditional independence are available, but so far both constraint-based procedures and Bayesian searches have been confined to the same few families of distributions. The sensitivities of policy predictions to specification errors in the outputs of the procedures are unknown. It has been argued that although the procedures are asymptotically correct under the assumptions, the posterior ratios of inequivalent models will not converge to zero if the number of possible latent variables increases sufficiently rapidly as a function of the sample size. In linear models there are sign constraints that can give information about structure without assuming faithfulness, but so far as I know they are little explored and uncharacterized. There is as yet no analytically useful theory of estimation errors for model search, and so on.

So there's lots of theoretical and computational work to do, but it is equally important to make use of the procedures that are available now. There is no excuse for analyzing a dataset only with the established tools.

Several of the constraint-based search methods have received extensive simulation tests, but perhaps more interesting, besides empirical applications in which they yield causal explanations whose truth or falsity is unknown, in several cases these procedures have been used to make causal forecasts that have been independently verified, for example, to correctly reconstruct the test batteries that made up an aggregate psychometric score, to predict the outcome of a randomized greenhouse experiment from uncontrolled plant samples (see Ref. 21), to predict the effects on college dropout rates of manipulating the average SAT scores of a class entering students,[26] and to recalibrate a misbehaving mass spectrometer aboard an orbiting satellite. In all but the last of these cases (to which regression did not apply), regression methods gave different, and incorrect or less informative, results. In the last case, principal components would have given the same result. In the plant experiments, the predictions of the search procedure, which experiment proved correct, contradicted those of the biologist.

The information that can be obtained from any of these searches depends on the strength of the assumptions; for example, usually more causal information can be obtained from data when it is known there are no unmeasured confounders than when there is no such prior knowledge. But the information that can be extracted also depends on the data, which in turn depends on the actual, usually unknown, causal structure. Suppose, for example, that in some empirical study a substantial part of the covariation of each pair of variables were due to an unrecorded common cause, with a distinct such cause for each measured pair. Then from the observed marginal, the methods I have just discussed (those allowing the possibility of latent variables) would give no information at all about the structure responsible for the data. That is the result that would be found if these methods were applied to Murray and Herrnstein's data: nothing can be inferred about causation. Arguably, that is the right conclusion, not only about *The Bell Curve* but also about many other social data sets.

Conclusions

The Bell Curve is an amalgam of several sorts of pseudo-science: the sort of pseudo-science of nineteenth century chemists who possessed and used everywhere rules for finding molecular formulae that worked only for special classes of molecules, mixed together with Leibniz's sort of pseudo-science, in which inquiry is driven by the desire for particular conclusions rather than the desire for knowledge. But Murray and Herrnstein are not unusual, only unpopular. Without offering a tedious survey, I claim that a great deal of empirical social science, without experimental control of the variables or features that are supposed to be "exogenous" or "explanatory" or "independent," suffers one or the other or both of the same defects. Leibniz thought each property of anything is a necessary property of that thing, but he was wrong: these methodological defects in current social science are accidental, eliminable, unnecessary.

Any pseudo–social science does harm, but so does the absence of any social science at all; in that vacuum policy will be made from myth or prejudice, and those who are policied will suffer the consequences. The most damaging—and, I think, malicious—pseudo-science is Pearson's, for it shatters hope that science can improve the human social condition. Only revolution remedies malice, and only science remedies ignorance. Time, therefore, for a social scientific revolution.

References

1. Herrnstein, R.J., and Murray, C. (1994), *The Bell Curve: Intelligence and Class Structure in American Life*, The Free Press, New York.
2. Taubes, G. (1993), *Bad Science: The Short Life and Wierd Times of Cold Fusion*, Random House, New York.
3. Pearson, K. (1911), *The Grammar of Science*, A. and C. Black, London,
4. Blau, P., and Duncan, O. (1967), *The American Occupational Structure*, Wiley, New York.
5. Gould, S.J. (1981), *The Mismeasure of Man*, Norton, New York.
6. Suppes, P., and Zanotti, M. (1981), "When Are Probabilistic Explanations Possible?" *Synthese*, 48, 191–99.

7. Thurstone, L. (1935), *The Vectors of Mind*, The University of Chicago Press, Chicago, IL.
8. Thurstone, L. (1947), *Multiple-Factor Analysis; a Development and Expansion of The Vectors of the Mind*, The University of Chicago Press, Chicago, IL.
9. Hayduk, L. (1996), *LISREL Issues, Debates, and Strategies*,. Johns Hopkins Press, Baltimore.
10. Dawes, R. (1988), *Rational Choice in an Uncertain World*, Harcourt Brace Jovanovich, San Diego, CA.
11. Thomson, G. (1939), *The Factorial Analysis of Human Ability*, Houghton Mifflin, Boston.
12. Mosteller, F., and Tukey, J.W. (1977), *Data Analysis and Regression*, Addison-Wesley, Reading, MA.
13. Spirtes, P., Meek, C., and Richardson, T. (1996), "Causal Inference in the Presence of Latent Variables and Selection Bias," P. Besnard and S. Hanks (Eds.), in *Proceedings of the Eleventh Conference on Uncertainty in Artificial Intelligence*,Morgan Kaufmann Publishers, San Francisco, CA, pp. 499–506.
14. Junker, B., and Ellis, J.L. (1996), *A Characterization of Monotone Unidimensional Latent Variable Models*, 2/95 (revised 3/96). `http://www.stat.cmu.edu/brian/bjtrs.html`
15. Madigan, D., Raftery, A.E., Volinsky, C.T., and Hoeting, J.A. (1996), *Bayesian Model Averaging*. AAAI Workshop on Integrating Multiple Learned Models. `http://bayes.stat.washington.edu/papers.html`
16. Heckerman, D. (1995), *A Bayesian Approach to Learning Causal Networks*, Technical Report MSR-TR-95-04, Microsoft Research. `http://www.research.microsoft.com/research/dtg/heckerma/heckerma.html`
17. Geiger, D., Heckerman, D., and Meek, C. (1996), *Asymptotic Model Selection for Directed Networks with Hidden Variables*, preprint, Microsoft Research Center.
18. Pearl, J., and Verma, T. (1990), *A Formal Theory of Inductive Causation*, Technical Report R-155, Cognitive Systems Laboratory, Computer Science Department, UCLA.
19. Pearl, J., and Verma, T. (1991), "A Theory of Inferred Causation," Principles of Knowledge Representation and Reasoning: Proceedings of the Second International Conference, Morgan Kaufmann, San Mateo, CA.
20. Spirtes, P., Glymour, C. and Scheines, R. (1990) "Causality From Probability," J. Tiles et al. [Eds.], *Evolving Knowledge in Natural Science and Artificial Intelligence*, Pitman, London, pp. 181–199.
21. Spirtes, P., Glymour, C., and Scheines, R. (1993), *Causation, Prediction and Search*, Springer Lecture Notes in Statistics. `http://hss.cmu.edu/html/departments/philosophy/TETRAD.BOOK/book.html`
22. Spirtes, P. (1996), *Discovering Causal Relations Among Latent Variables in Directed Acyclic Graphical Models*, Technical report, CMU-Phil-69, Department of Philosophy, Carnegie Mellon University.
23. Richardson, T. (1996), *Discovering Cyclic Causal Structure*, Technical Report CMU Phil 68.
24. Pearl, J., and Dechter, R. (1996), *Identifying Independencies in Causal Graphs with Feedback*, Technical Report (R-243), Cognitive Science Laboratory, UCLA. `http://singapore.cs.ucla.edu/csl-papers.html`

25. Scheines, R. (1994), "Inferring Causal Structure Among Unmeasured Variables," in *Proceedings of the Fourth International Workshop on Statistics and AI*, Springer-Verlag, Ft. Lauderdale, FL.
26. Druzdzel, M., and Glymour, C. (1994), "Application of the TETRAD II Program to the Study of Student Retention in U.S. Colleges," in *Working Notes of the AAAI-94 Workshop on Knowledge Discovery in Databases (KDD-94)*, Seattle, WA, pp. 419–430.

PART V

THE BELL CURVE AND PUBLIC POLICY

The Bell Curve is an essay about public policy as well as a report on the nature and distribution of American intelligence. Herrnstein and Murray's sense of excitement is real as they tell their story: "... there is no major domestic issue for which the news we bring is irrelevant" (p.387). But the news that they bring does not rest on the authority of gospel. It depends on sound interpretation of a body of data and is subject to challenge by the same scientific procedures.

In their zeal to confirm that unequal distribution of intelligence is fixed and unalterable, Herrnstein and Murray attack the project of compensatory education for three- to five-year-olds in poverty. The Head Start program, the largest provider of health services to poor children, is condemned for having failed to raise IQ scores. But is it realistic or appropriate to judge the success of such programs by their impact on IQ scores?

Zigler and Styfco agree with Herrnstein and Murray's judgment that Head Start effects no lasting changes in IQ. They question, however, why evaluators would seek to judge the program on so narrow a criterion. There have been significant results reported for the program in school adjustment, social competence, and parental involvement. These are cost-effective as an investment for the nation and meet the Head Start mandate to increase school readiness. Zigler and Styfco review the history and operation of Head Start, offering the best chapter-length study of the program now available.

Does America have a meritocracy that makes the most intelligent the most highly rewarded and the most influential? Lemann sets out, in a review of college admissions, business recruitment, survey data, and tax returns, to establish the relationship between IQ and the rewards of money, employment, and status. He finds that there is no convincing evidence that today's elites are more cognitively distinguished than their predecessors (and it would be difficult to generate such data) or that wealth is intimately tied to IQ. There are other problems as well. Herrnstein and Murray treat test scores and their increasing use as independent data, without reference to the marketing efforts of test publishers or the reliance on test scores by degree-granting institutions. It is important not to confound the power of credentials with esteem for intelligence, nor to assume a greater selectivity for our post-secondary institutions than they in fact possess.

Resnick and Fienberg, after reviewing the critiques made in earlier chapters of this book, turn to consider how to address the problem of inequality. *The Bell Curve* found that high levels of inequality is a price that the United States has to pay for creating and maintaining a competent business and professional leadership in the face of a society with limited intellectual capacity.

Resnick and Fienberg find Herrnstein and Murray far too complacent. Inequality on the scale we experience is neither necessary nor desirable, they argue. Instead they recommend remedies already operative in our society for reducing discrimination in the workplace, unequal access to good education and health services, and unacceptable levels of poverty. They see no warrant, in particular, for Herrnstein and Murray's attack on affirmative action and Head Start, or for their broader attack on the legitimacy of government.

chapter thirteen

A "Head Start" in What Pursuit? IQ Versus Social Competence as the Objective of Early Intervention

EDWARD ZIGLER AND SALLY J. STYFCO

"Project Head Start, which began as an experiment, is now battle tested and it has been proven worthy." So spoke President Lyndon Johnson in August 1965, when he announced that Head Start would become a year-round program after its inaugural summer session.[1] "Compensatory education has been tried and it apparently has failed," wrote educational psychology professor Arthur Jensen just 4 years later.[2] By the time Head Start celebrated its twenty-fifth birthday, it was lauded as "the nation's pride"[3] and "the one program everyone wants to help."[4] Soon, however, the public was reading about "The Head Start Scam" and Herrnstein and Murray's blunt opinion on "the disappointment of Head Start" in *The Bell Curve*.[6]

What caused such vacillations in the esteem accorded our nation's oldest and largest early childhood intervention program? The basic Head Start format has not changed greatly since its inception. The program still holds to its original goals of boosting children's school readiness and behavioral competence. Its methods still entail the provision of health and social services, preschool education, and family support. Parent involvement remains the bedrock of program operations. And the 15 million children and families served by Head Start over the years form a stalwart group of believers and supporters. What, then, has repeatedly swayed judgments from "worthy" to "disappointment"?

The answer resides not in the reality of Head Start but in the expectations scientists and policy makers hold for it and other early intervention efforts. Changes in these expectations have curiously followed changes in prevailing beliefs concerning the malleability of intelligence. Debate over whether intelligence is governed by genes, and thus is relatively constant, or whether it is environmentally determined and therefore mutable has droned throughout modern history, with proponents of either side periodically gaining then losing the lead. The perpetual swing between hope and disillusionment over the value of early intervention is driven by what particular position is the front-runner at any given time. When the *zeitgeist* is that intellectual ability can be raised appreciably, such programs enjoy a surge in approval. When the mood is that intelligence cannot be nudged much from its genetic underpinnings, they are viewed as a waste of effort and money. Experience has shown that one such extreme actually leads to the other. When we allow ourselves to become overly optimistic about the potential of intervention programs to increase intelligence, our hopes are dashed and we revert to pessimism about whether they have any worth at all.

What is forgotten in these rash judgments is that most intervention programs were never intended to raise intelligence. For example, we mentioned earlier that Head Start's goal is to promote children's school readiness and behavioral competence. Nowhere in its mission statement is there a reference to IQ scores or a hint about making children smarter. The same is true for most other preschool interventions (with the exception of a few that were mounted to prevent mental retardation). Yet even scholars such as Herrnstein and Murray—schooled at the prestigious intellectual training grounds of Harvard and MIT—erroneously describe Head Start as "designed to foster intellectual development ... [by] providing classes for raising preschoolers' intelligence" (p. 403). From this basis in belief, it appears logical for them to summarize the evidence on the program's efficacy with the statement, "no lasting improvements in intelligence have ever been statistically validated with any Head Start program" (p. 404). From there, it is not a *non sequitur* to conclude that Head Start does "not affect cognitive functioning much if at all" so "we need to look elsewhere for solutions ... " (pp. 414, 416).

How exactly did Head Start and similar programs get snared in the IQ trap, where their success is measured by impertinent criteria? While it might seem surprising that learned analysts like Herrnstein and Murray (and Jensen before them) would hold the false assumption that such programs exist to raise children's measured intelligence, this misconception has long existed in the annals of early intervention. It can be traced to the spirit of the times when the first intervention efforts were launched. The early 1960s were characterized by unbridled social and scientific optimism. The nation had declared an all-out war against poverty, and expectations were high that the war would be won. At the same time, social scientists thought they had discovered the means to push IQ scores way beyond expected limits. The two agendas converged at about the time early intervention debuted, so preschool programs were assumed to share the goals of ending poverty and turning children into geniuses. The enthusiasm of the sixties has long since waned, and more realistic views of the potential of intervention have been shaped. Yet the field has never been able to put to rest the impossible claims once made for it.

In this chapter we explain the political and scientific environments that created, influenced, and ultimately threatened early intervention efforts. We also examine the wealth of evidence that has now accrued to show what such programs can and—of equal importance—cannot accomplish. We then discuss ideas for strengthening early intervention programs to maximize their potential benefits to young children and families in poverty.

The Nation Declares War on Poverty

The early 1960s were a time of national pride and prosperity that are worth remembering in today's troubled times. Americans had recovered from the embarrassment of seeing the U.S.S.R. launch the first unmanned and manned space flights, and were impressing the world with their own successful space program. They were reforming their educational system to emphasize traditional academics so that no nation would outperform us again. The economy was healthy, and at the beginning of the decade there was actually a surplus in the national budget. The Vietnam War had not yet begun to drain America's spirits and resources, so ample amounts of both were available to devote to domestic problems.

Although poverty had always been a concern, there was now a growing realization about how grave a problem it had become. Government estimates showed that nearly one fourth of Americans were poor, and that half of these were children. The heads of many poor families had only a grade-school education, making personal economic improvement difficult. Poverty was blamed for the rising crime rate and increasing number of individuals who were not capable of assuming productive roles in the military service and private industry. On another level, Michael Harrington's 1962 book, *The Other America*, highlighted the devastating personal effects on the individual living in poverty, isolated from the mainstream of American culture.[7] This image troubled President John Kennedy, who was assassinated before he could act on his desire to help. Kennedy's successor, Lyndon Johnson, acted in the big manner that was his fashion. He would not only end poverty but would create a "Great Society" in the process. He thus declared the War on Poverty—a national effort to transform the lives of poor Americans through education and self-help activities.

The 1964 Economic Opportunity Act opened the War with three programmatic weapons: the Job Corps, the Community Action Programs (CAP), and VISTA (a domestic Peace Corps). The CAP were designed to assist local communities in establishing and administering their own antipoverty programs. However, some local governments opposed the CAP's proposed placement of administrative control and resources in the hands of poor people and refused to apply for program grants. In an effort to make the CAP more palatable to local officials, while using what would have been an embarrassing budget surplus, the Head Start project was born. Sargent Shriver, head of the Office of Economic Opportunity, envisioned an early intervention program as a way to "overcome a lot of hostility in our society against the poor" because children could not be blamed for their economic predicament (quoted in Ref. 8).

Head Start Is Born

A program for poor children was vaguely authorized under the Economic Opportunity Act, which mentioned that CAP could provide "a balanced program of educational assistance [that] might include ... creation of, and assistance to, preschool day care, or nursery centers for 3- to 5-year-olds."[9] But, with the exception of a few experimental projects, there was very little experience to suggest how to meet the needs of poor preschoolers. Therefore issues of program content, size, and duration were open questions for Shriver as he began piecing together a national intervention project. His first step was to appoint a planning committee of fourteen experts whose backgrounds were in education, child development, mental health, and pediatrics. This diversity ensured that Head Start would become more than an educational program. (For the history of the development of Head Start, see Refs. 10 and 11.)

Dr. Robert Cooke, a pediatrician, chaired the group, which had only 3 months to develop their plans. The committee's recommendations, presented to Shriver in February 1965, were based on a "whole child" philosophy that embraced a variety of goals. Unfortunately, these goals were not always presented clearly or in their entirety by policy makers and the media, creating confusion over Head Start's mission that continues to this day. Herrnstein and Murray are not the first to think that Head Start's purpose was to raise IQ scores and, when it failed at that, advocates' response was to "redefine its goals" to save face and the program (p. 404). To set the record straight, we quote the goals of Head Start from the original planning document:

The objectives of a comprehensive program should include:

A. Improving the child's physical health and physical abilities.

B. Helping the emotional and social development of the child by encouraging self-confidence, spontaneity, curiosity, and self-discipline.

C. Improving the child's mental processes and skills with particular attention to conceptual and verbal skills.
D. Establishing patterns and expectations of success for the child which will create a climate of confidence for his future learning efforts.
E. Increasing the child's capacity to relate positively to family members and others while at the same time strengthening the family's ability to relate positively to the child and his problems.
F. Developing in the child and his family a responsible attitude toward society, and fostering constructive opportunities for society to work together with the poor in solving their problems.
G. Increasing the sense of dignity and self-worth within the child and his family.[12]

Comprehensive Services to Meet Comprehensive Goals

The objectives of Head Start obviously went far beyond cognitive benefits, so the program's methods had to go beyond traditional preschool education. Instead of being modeled after typical middle-class nursery schools designed to give children learning and social experiences, each Head Start center was to have six components: health screening and referral, mental health services, early childhood education, nutrition education and hot meals, social services for the child and family, and parent involvement. The latter component was the most daring and controversial proposal in the planning document. Up to this time, poor people were treated as passive recipients of services dispensed by professionals. School involvement of parents—poor or not—was generally limited to bake sales and child discipline conferences. But Head Start parents were to be involved in the planning, administration, and daily activities of their local centers.

One reason for this decision derived from Head Start's origins in the CAP, which were mandated to allow poor citizens "maximum feasible participation" in running antipoverty efforts. The planning committee was also influenced by the ideas of one of its members, Urie Bronfenbrenner, who was just beginning to develop his ecological approach to human development. Bronfenbrenner argued that the course of development is influenced by a complex interrelationship of the child's traits, the family, and the community, so intervention must touch all of these areas to be effective.[13] This insight was an astute one: today, parent participation is recognized as crucial to the success of early intervention as well as later education.[14–16]

Although Head Start has grown in the 30 years since its birth, the basic format remains the same. In Fiscal Year 1994–95 there were 1,400 Head Start grantees serving about 750,000 children and their families. The majority of enrolled children are between 3 and 5 years old. Most attend a half-day program for one academic year. A small percentage are served before the age of 3 and/or for two years. Full-day services and home-based programs are provided in some areas.

Head Start is a federal-to-local effort. States may contribute to their Head Start programs (many do), but they have no voice in administration. By law, grantees receive 80% of their funding from the federal government and the rest from other, usually community sources, which may be in the form of in-kind services. Local programs are overseen by Policy Advisory Councils, composed of parents, staff, and community representatives, who are responsible for operating and staffing decisions. Each program is required to provide the six components listed earlier. Although they must adhere to a body of national performance standards in these areas, centers are encouraged to adapt the program in response to local needs and resources. Thus it is somewhat misleading to speak of Head Start as a singular intervention because of the variety in local programming.

Federal guidelines require that at least 90% of Head Start families have incomes below the poverty line, presumably leaving some slots open to other income groups. This allowance was meant as a signal that the program should eventually move toward full socioeconomic integration (perhaps charging those above poverty tuition on a sliding-fee scale based on family income). The planners believed that children of all income levels could benefit from interacting with one another, and they hoped to avoid the socioeconomic segregation within the program that Herrnstein and Murray see as coming to characterize society. Yet Head Start was never funded to allow expansion toward mixed-income enrollment. In fact, only 29% of eligible (below poverty) preschoolers are currently being served.[17] This group receives priority for available slots. If any are remaining, they are offered to families who are eligible for other types of federal assistance, namely, Aid to Families with Dependent Children (now Temporary Assistance for Needy Families). Current enrollment practices thus are meant to ensure that Head Start serves the children and families who are neediest and can benefit the most from the program.[18]

This description of the Head Start model hopefully puts into perspective the misplaced hopes and condemnations directed toward the program and other early interventions. Head Start was developed as part of the War on Poverty, an admirable but overly optimistic attempt to overcome the social, economic, and racial inequalities that had long divided the nation. The preschool project was only one small piece in an arsenal of programmatic weapons to combat poverty, all of the others targeting adults. Head Start was designed with a variety of goals and a variety of services to accomplish them. It was the pioneer of comprehensive, two-generation strategies to better some aspects of the lives of young children and families in poverty. To judge it on the basis of how much participants' IQ scores have gone up is absurd when its purpose, methods, and resources are taken into consideration. Understanding how IQ effects came to be expected requires understanding another history—that of changes in social scientists' attitudes about human development, from strict determinism to rampant environmentalism.

The Environmental Mystique

If policy makers in the early 1960s seemed too hopeful about the prospects of ending poverty, their enthusiasm was dwarfed by social scientists' dreams to raise intelligence. The decade was characterized by unabashed promises that, given the "right" experiences at the "right" time, all children could develop into great intellects. Many psychologists devoted themselves to discovering just what those right experiences might be. Their efforts resulted in a growth industry for crib mobiles (to keep baby's brain stimulated), talking typewriters (to teach toddlers what they might otherwise not be interested in knowing), and "educational" toys (to put an eruditional spin on play time). Although early childhood intervention programs also experienced growth at this time, few of them were designed explicitly to deliver higher IQ scores. Yet considering the wild beliefs of the day concerning intelligence, it is easy to see how their purpose became confused with that of the educational gimmickry crowd.

Fixation on Intelligence

Up until about the middle of this century, American psychologists generally concentrated on studies of observable behavior and the role of conditioning and external reinforcements in learning. Somewhat belatedly, they discovered the sizable body of work on cognition being produced by prominent European scholars, particularly Jean Piaget. They became so captivated with his classic efforts in charting the sequence of cognitive development that they began to focus on intelligence to the exclusion of other human behaviors. At the same time, they broke from the traditional views of Gesell and others that physical and cognitive development were predetermined processes governed by hereditary and maturational factors. Instead of building new theories on the wisdom of old ones, scientists quickly abandoned predeterminism in favor of environmentalism, and development came to be viewed as wide open to the manipulation of experts.

The excitement with raising intelligence was fed by the writings of some respected theorists whose ideas were exactly what newly "enlightened" scientists wanted to hear. For example, J.McV. Hunt argued against the notion of genetically fixed intelligence by taking the opposite stance, that intelligence is essentially a product of the environment. He theorized that it is possible to promote a faster rate of intellectual development and higher level of adult intelligence by "governing the encounters that children have with their environments, especially during the early years of their development."[19] Hunt continued into the 1970s to argue that IQ changes of 50 to 70 points could be produced by appropriate intervention.[20]

The link between early environment and later intelligence was given credence by another notion gaining momentum at the time—that of critical or "magical" periods in development. This concept was popularized in Benjamin Bloom's work, *Stability and Change in Human Characteristics*.[21] Bloom argued that the human organism is most sensitive to environmental inputs during periods of rapid growth. He examined the ages when various human traits appear to stabilize and argued that proportionally more development occurs during the early years.

From patterns of correlations between IQ test scores given at different ages, Bloom noted that half of the variance of adult IQ scores could be statistically explained by IQ scores obtained at age 4. Hence, there emerged what came to be a cliché of the 1960s: half of the child's learning is over by the age of 4. Of course, we did not know then and still do not know when the complete learning process is over, so how could we possibly determine when half of it has been achieved? Yet, the idea that much learning was over before children began school was believed by many and fed the emphasis on early intervention.

The Deficit Model

To some extent, the enthusiasm with intellectual improvement was spurred by animal research on early deprivation. For example, various studies involving rats, monkeys, and other creatures showed that those reared in unstimulating conditions, such as darkness or isolation, did not become very good learners or problem solvers. Because these findings imply that early deprivation creates a permanent deficit in intellectual ability, the concept of early enrichment programs to counter environmental deprivation was a logical step, if one can equate the deprivation experienced by animals in laboratory cages with that experienced by economically disadvantaged children. This analogy, however, is obviously far fetched.

Remember, though, that these were the War on Poverty years—years when policy makers were finally confronting the fact that children reared in poverty generally did not do as well in school as did their wealthier classmates. Apparently whatever middle-class parents did for their children at home worked, but that something was missing in the rearing environments of poor children. Some hoped that early enrichment could fill the presumed deficits in poor children's lives so that class differences in performance could be eliminated by the time of school entry.

The deficit model embodied some paternalistic ideas that have had a long life in civilized societies. The notion of "social" or "cultural" deprivation was a blatant assumption that the culture of the lower classes was inferior to that of the middle class. Some of the earliest intervention programs were intended to provide poor children with the values and experiences common in wealthier homes. Others, such as the Abecedarian and Milwaukee projects noted by Herrnstein and Murray, were so lengthy and intensive that they took over from poor parents much of the responsibility for child rearing. Head Start's planners were the first to design a program that avoided the deficit model. Parents were not seen as the problem but as part of the solution, and they were welcomed to fill a variety of roles in Head Start centers. The success of parent involvement in Head Start seeded the current recognition that parents, regardless of their material wealth, are the primary educators and socializers of their children and need to be supported in this role.

The practice of involving parents in running Head Start was way ahead of its time. As late as 1984, debate still raged over whether poor (and by this time, minority) children suffered intellectual defects as a result of socioeconomics and parenting (see Refs. 22 and 23). Eventually, though, the inferiority–superiority tenet of the deficit model came to be replaced by approaches emphasizing differences among groups rather than deficiencies.[24] In education programs, the goal is no longer to inculcate middle-class values (as in Riessman's 1962 book, *The Culturally Deprived Child*[25]) but to build on the unique strengths that each child brings to the classroom (as in Riessman's 1976 subsequent book, *The Inner-City Child*[26]). Yet despite this more egalitarian view, preschool programs are still judged by how well graduates fare according to middle-class norms. Indeed, the "failure" of preschool education to turn poor children into middle-class ones (and the depressed scores of poor adults and minorities on the "Middle Class Values Index") premised Herrnstein and Murray's resurrection of the deficit model.

Evaluations of Cognitive Benefits

Given that much of the social science community had embraced both naive environmentalism and the notion of magical periods in development, it is not surprising that early research focused on how much intervention programs could raise children's intelligence and achievement test scores. In the case of Head Start, the project's broad scope, multiple (if ill-defined) goals, and local diversity complicated evaluation because the standard design and measurement procedures available to researchers were not readily applicable; reliable commercial tests were an easy alternative. The fact that there were no established measures of some program aims such as parent involvement and social relationships is another reason why IQ and achievement tests came to dominate assessments. Perhaps the most compelling reason that such tests were used nearly exclusively in intervention research is that the results were so positive. Program designers, of course, wanted their efforts to succeed, and it was very tempting to employ measures—however inappropriate—that were sure to show success.

Early reports of IQ gains following almost any program ignited soaring hopes about the potential of intervention. For example, Eisenberg and Conners reported an amazing 10-point IQ increase for graduates of the first Head Start short summer program.[27] Results often reached the public in dramatic fashion, as when the Deutschs' preliminary report of 10-point IQ increases after their 10-month program was published in New York newspapers under the headline, "Program raises children's IQs a point a month." No wonder there was optimism about the potential of intervention to enhance cognitive ability. And no wonder there was disappointment when the bubble burst.

The first shoe to drop was a study on Head Start by the Westinghouse Learning Corporation.[28] The researchers did not administer IQ tests but achievement tests; they evaluated children who attended Head Start summer sessions, sampling only a small group who had been in full-year programs.[29] These should have been cautionary flags to even cursory interpretations, but the Westinghouse findings, showing the achievement gains made in preschool purportedly faded away in the early grades of school, did much to cast doubt on the efficacy of early intervention.

Experts in statistics and evaluation eventually discredited the Westinghouse report after questioning the sampling procedures, data analyses, and appropriateness of the outcome measures.[30,31] Yet, subsequent studies of Head Start and almost every other early intervention program led to the same conclusion: preschool graduates generally do not continue to do better on cognitive or achievement tests. There are certainly many exceptions, but the more common findings of fade-out dashed hopes that brief preschool experiences could guarantee higher IQs or academic success for poor children.

As we discuss shortly, other worthwhile benefits of early intervention have since been revealed. Nonetheless, preschool programs continue to be judged largely on the basis of sustained increases in IQ—the approach taken in *The Bell Curve*. This is puzzling because by now it has been shown that it is extremely difficult to effect true gains in measured intelligence.[32,33] The fact is that only a few programs appear to have had lasting effects on IQ.[34,35] But looking at why so many others at least temporarily affected IQ scores tells us much about the degree of malleability of different human systems.

The developing child is a complex combination of socioemotional, physical, and cognitive subsystems, each differentially sensitive to environmental input. The brain appears to be the most buffered against external events and stressors, making intelligence a highly stable trait. This does not mean that it is predetermined, but scholars' opinions on the practical reaction range for IQ cluster between about 10 and 25 points; this is a defensible expectation of what intervention can in fact accomplish.[36–38] The physical and socioemotional aspects of development are more strongly controlled by the environment and, therefore, more effectively targeted by intervention. Indeed, improvements in these aspects are suspected to underlie the initial gains in IQ following preschool experience. For example, several studies have traced higher scores to factors such as better familiarity with the test content, improved motivation, and comfort and self-confidence in the testing situation.[39–41] There is no doubt that the child who enjoys good health and has the desire and confidence to do well will try harder on an IQ test and later in school than a child who suffers hearing or vision problems and has an "I can't do it" attitude.

Judging early intervention programs by the sole criteria of elevated IQ scores not only shortchanges the scope of their potential impact, but gives the trait of intelligence unwarranted import. IQ tests were designed to predict school performance, and this they do fairly well. Scores on IQ and academic measures correlate a respectable .70. However, this correlation indicates that only about half [(.70)2 = .49] the variance in school performance is accounted for by children's IQ scores. What, then, accounts for the remainder? The answer most likely includes some collection of personal attributes that standard IQ tests do not assess very well. These attributes are even more important to success outside of school and explain why—contrary to Herrnstein and Murray's misreading of the evidence—IQ scores are not as robust predictors of life outcomes.[42–43] This is not to say that IQ has no bearing on adult occupational and social status, because it certainly does. Respondents to *The Bell Curve* are in agreement that life success is related to a *combination* of intelligence and character, judgment, common sense, social intelligence, morality, creativity, and other personal skills.[44,45] Expecting intervention to raise IQs is therefore a narrow, incomplete goal. Broader and richer outcomes are to be desired and, as we now describe, may in fact result from quality intervention.

The Broader Picture

Hopes that preschool enrichment can make children smarter have never been extinguished but have been joined by more exaggerated promises. Recent popular and political support of early intervention has assumed that it will improve high school graduation rates, stop teen pregnancy and juvenile delinquency, end welfare usage, and save a great deal of taxpayers' money. Only a small number of studies have produced evidence that these outcomes might be possible, so such conclusions are vastly premature. They also overshadow the less idolized but hardly trivial benefits of intervention that have been substantiated by over three decades of research in the field. Positive effects in the areas of academic and prosocial behavior are among them.

School Adjustment

In a welcome break from the narrow focus on IQ that characterized early research, the Consortium for Longitudinal Studies brought to light some of the noncognitive benefits of early intervention.[46] The Consortium consisted of researchers who had evaluated eleven preschool programs during the 1960s and early 1970s. (Although Herrnstein and Murray claim that "None of them was a Head Start program" [p. 405], in actuality two of them were of Head Start.) The researchers located original program participants and collected a uniform set of information about their current status. The findings confirmed that children who attend quality preschool programs do gain an initial boost in IQ and achievement scores that lasts for some years but appears to fade. However, lasting effects were found in other areas: participants were less likely to be assigned to special education classes and were somewhat less likely to be held back a grade in school. The rigor of the Consortium methodology, and the findings of benefits that persisted until many children had reached 12 or more years of age, did much to restore public and scientific faith in the value of early intervention.

More recent studies and research reviews have yielded similarly encouraging results. Among the most comprehensive analyses of the literature are reviews by Barnett,[47] Haskins,[48] and Woodhead.[49] All were careful to separate experimental preschools (typically small and university based; carefully designed, implemented, and evaluated; and of high quality) from Head Start (which serves some 750,000 children and families each year in programs across the nation that range from excellent to poor quality). The reviewers noted that Head Start and the smaller programs produced immediate gains in intelligence and achievement that generally dissipated within a few years. Haskins concluded that after that, there was very strong evidence that the experimental models improved school performance, including less grade retention and special education class placement. Although he described the evidence for Head Start in this domain as only modest, Barnett found it to be "significant" for large-scale programs, including Head Start. Life success indices (avoidance of delinquency, teen pregnancy, welfare, etc.) produced some evidence for the smaller projects but "virtually no evidence for Head Start" (see Ref. 48). This does not mean that Head Start has no such benefits, but that little or no data have been collected about them. In fact, because Head Start offers a wider variety of services than most of the comparisons, it seems reasonable to expect greater benefits if the services are of high quality.[50]

We should note that, while there is little doubt children eventually appear to lose the IQ gains made in preschool, the issue of whether academic benefits also dissipate is not yet decided. A problem in many reviews, particularly the most inclusive, is that the results of good and poor programs (and sound and unsound studies) are combined so that robust results from some investigations are statistically diluted by null effects from others.[51] For example, a glance at the individual findings in Barnett's review and in a meta-analysis of Head Start studies[52] reveals a number that do not show fade-out. Other studies have found retention of benefits for some participant groups but not others.[53]

Further arguments against the reality of fade-out are that the paper-and-pencil, standardized group achievement tests used in many studies are inappropriate for children in the early primary grades and are not reliable indicators of their academic progress.[54–56] Barnett, who holds the position that the achievement gains made in preschool probably do not fade out, suggests that their apparent loss is an artifact not only of measurement but of statistical analysis and sampling procedures.[57] For example, poor achievers in both preschool graduate and comparison groups could be retained in grade and thus dropped from the evaluation, thereby minimizing the differences among the remaining students.

In sum, the sizeable body of research on varying intervention programs tentatively shows enduring effects on school adjustment and other aspects of social competence.[58] The persistence of academic benefits is a possibility that warrants further review. Findings concerning immediate program effects are much more definite: When children leave preschool they have better IQ test scores and school readiness skills. In other words, they are better prepared for school. In lamenting the loss of initial cognitive benefits, many critics have overlooked the contribution of early intervention toward realizing the first national education goal for the year 2000—that all children will enter school ready to learn. If this advantage later fades, the fault lies beyond the preschool.

Prosocial Behavior

Because most early intervention programs were created to improve school performance, few evaluators have looked for outcomes beyond the realm of school. Those who have done so have discovered some surprising and welcome benefits, particularly in the area of delinquency—a behavior that is negatively correlated with success in school. (For reviews of the literature on early intervention and later criminality, see Refs. 59–61).

The most frequently cited study of the effects of early intervention on delinquency is that of the Perry Preschool.[62] The program (which predated Head Start and ended in 1967) was a high-quality preschool that most of the 58 participants attended for 2 years. The well-trained teachers also conducted weekly home visits to teach the children's mothers to reinforce the curriculum at home. As in most studies, the Perry graduates showed no permanent gains in IQ, but as a group they did better in school and beyond than controls. They were retained in grade less often, received fewer special education services, were more likely to graduate from high school, and had higher employment rates and lower welfare usage. They were also less likely to be chronic troublemakers. Although over 60% of the program men had been arrested between one and four times, only 12% of them were arrested more times than that. Nearly half of the men in the control group had five or more arrests.

These positive program effects saved a great deal of taxpayer money. An economist on the research team, W. Steven Barnett, estimated that every dollar spent on the preschool saved $7.16 in fewer special school services, less welfare use, higher employment, and of course, reduced crime.[63] In fact, over 90% of the savings were in this area. Most of these benefits were projected savings to crime victims who did not become victims because the crime was not committed.

No other study has followed preschool graduates for such a long time and for so many possible outcomes, but research on several other programs has produced results suggestive of reduced delinquency. For example, the Family Development Research Program at Syracuse University offered family support and child care and education services to young, poor single mothers for several years.[64] Although the children did not do much better in school than controls, they were less likely to be seen at the county probation department when they got older. By the age of 13 to 16, only 6% of the Center children had been processed as probation cases, compared with 22% of the controls. Court and penal costs for each child in the program group averaged $186. These costs were 10 times as great for each child in the control sample.

Another example is the Yale Child Welfare Research Program, which was designed more for parents than for children.[65] Participants were young mothers raising children in high-risk environments. Services began prenatally and lasted until the children were $2\frac{1}{2}$ years old. These included parenting education and practical supports, such as help in accessing housing and food programs. The children received pediatric care, and most attended child care, which was an option provided as a family support service. At the 10-year follow-up, the intervention mothers had obtained more education than controls and had fewer children.[66] Almost all of the program families were self-supporting by this time. Although the researchers did not assess delinquent behavior, teacher ratings showed that control boys were more likely to skip school and to show aggressive, acting-out, predelinquent behavior. The program boys required fewer remedial and supportive services, including court hearings, at an average savings of over $1,100 each academic year.

How can programs designed to help children succeed in school or strengthen their families have such an effect on behaviors associated with delinquency? In a 1994 review of the risk factors in chronic delinquency, Yoshikawa grouped risks into those associated with the child and those occurring within the child's family. Child correlates of delinquency include gender, low intelligence, poor school achievement, aggressiveness, and poor intersocial skills. Family correlates involve large family size, low socioeconomic status, low educational attainment among parents, and poor parenting skills. A third source of risks that can be added to Yoshikawa's model involves community factors, including poor schools, poor housing, lack of employment opportunities, and a lack of positive role models. The more of these risk factors that affect a child, the more likely she, or probably he, is to engage in delinquent behavior. This analysis is much richer than that of Herrnstein and Murray, who, although acknowledging that there may be more correlates of criminality, explain criminal behavior by low IQ alone. Their perspective closes the door on prevention efforts, while Yoshikawa's opens it wider.

Yoshikawa described two pathways through which intervention might mediate the effects of two layers of risk factors on delinquency. Programs aimed at reducing a child's risk factors might work to enhance cognitive and social development and school achievement. Programs that focus on family risks might attempt to improve, for example, socioeconomic status and parenting skills. (The latter are particularly important to nondelinquent outcomes in light of strong evidence presented by Rothbaum and Weisz linking child externalizing behavior to parental caregiving.)[67] Yoshikawa argued that interventions that take both child and family pathways will be the most effective in preventing chronic delinquency and later adult criminality (see Ref. 60). Comprehensive, early childhood programs that provide preschool education as well as family support fill this bill. More research, however, is needed to substantiate the link between early intervention and prosocial behavior.

Lessons to Be Learned

We are in complete agreement with Herrnstein and Murray's assertion that "Next to nothing is to be learned about how to raise IQ by more evaluations of Head Start, or even by replicating much better [sic] programs such as Perry Preschool or Abecedarian" (p. 413). Decades of research on dozens of early intervention programs have by now proved that most of them do not permanently raise IQ scores. Yet even the most recent studies of Head Start have searched for lasting cognitive benefits and, to no one's surprise, turned in less than robust results.[53,68] Empiricists' IQ habit has not only created mounting disappointment in the value of early intervention, but it has stymied the discovery of other possible effects.

We must disagree with the disparaging intent of Herrnstein and Murray's remark, "Nothing is more predictable than that researchers will conclude that what is most needed is more research" (p. 413). A great deal more research *is* needed, but not of the cognitive-test variety, and certainly not limited to the physiological basis of intelligence (the agenda promoted by Herrnstein and Murray). Programs that offer comprehensive services to young children have the potential to impact more than one aspect of their development. As described earlier, children's adaptation to school and prosocial behavior are among them. Here we discuss potential benefits of some of the under-researched components of the Head Start model because at least some of these features are to be found in other good preschool programs.

Health

There is no argument that children who are healthy and adequately nourished are better learners than those who are hungry and frequently absent from school because of illness. Head Start performance standards require that all enrolled children receive medical screenings, immunizations, dental exams, and corrective treatment if needed. A high percentage of them do, making Head Start the nation's largest single provider of health services to poor children.[69] This role will expand now that the program is authorized to serve the health needs of participants' siblings. Head Start is also becoming a major provider of screening and diagnostic services required for children covered under Medicaid, and it delivers many of the services to handicapped preschoolers and their families mandated under the Individuals with Disabilities Education Act. In fact, children with disabilities comprise 13% of enrollment (higher than the 10% mandated by law), so the program has done well in meeting the needs of this underserved population.

Another health benefit derives from Head Start's nutrition component. Centers are required to serve hot meals and healthy snacks, and to provide nutrition counseling to parents. The purpose of these services is the obvious one: to improve children's health and, ultimately, their ability to attend to learning tasks. In their preconceived dismissal of Head Start, Herrnstein and Murray failed to include the program in their discussion of efforts that they see as holding the most promise for raising intelligence—nutrition and vitamin supplementation. Improved nutrition is in fact thought to be responsible for the Flynn effect, which is the gradual increase in IQ over succeeding generations.[70] Among children at high risk for school failure, substantial increases in IQ have been documented after vitamin supplement regimens.[71] Such findings have been questioned, however, and not all studies have shown cognitive gains as a result of better nutrition.[72] The mixed evidence notwithstanding, no component of Head Start should be evaluated on the basis of IQ effects. The nutrition component must be appreciated for its salutary effects on children's health, particularly children who are at risk of inadequate nourishment.

Perhaps because Head Start's health benefits are so obvious (or because psychologists rather than physicians have conducted most of the research), not much work has been done to highlight their value. One study did show that Head Starters received more health screenings and dental exams than did middle-class children.[73] Similar physical exam histories were also found for the two groups' siblings. The Synthesis Project also revealed that the health status of Head Start participants was comparable with that of more advantaged children (see Ref. 52). Even a recent study showing that white children purportedly benefit more from Head Start than do blacks found equal and substantial effects in the area of health (see Ref. 53).

Families

Head Start's impact on families is another outcome that has not received the research attention it merits. In 1992, 94% of enrolled families identified as needing supportive services received them.[74] Each year thousands of low-income parents benefit financially by obtaining jobs and training through Head Start. Over 35% of program staff are parents of enrolled children or graduates, and many have earned Child Development Associate credentials and have entered careers in early childhood education.[75] Parents have also reported improved relationships with their children (see Ref. 3) and greater life satisfaction and psychological well-being resulting from the supportive social network of the preschool community.[76] Parents' involvement in Head Start might also benefit their other children, as has been found in studies of different programs (reviewed in Ref. 77). It makes sense that if parents are respected for their roles and relieved of some of life's burdens, they may become better socializers of all their children. If so, the intervention's impact and cost effectiveness are much broader than what is suggested by the research, nearly all of which has been limited to children directly participating in the program.

Communities

Another under researched area is the impact of Head Start on local communities. A few studies have shown that the presence of a Head Start program enhances a community's capacity to meet local needs (see Ref. 52). In one survey, almost 1,500 institutional changes in the educational and health systems were identified in 48 communities that housed Head Start centers.[78] A more recent General Accounting Office report praised Head Start's methods of linking families with local services, judging this approach far more successful than efforts to create new services or delivery mechanisms.[79] The impact of such efforts to better the living environments of families deserves further scrutiny.

There is clearly a vast amount of knowledge that we do not yet have about the scope of the effects of Head Start and other comprehensive interventions. This void leaves the door wide open for judgments of programs on the basis of the information that we do have—that they do not increase intelligence over the life span. Only if we value intelligence over all other human traits, as Herrnstein and Murray appear to do, can we conclude from this incomplete dataset that early intervention is a total failure. The literature is bulging with suggestions of broader benefits to individuals, families, and communities that are far more important to the nation's quality of life than psychometrics.

Attempts to Measure Social Competence

Many of the salutary effects accruing from early intervention are in behavioral areas such as meeting school and social expectancies and relating well to family and others. These abilities fall under the rubric of social competence. This construct may make complete sense to anyone who hears it, but it has long defied scientific definition and measurement.

The need to assess outcomes other than cognitive ones became apparent early in the history of intervention programs. Psychologists and educators had begun to question the cultural-fairness of IQ tests, which seemed biased in favor of children with white, middle-class experiences. The value of the IQ construct as a predictor of performance outside of the realm of school also became suspect. For example, Mercer referred to the "6-hour" retarded child—one whose IQ and school test scores are low but who functions perfectly adequately before and after the school day.[80] By the same token, there are many individuals who achieve very high IQ scores but do not behave competently at home, work, or in social settings. IQ, then, was just not measuring what early intervention specialists hoped to achieve.

Not long after the release of the Westinghouse report, the Office of Child Development (OCD)—the government agency then responsible for Head Start—began a concerted effort to redefine ways to measure the success of the program. Social competence was formally adopted as an overarching goal of Head Start, and therefore as the critical dependent variable for any future evaluations. Social competence included, but was not limited to, intellectual functioning, and was defined as "the child's everyday effectiveness in dealing with his environment and later responsibilities in school and life."[81]

In 1973, OCD commissioned the Educational Testing Service to construct a comprehensive definition for use in Head Start evaluations. The ETS assembled a conference of child development experts who eventually identified 29 variables as "facets" of social competence.[82] This unworkable number left OCD no closer to an operating definition that could be used in evaluation. Subsequently, the Rand Corporation was contracted to design a national assessment of social competence in Head Start children.[83] Again, the effort proved futile.

The last attempt began in 1977, when Mediax Associates was commissioned to develop and field test a battery of social competence measures for use in Head Start. Mediax (Ref. 81) proposed a model of competence that included four domains: (1) health and physical, (2) cognitive, (3) social-emotional, and (4) "applied strategies." This fourth domain involved such characteristics as motivation, curiosity, initiative and persistence, and task orientation. Mediax then subcontracted the development of three of the four domains to prominent research groups.

In 1982 the project came to an abrupt halt when funding for all but one of the parts was terminated. It is difficult to pinpoint exactly why the contracts were canceled, but one reason is that the new Reagan Administration felt that the development of competence measures for Head Start was of little value and should be cut from the budget. Ironically, the only part of the project that was salvaged was the part devoted to cognitive measures. Not only did such measures already exist, but their use had yielded little in previous Head Start evaluations. Early childhood specialists were disappointed and frustrated, not only because they still had no tool to assess competence, but because so much concern for the physical, social, and emotional development of children had been abandoned (see Ref. 55).

Despite the staggering amount of work put into the development of measures of social competence, no practical instrument has been forthcoming. In 1978, Zigler and Trickett warned, "the hour grows late, and unless the social sciences develop a practical and coherent measure, social competence will never replace the IQ as our primary measure of the success of intervention programs."[84] Their prediction has unfortunately come true for Head Start. The program's failure to produce permanent increases in IQ has been proven over and over again, as Herrnstein and Murray remind us. But the possibility that it has positive effects in other areas has been largely ignored.

The lack of standardized measures has left evaluators of different programs to pursue their own ideas about what might indicate social competence. Researchers studying graduates of the Perry Preschool in adulthood, for example, looked at factors such as whether they owned a second car and how often their children used the library (see Ref. 61). The use of such varied outcome measures is valuable in imparting a sense of the breadth of the effects of intervention. They contribute little to the knowledge base, however, because findings across studies are not easily compared and do not build upon one another.

By now there does seem to be a consensus about what features enter into social competence and are appropriate to include in evaluations. These elements are close to those proposed by Mediax (Ref. 81) and Zigler and Trickett

(Ref. 82), whose papers should be consulted for more complete descriptions. They include: (1) physical health, assessed by measures such as growth rates and immunization histories; (2) cognition, tapped by standardized intelligence tests; (3) achievement, also indicated by standard tests; and (4) socioemotional factors. These cluster around motivation, self-image, attitudes, and social relationships and can be evaluated using a variety of measures with established reliability and validity. Academic histories including grades, retention rates, and need for special services are good indicators of school adjustment. Because of the stability of traits related to predelinquent and delinquent behavior,[59,85] and emerging evidence about traits characteristic of resilient children,[86] personality inventories can be used to project later social adaptation. Other contributors to social competence might be examined by looking at program effects on families, for example, parents' subsequent education, income, employment status, family size, child-rearing skills, involvement in their child's education, mental health, and family functioning.

Not every study can look into all of the possible results on behavioral competence of quality intervention programs, but hopefully this discussion has made clear that studies limited to cognitive effects miss the point. They also offer no constructive criticism. The purpose of evaluation is to guide improvements in services to make them more effective as well as to justify programs to their funders. Interpretations of myopic evaluations, particularly those focused on intelligence, historically have been used instead to advance political or social agendas. Herrnstein and Murray's is no exception.

Improving Programs for At-Risk Children

The field of early childhood intervention has matured since the 1960s and has generated a wealth of knowledge to inform directions for change. As we discuss in turn, it is now clear that programs must be comprehensive and of high quality to achieve their desired ends. Second, because there are no magical periods in human development, programs must begin earlier and last longer than the year or two before a child who lives in poverty enters school. Third, programs that address the broad ecology in which the child develops will be the most effective. The final and perhaps the hardest lesson is not to expect miracles. Early intervention can help prepare children for school and enhance some aspects of their families' functioning, but it alone cannot end poverty, crime, and school failure.

Elements of Effective Intervention

The literature on early intervention has generated some guiding principles that underlie the most salutary efforts.[3,24,87–89] One is that programs must be comprehensive in scope, attending to the many factors that underlie the complex phenomenon of school performance. All children need certain learning experiences to be ready for school, but poor children often have myriad other needs as well. Children who are hungry, homeless, witnesses or victims of abuse, or live in neighborhoods governed by gangs and guns will not be able to give their full attention to school, regardless of whether they attended preschool. The services poor children need may go beyond the traditional mission of the education establishment, but their absence makes that mission unattainable.

Preschool programs have proliferated in recent years, but few of them provide comprehensive services. Of particular concern are the public prekindergartens implemented in many states after the adoption of the national education goals. These programs typically serve 4-years-olds and, like Head Start, most are for children deemed at risk of having problems in elementary school. However, only half of the public prekindergartens are required to provide services that go beyond education.[90] Few do so, and none approaches the level of services in Head Start.[17,91] Further, the quality of the programs may not be sufficiently high to meet the needs of very young, at-risk children. Quality involves features such as good teacher/child ratios, staff trained in early childhood, small group sizes, and developmentally appropriate curriculum. Yet in many public preschools, the number of children per teacher and the curriculum used often mirror typical kindergartens and are simply inappropriate for preschoolers.[92,93] Finally, parental involvement—particularly important in intervention with young children—is a practice that not all public schools have implemented or even fully accepted.[94] Public preschools that attend only to children's educational needs may give the impression that they are preparing children for school, but they are unlikely to be very successful.

Both the state and federal preschool efforts can be enhanced by ensuring that services for children are of sufficient quality to deliver the intended benefits. This point deserves emphasis because in the recent Head Start expansion, efforts to serve more children proceeded more rapidly than efforts to serve them well. Years of inadequate funding, oversight, and training and technical assistance have left many Head Start centers struggling to provide adequate services and some unable to do so.[3,95–99] This is not to say that all Head Start centers conform to Herrnstein and Murray's description as places where "a few modestly trained adults who enjoy being with children watch over a few dozen children in a pleasant atmosphere" (p. 414). Even the worst of centers provide some degree of preschool education and health and social services. Yet the existence of some poor centers jeopardizes the reputation of the many good ones.

The need for uniformly high quality cannot be overstated. The literature makes clear that "only high quality programs consistently show success."[100] Zigler told a U.S. Senate committee that "Head Start is effective only when quality is high.... Below a certain threshold of quality, the program is useless, a waste of money regardless of how many children are enrolled."[101] Quality improvement plans have now been drawn and are starting to be implemented (see Ref. 95). Reauthorization of Head Start in 1994 continues the practice begun in 1990 of putting aside 25% of budget increases to upgrade quality. Without increased funding for expansion, however, the quality set-aside will disappear and the efforts already begun will not reach fruition.

Dovetailed Programming

The first major disappointment over Head Start came from the findings of the Westinghouse report that the academic gains children made in preschool appeared to be short lived. While many were quick to dismiss Head Start as a failure, others suggested that the blame belonged to the elementary schools the children later attended. They were prepared for school when they arrived, but the education establishment failed to build upon this advantage.[102] Still others argued that Head Start came too late—that by the time children began preschool, their potential had already been dimmed by the harmful effects of poverty. Today it is widely recognized that both sides were right. There are no magical periods in development. All stages of growth are important and require appropriate environmental nutrients.

Head Start's planners had never fallen prey to the "inoculation model." They never believed that there was a quick fix that would erase the past and prevent the future effects of growing up in poverty. They encouraged the development of model programs to serve children both before and after the preschool years. Now that social scientists have accepted the view that development is a continuous process—and that no brief program can dissolve the risks associated with poverty—these models may become the basis for a national system of extended intervention.

The need for preventive intervention beginning very early in the child's life has now been firmly established. Recognizing that preventive services can be more effective, and more cost effective, than remedial ones, Congress has bolstered efforts to serve Head Start's pioneer program for very young children and their families. An exciting new 0-to-3 effort called Early Head Start has also been initiated with the Human Services Reauthorization Act of 1994. The Clinton Administration followed the advice of the Carnegie Task Force on Meeting the Needs of Young Children, the Children's Defense Fund, and other advocacy and professional groups and convened a multidisciplinary committee to plan and implement this younger version of Head Start.[103] Services for at-risk families will begin prenatally and include nutrition, health care, parenting education, and family support. The rationale for the project is clear: Waiting until a child is 3 or 4 years old is waiting too long. Children who are healthy, have sound relationships with their primary caregivers, and who have received adequate nurturing and stimulation will have the socioemotional foundations needed for learning in preschool and beyond.[104]

For children who may still need services after infancy and preschool, the Head Start-Public School Early Childhood Transition Project (described in Ref. 105) has been reauthorized by the 1994 Act. The project follows Head Start graduates from kindergarten through grade 3. Local Head Start and public school personnel work to introduce each child and family to the new school experience and to familiarize kindergarten teachers with the child's progress, program, and needs. Comprehensive services, parental involvement, and family support are continued for the next 4 years.

A small but convincing body of evidence supports the premise of the Transition Project, that is, that longer, coordinated intervention produces longer lasting gains. Results (reviewed in Ref. 89) of the Abecedarian Project, Success for All, the Chicago Child-Parent Centers, Follow Through, and the Deutschs' early enrichment program confirm that the advantages derived from preschool can be sustained with dovetailed, school-age programming.

This research offers promise that the Transition Project will be effective in helping Head Start graduates be more competent students throughout school. If evaluation supports this expectation, it will be compelling to move the project into the educational mainstream. At the Yale Bush Center, we have developed a plan to do so using current federal education expenditures (see Ref. 89). A large part of the Department of Education's budget (over $7 billion annually) is spent on Title I of the Elementary and Secondary Education Act of 1965 (ESEA). Title I is a compensatory education program for economically and "educationally" deprived children in preschool through grade 12. Originally intended to enhance the educational services of impoverished school districts, the program now operates in the majority of the nation's schools—mostly as a pull-out program offering remedial instruction to children who have fallen behind the academic expectations of their grade level.

There has been little evaluation of Title I considering the size of the program, but what there is shows that most students do not exhibit a meaningful gain in achievement.[106] The reasons for the lackluster results may be in the program's design: services are remedial rather than preventive, and narrow instead of comprehensive; parental involvement is minimal; and health and family problems that can interfere with school performance are not addressed. Debate during reauthorization of the ESEA in 1994 focused on suggestions to revamp the program by training teachers and narrowing the target population.[107] Although such efforts may do some good, they do not bring the elements of effective intervention to Title I. To make a difference in the education of low-income children, we must put aside the ineffectual educational model of Title I and adopt on a large scale the proven model of comprehensive, family-focused services.

Although the Improving America's School Act of 1994 mandates that Title I do more to provide children with access to health and social services and to include their parents in the program, such stipulations have been poorly implemented in the past (see Ref. 106). Our proposal is instead for Title I to follow Transition project plans and to become the school-age version of Head Start. As Head Start eventually expands to serve all eligible children, Title I can continue their intervention in grammar school. Coordinated curricula and continued parental involvement and comprehensive services will then be firmly placed in schools that serve populations below the poverty line.

A national system of extended intervention, beginning with Early Head Start and followed by preschool Head Start and a Title I transition, will do more than make a meaningful impact on the lives of at-risk children and their families. Together, the three programs will form a coherent federal policy to meet the needs of poor children beginning prenatally. Instead of funding a hodgepodge of programs with similar goals, tax dollars will be more efficiently spent on a system that can produce benefits greater than the sum of its parts.

Ecological Approach

Today it is generally agreed that successful intervention programs target not only the child but the family who rears the child. It makes sense that parents who have some of their basic needs met, feel a degree of social support, have some child-rearing skills, and have a sense of control over their own and their children's future can do a better job of parenting. The family systems approach provides a viable hypothesis to explain the long-term effects that have been found for early intervention. Several theorists have advanced the view that persistent benefits cannot be due to a half-day program experienced by the child during 1 year of preschool, but are rather due to the parents.[16,38,64] As a result of their involvement with the intervention, parents become more optimal socializers of the child throughout the rest of the day, and hopefully throughout the child's course of development.

Two-generation intervention has now become the latest emphasis in children's programming, not only because of scientific support but for political reasons. Today public sentiments are strong to reform welfare and move recipients from the public dole into jobs. Early childhood programs can provide their children with needed child care. Two-generation programs can also supply the adults with training in literacy and basic skills, work experience, and job networks within the community. Parental involvement may have become an accepted practice in effective intervention for children, but it also fits the needs of a nation that has lost patience with expensive social entitlements.

Two-generation efforts, of course, respond to only part of the ecological model, which holds that the child's development is influenced by the family system as well as by other systems far removed from the family's control.

These include work, neighborhoods, the school, the media, health services, and child care. The more responsive these systems are to meeting the family's needs, the more responsive the family can be in tending to the child's needs. Obviously, no early childhood program can tackle all of these influences. The best of them can strive to achieve collaborations with other human services providers, but the strength of their impact will weaken the further they move from their child and family core.[108]

Realistic Expectations

The ecological model has clear implications that we not oversell what we can realistically accomplish with early intervention. These programs simply cannot guarantee that children will develop optimally when their larger rearing environment is not conducive to optimal development. Many people have been led to believe that graduates of preschool programs will turn out to be model citizens. Yet even for the highly acclaimed Perry Preschool, not all of the participants had good outcomes. Over half of the Perry graduates were arrested at least once by the time they were 27 years old, and 44% of the program females had been teen mothers (see Ref. 62). Longitudinal studies of other preschool projects have also shown that although children do better than they would have without the experience, they still do not approach the achievements of middle-class students.[109,110]

Such findings lead to the sobering conclusion that early childhood interventions alone cannot transform lives. Herrnstein and Murray blame this on their failure to produce higher IQs. Looking mostly at the evidence relating to cognitive ability, they dismiss Head Start and the experimental projects they describe by concluding, "to think that the available repertoire of social interventions can do the job [of raising IQ] is illusory" (p. 416). They carry their dismal judgment a step further by speaking against the K through 12 educational system (which they view as an "attempt to raise intelligence"): "school is not a promising place to try to raise intelligence, or to reduce intellectual differences" (p. 414). They do not recommend beginning intervention earlier than the preschool or school years. Instead, they point to the "cognitive benefits of adoption" at birth of children born to the "wrong women" (those who have low IQs) (pp. 414, 548).

In our chapter we have railed against this narrow focus because we do not believe that intelligence is the only important human trait. Nor do we believe it is primarily responsible for most individuals' successes and failures. We therefore think that early intervention should focus on other features of human development—the social, learning, and physical activities that define and enrich the course of life. We know that this focus will not permanently enhance participants' IQs to any huge extent. We also know that it will not end school failure, poverty, or crime and, in Herrnstein and Murray's words, make "it easier for people to live a virtuous life" (p. 543). They state the reason why themselves: "It is tough to alter the environment ... an unimaginably complex melange of influences and inputs.... No meaningful proportion of that melange can reasonably be expected to be shaped by any outside intervention.... " (p. 413). They are, of course, referring to the environment "for the development of general intellectual ability," but that is the same environment in which all human systems develop.

If intervention cannot improve the total rearing environment—the family, the neighborhood, the nation, and indeed, the whole world—does this mean it is not worth undertaking? If intervention cannot improve IQ, does this mean it cannot improve anything worthwhile? We answer no on both counts. Early childhood programs cannot change the universe, but they can alleviate some of the risks to development faced by young children reared in poverty (low IQ is only one of them). Because these risks have a cumulative effect, reducing their numbers by even a small amount may be enough to guard the child from those remaining. The same is true for protective factors (high IQ is only one of them). Intervention cannot give poor children the smart, rich, nurturing parents esteemed by Herrnstein and Murray, who will help them grow up to be smart and rich themselves. Nor can it give them wonderful schools that ensure achievement. But comprehensive programs can strengthen families to some extent and ease the transition to school, thus bolstering these protections against the risks.

Early childhood intervention thus has more modest goals than those assumed by Herrnstein and Murray. Comprehensive programs such as Head Start can prepare poor children for school and enable them to meet social expectancies during the school years and perhaps beyond. However, eliminating the problems of socioeconomic disadvantage must entail broader efforts to improve the climate of education, health care, employment, and community supports for all Americans. These efforts will not create the nostalgic society depicted by Herrnstein and Murray, nor will they give anyone an elite IQ score. They will accomplish much more.

References

1. Califano, J.A., Jr. (1979), "Head Start, A Retrospective View: The Founders, Section 1: Leadership Within the Johnson Administration," in E. Zigler and J. Valentine (Eds.), *Project Head Start: A Legacy of the War on Poverty*, Free Press, New York, pp.43–72.
2. Jensen, R.J. (1969), "How Much Can We Boost IQ and Scholastic Achievement?" *Harvard Educational Review*, 39, 1–123.
3. National Head Start Association. (1990), *Head Start: The Nation's Pride, a Nation's Challenge*, Report of the Silver Ribbon Panel, Alexandria, VA.
4. Rovner, J. (1990), "Head Start Is One Program Everyone Wants to Help," *Congressional Quarterly*, 48(16), 1191–1195.

5. Hood, J. (1992, December), "Caveat Emptor: The Head Start Scam," *Policy Analysis*, #187, Cato Institute, Washington, D.C.
6. Herrnstein, R.J., and Murray, C. (1994), *The Bell Curve: Intelligence and Class Structure in American Life*, The Free Press, New York.
7. Harrington, M. (1962), *The Other America: Poverty in the United States*, Macmillan, New York.
8. Zigler, E., and Anderson, K. (1979), "An Idea Whose Time Had Come: The Intellectual and Political Climate for Head Start," in E. Zigler and J. Valentine (Eds.), *Project Head Start: A Legacy of the War on Poverty*, Free Press, New York, pp. 3–19
9. *Senate Reports*. (1964), No. 23620, p. 20.
10. Zigler, E., and Muenchow, S. (1992), *Head Start: The Inside Story of America's Most Successful Educational Experiment*, Basic Books, New York.
11. Zigler, E., and Valentine, J. (Eds.). (1997), *Project Head Start: A Legacy of the War on Poverty*, National Head Start Association, Alexandria, VA.
12. Recommendations for a Head Start Program by a Panel of Experts. (1965, February), U.S. Department of Health, Education and Welfare, Office of Child Development, Washington, D.C.
13. Bronfenbrenner, U. (1974), "Is Early Intervention Effective?" *Day Care and Early Education*, 44, 14–18. Bronfenbrenner, U. (1979), *The Ecology of Human Development*, Harvard University Press, Cambridge, MA.
14. Bowman, B.T. (1994), "Home and School: The Unresolved Relationship," in S. L. Kagan and B. Weissbourd (Eds.), *Putting Families First: America's Family Support Movement and the Challenge of Change* Jossey-Bass, San Francisco, pp. 51–72.
15. Comer, J.P. (1991), "Parent Participation: Fad or Function?" *Educational Horizons*, 69, 182–188.
16. Seitz, V. (1990), "Intervention Programs for Impoverished Children: A Comparison of Educational and Family Support Models," *Annals of Child Development*, 7, 73–103.
17. General Accounting Office. (1995, March), *Early Childhood Centers, Services to Prepare Children for School Often Limited*, (GAO/HEHS-95-21), Washington, D.C.
18. Head Start Program, Final Rule, 57 *Fed. Reg.* 46725 (1992), (45 C.F.R. Part 1305).
19. Hunt, J.McV. (1961), *Intelligence and Experience*, Ronald Press, New York.
20. Hunt, J.McV. (1971), "Parent and Child Centers: Their Basis in the Behavioral and Educational Sciences," *American Journal of Orthopsychiatry*, 41, 13–38.
21. Bloom, B.S. (1964), *Stability and Change in Human Characteristics*, Wiley, New York.
22. Humphreys, L.G. (1984), "The Orwellian Difference Model," *American Psychologist*, 39, 916.
23. Zigler, E. (1984), "Meeting the Critics on their own Terms," *American Psychologist*, 39, 916–917.
24. Zigler, E., and Berman, W. (1983), "Discerning the Future of Early Childhood Intervention," *American Psychologist*, 38, 894–906.
25. Riessman, F. (1962), *The Culturally Deprived Child*, Harper, New York.
26. Riessman, F. (1976), *The Inner-City Child*, Harper and Row, New York.
27. Eisenberg, L., and Conners, C.K. (1966, April 11), "The Effect of Head Start on the Developmental Process", Paper presented at the Joseph P. Kennedy, Jr., Foundation Scientific Symposium on Mental Retardation, Boston.

28. Cicirelli, V.G. (1969), *The Impact of Head Start: An Evaluation of the Effects of Head Start on Children's Cognitive and Affective Development*, Report presented to the Office of Economic Opportunity (Report No. PB 184 328), Westinghouse Learning Corporation, Washington, D.C.
29. Datta, L. (1979), "Another Spring and Other Hopes: Some Findings From National Evaluations of Project Head Start," in E. Zigler and J. Valentine (Eds.), *Project Head Start: A Legacy of the War on Poverty*, Free Press, New York, pp. 405–432.
30. Campbell, D.T., and Erlebacher, A. (1970), "How Regression Artifacts in Quasi-Experimental Evaluations Can Mistakenly Make Compensatory Education Look Harmful," in J. Hellmuth (Ed.), *Compensatory Education: A National Debate*, Brunner/Mazel, New York, pp. 185–210.
31. Smith, M., and Bissell, J.S. (1970), "Report Analysis: The Impact of Head Start," *Harvard Educational Review*, 40, 51–104.
32. Spitz, H.H. (1986), *The Raising of Intelligence: A Selected History of Attempts to Raise Retarded Intelligence*, Erlbaum, Hillsdale, NJ.
33. Zigler, E. (1988), "The IQ Pendulum," (Review of H. Spitz, *The Raising of Intelligence*,) *Readings*, 3, 4–9.
34. Garber, H.L., (1988), *The Milwaukee Project: Preventing Mental Retardation in Children at Risk*, American Association on Mental Retardation, Washington, D.C.
35. Ramey, C.T., and Campbell, F.A. (1991), "Poverty, Early Childhood Education, and Academic Competence: The Abecedarian Experience," in A.C. Huston (Ed.), *Children in Poverty. Child Development and Public Policy*, Cambridge University Press, New York, pp. 190–221.
36. Cronbach, L.I. (1971), "Five Decades of Public Controversy Over Mental Testing," *American Psychologist*, 30, 1–14.
37. Devlin, B., Fienberg, S.E., Resnick, D.P., and Roeder, K. (1995), "Galton Redux: Eugenics, Intelligence, Race, and Society," *Journal of the American Statistical Association*, 90:1483–1488.
38. Zigler, E., and Seitz, V. (1982), "Social Policy and Intelligence," in R. Sternberg (Ed.), *Handbook of Human Intelligence*, Cambridge University Press, New York, pp. 586–641.
39. Seitz, V., Abelson, W. D., Levine, E., and Zigler, E. (1975), "Effects of Place of Testing on the Peabody Picture Vocabulary Test Scores of Disadvantaged Head Start and Non-Head Start Children," *Child Development*, 46, 481–486.
40. Zigler, E., and Butterfield, E.C. (1968), "Motivational Aspects of Changes in IQ Test Performance of Culturally Deprived Nursery School Children," *Child Development*, 39, 1–14.
41. Zigler, E., Abelson, W.D., Trickett, P.K., and Seitz, V. (1982), "Is an Intervention Program Really Necessary to Raise Disadvantaged Children's IQ Scores?" *Child Development*, 53, 340–348.
42. Gardner, H. (1995), "Cracking Open the IQ Box," in S. Fraser (Ed.), *The Bell Curve Wars: Race, Intelligence, and the Future of America*, Basic Books, New York, pp. 23–35.
43. Sternberg, R.J., Wagner, R.K., Williams, W.M., and Horvath, J.A. (1995), "Testing Common Sense," *American Psychologist*, 50, 912–927.
44. Allman, W.F. (1994, October 24), "Why IQ Isn't Destiny," *U.S. News and World Report*, pp. 73–80.

45. Fraser, S. (Ed.). (1995), *The Bell Curve Wars: Race, Intelligence, and the Future of America*, Basic Books, New York.
46. Consortium for Longitudinal Studies (Ed.). (1983), *As the Twig is Bent: Lasting Effects of Preschool Programs*, Erlbaum, Hillsdale, NJ.
47. Barnett, W.S. (1995), "Long-Term Effects of Early Childhood Programs on Cognitive and School Outcomes," *Future of Children*, 5(3), 25–50.
48. Haskins, R. (1989), "Beyond Metaphor, The Efficacy of Early Childhood Education," *American Psychologist*, 44, 274–282.
49. Woodhead, M. (1988), "When Psychology Informs Public Policy: The Case of Early Childhood Intervention," *American Psychologist*, 43, 443–454.
50. Zigler, E., and Styfco, S.J. (1994b), "Is the Perry Preschool Better than Head Start? Yes and No," *Early Childhood Research Quarterly*, 9, 269–287.
51. Gamble, T., and Zigler, E. (1989), "The Head Start Synthesis Project: A Critique," *Journal of Applied Developmental Psychology*, 10, 267–274.
52. McKey, R.H., Condelli, L., Ganson, H., Barrett, B., McConkey, C., and Plantz, M. (l985), *The Impact of Head Start on Children, Family, and Communities: Final Report of the Head Start Evaluation, Synthesis and Utilization Project*, U.S. Government Printing Office, Washington, D.C. (DHHS Pub. No. OHDS 85-31193).
53. Currie, J., and Thomas, D. (1995), "Does Head Start Make a Difference?," *American Economic Review*, 85, 341–364.
54. Meisels, S. J. (1992), "Doing Harm By Good: Iatrogenic Effects of Early Childhood Enrollment and Promotion Policies," *Early Childhood Research Quarterly*, 7, 155–174.
55. Raver, C.C., and Zigler, E. (1991), "Three Steps Forward, Two Steps Back: Head Start and the Measurement of Social Competence," *Young Children*, 46, 3–8.
56. Shepard, L.A. (1994), "The Challenges of Assessing Young Children Appropriately, *Phi Delta Kappan*, 76, 206–212.
57. Barnett, W.S. (1992), "Benefits of Compensatory Preschool Education," *Journal of Human Resources*, 27, 279–312.
58. McCall, R. (1993), *Head Start: Its Potential, its Achievements, its Future, a Briefing Paper for Policymakers*, University of Pittsburgh Center for Social and Urban Research, Pittsburgh, PA
59. Yoshikawa, H. (1994), "Prevention as Cumulative Protection: Effects of Early Family Support and Education on Chronic Delinquency and Its Risks," *Psychological Bulletin*, 115, 28–54.
60. Yoshikawa, H. (1995), "Long-Term Effects of Early Childhood Programs on Social Outcomes and Delinquency," *Future of Children*, 5(3), 51–75.
61. Zigler, E., Taussig, C., and Black, K. (1992), "Early Childhood Intervention: A Promising Preventative for Juvenile Delinquency," *American Psychologist*, 47, 997–1006.
62. Schweinhart, L.J., Barnes, H.V., and Weikart, D.P. (1993), *Significant Benefits: The High/Scope Perry Preschool Study Through Age* 27, Monographs of the High/Scope Educational Research Foundation, 10, High/Scope Press, Ypsilanti, MI.
63. Barnett, W.S. (1993), "Benefit-cost Analysis of Preschool Education: Findings From a 25-Year Follow-up," *American Journal of Orthopsychiatry*, 63, 500–508.
64. Lally, R.J., Mangione, P.L., and Honig, A.S. (1988), "The Syracuse University Family Development Research Program: Long-range Impact of an Early Intervention with Low-Income Children and their Families," in D. Powell (Ed.), *Parent*

Education as Early Childhood Intervention: Emerging Directions in Theory, Research and Practice, Ablex, Norwood, NJ, pp. 79–104.

65. Provence, S., and Naylor, A. (1983), *Working with Disadvantaged Parents and Children: Scientific Issues and Practice*, Yale University Press, New Haven, CT.
66. Seitz, V., Rosenbaum, L.K., and Apfel, N.H., (1985), "Effects of Family Support Intervention: A Ten-Year Follow-Up," *Child Development*, 56, 376–391.
67. Rothbaum, F., and Weisz, J.R. (1994), "Parental Caregiving and Child Externalizing Behavior in Nonclinical Samples: A Meta-Analysis," *Psychological Bulletin*, 116, 55–74.
68. Lee, V.E., Brooks-Gunn, J., Schnur, E., and Liaw, F.R. (1990), "Are Head Start's Effects Sustained? A Longitudinal Follow-Up Comparison of Disadvantaged Children Attending Head Start, No Preschool, and Other Preschool Programs," *Child Development*, 61, 495–507.
69. Zigler, E., Piotrkowski, C., and Collins, R. (1994), "Health Services in Head Start," *Annual Review of Public Health*, 15, 511–534.
70. Lynn, R. (1990), "The Role of Nutrition in Secular Increases in Intelligence," *Personality and Individual Differences*, 11, 273–285.
71. Schoenthaler, S.J., Amos, S.P., Eysenck, H.J., Peritz, E., and Yudkin, J. (1991), "Controlled Trial of Vitamin-Mineral Supplementation: Effects on Intelligence and Performance," *Personality and Individual Differences*, 12, 351–362.
72. Pagliari, H.C. (1993), "Effects of Nutritional Supplements on Intelligence: Comment on Schoenthaler et al.'s Paper," *Personality and Individual Differences*, 14, 493.
73. Hale, B.A., Seitz, V., and Zigler, E. (1990), "Health Services and Head Start: A Forgotten Formula," *Journal of Applied Developmental Psychology*, 11, 447–458.
74. Brush, L., Gaidurgis, A., and Best, C. (1993), *Indices of Head Start Program Quality*, Report prepared for the Administration on Children, Youth and Families, Head Start Bureau, Pelavin Associates, Washington, D.C:
75. Collins, R.C. (1990), *Head Start Salaries: 1989-90 Staff Salary Survey*, National Head Start Association, Alexandria, VA.
76. Parker, F.L., Piotrkowski, C.S., and Peay, L. (1987), "Head Start as a Social Support for Mothers: The Psychological Benefits of Involvement," *American Journal of Orthopsychiatry*, 57, 220–233.
77. Seitz, V., and Apfel, N.H. (1994), "Parent-Focused Intervention: Diffusion Effects on Siblings," *Child Development*, 65, 677–683.
78. Kirschner Associates, (1970), *A National Survey of the Impacts of Head Start Centers on Community Institutions*, Albuquerque, NM.
79. General Accounting Office. (1992), *Integrating Human Services*, (Report No. GAO/HRD-92-108), Washington, D.C.
80. Mercer, J.R. (1973), *Labeling the Mentally Retarded*, University of California Press, Berkeley.
81. Mediax Associates. (1980), *Accept My Profile: Perspectives for Head Start Profiles of Program Effects on Children*, Westport, CT.
82. Anderson, S., and Messick, S. (1974), "Social Competency in Young Children," *Developmental Psychology*, 10, 282–293.
83. Raizen, S., and Bobrow, S.B. (1974), *Design for a National Evaluation of Social Competence in Head Start Children*, Prepared for OCD, Dept. of Health, Education and Welfare, Rand Corporation, Santa Monica, CA.

84. Zigler, E., and Trickett, P. (1978), "IQ, Social Competence, and Evaluation of Early Childhood Intervention Programs," *American Psychologist*, 33, 789–798.
85. Farrington, D.P. (1994, April), "Delinquency Prevention in the First Few Years of Life", Plenary address delivered to the Fourth European Conference on Law and Psychology, Barcelona, Spain.
86. Garmezy, N. (1991), "Resilience in Children's Adaptation to Negative Life Events and Stressed Environments," *Pediatric Annals*, 20, 459-466.
87. Price, R.H., Cowen, E., Lorion, R.P., and Ramos-McKay, J. (Eds.) (1988), *Fourteen Ounces of Prevention: A Casebook for Practitioners*, American Psychological Association, Washington, D.C.
88. Schorr, L.B. (1988), *Within Our Reach: Breaking the Cycle of Disadvantage*, Doubleday, New York.
89. Zigler, E., and Styfco, S.J. (1993), "Strength in Unity: Consolidating Federal Education Programs for Young Children," in E. Zigler and S. J. Styfco (Eds.), *Head Start and Beyond: A National Plan for Extended Childhood Intervention.*, Yale University Press, New Haven, CT, pp. 111–145.
90. Mitchell, A., Seligson, M., and Marx, F. (1989), *Early Childhood Programs and the Public Schools*, Auburn House, Dover, MA.
91. Adams, G., and Sandfort, J. (1994), *First Steps, Promising Futures. State Prekindergarten Initiatives in the Early 1990s*, Children's Defense Fund, Washington, D.C.
92. Bauch, J.P. (Ed.). (1988), *Early Childhood Education in the Schools*, National Education Association, Washington, D.C.
93. Kagan, S.L., and Zigler, E. (Eds.), (1987), *Early Schooling: The National Debate*, Yale University Press, New Haven, CT.
94. Kagan, S.L. (1991), "Moving From Here to There: Rethinking Continuity and Transitions in Early Care and Education," in B. Spodek and O. Saracho (Eds.), *Yearbook in Early Childhood Education* 2, Teacher's College Press, New York, pp. 132–151.
95. Advisory Committee on Head Start Quality and Expansion. (1993), *Creating a 21st Century Head Start*, U.S. Department of Health and Human Services, Washington, D.C.
96. Chafel, J.A. (1992), "Funding Head Start: What are the Issues?" *American Journal of Orthopsychiatry*, 62, 9–21.
97. Office of the Inspector General. (1993), *Evaluating Head Start Expansion Through Performance Indicators*, (OEI-09-91-00762), U.S. Department of Health and Human Services, Washington, D.C.
98. U.S. Department of Health and Human Services. (1980), *Head Start in the 1980's, Review and Recommendations*, Washington, D.C.
99. Zigler, E., and Styfco, S.J. (1994a), "Head Start: Criticisms in a Constructive Context," *American Psychologist*, 49, 127–132.
100. Weikart, D.P., and Schweinhart, L.J. (1991), "Disadvantaged Children and Curriculum Effects," *New Directions for Child Development*, 53, 57–64.
101. U.S. Senate. (1990, August 3), *Human Services Reauthorization Act of 1990*, Report to accompany H.R. 4151, Report 101-421, Washington, D.C.
102. Lee, V.E., and Loeb, S. (1995), "Where do Head Start Attendees End Up? One Reason Why Preschool Effects Fade Out," *Educational Evaluation and Policy Analysis*, 17, 62–82.

103. Advisory Committee on Services for Families with Infants and Toddlers. (1994), *Statement of the Advisory Committee on Services for Families with Infants and Toddlers*, U.S. Department of Health and Human Services, Washington, D.C.
104. Zero to Three: National Center for Clinical Infant Programs. (1992), *Heart Start: The Emotional Foundations of School Readiness*, Arlington, VA.
105. Kennedy, E.M. (1993), "The Head Start Transition Project: Head Start Goes to Elementary School," in E. Zigler and S. J. Styfco (Eds.), *Head Start and Beyond: A National Plan for Extended Childhood Intervention*, pp. 97-109, Yale University Press, New Haven, CT.
106. Arroyo, C.G., and Zigler, E. (1993), "America's Title I/Chapter 1 programs: Why the Promise Has Not Been Met," in E. Zigler and S. J. Styfco (Eds.), *Head Start and Beyond: A National Plan for Extended Childhood Intervention*, Yale University Press, New Haven, CT, pp. 73–95.
107. Commission on Chapter 1. (1992), *Making Schools Work for Children in Poverty*, U.S. Department of Education, Washington, D.C.
108. Melaville, A.I., with Blank, M.J.. (1991), *What it Takes: Structuring Interagency Partnerships to Connect Children and Families with Comprehensive Services*, Education and Human Services Consortium, Washington, D.C.
109. Fuerst, J.S., and Fuerst, D. (1993), "Chicago Experience with an Early Childhood Program: The Special Case of the Child Parent Center Program," *Urban Education*, 28, 69–96.
110. Hebbeler, K. (1985), "An Old and a New Question on the Effects of Early Education for Children from Low Income Families," *Educational Evaluation and Policy Analysis*, 7, 207–216.

chapter fourteen

Is There a Cognitive Elite in America?

NICHOLAS LEMANN

Even before *The Bell Curve* was published, the idea of what Richard Herrnstein and Charles Murray call "the cognitive elite"—a dominant and rich group of well-educated smart people—was taking form.[1] This elite has enemies across the political spectrum, but usually it is described as liberal and most criticism of it comes from conservatives. This chapter argues that the dimensions of the cognitive elite have been wildly exaggerated. After examining the original historical documents from which *The Bell Curve* makes its case for the rise of the cognitive elite, it finds that the case is much flimsier than Herrnstein and Murray let on. The chapter also points out that people with elite educational backgrounds control some but by no means all of the turf in the United States; they don't control the central American institution, business, contra *The Bell Curve*. An elite education neither guarantees money and power, nor provides the only route to them. Therefore the cognitive elite should be understood as a sociological cartoon with political uses, not a phenomenon to be accepted at face value.

The Evil Elite

Populists used to hate the rich, but now they hate the elite—a shift that has made possible the migration of populism from the Democratic to the Republican Party and the rise of the Republicans to control of the Congress. The conservative notion of the elite has some internal variations—for former Vice President Dan Quayle's "cultural elite" isn't exactly the same thing as House Speaker Newt Gingrich's "corrupt elite"—but the basic concept is descended from two books published on the other side of the Atlantic in the late 1950s by left-wing politicians: Milovan Djilas's *The New Class*[2] and Michael Young's *The Rise of the Meritocracy*.[3]

From Young comes the idea that not long after a society institutes mass educational sorting based on the results of IQ tests, a distinct high-IQ ruling class will begin to emerge. Because of the tendency of people in this class to marry fellow students at highly selective universities and to pass on their IQ-rich genes to their offspring, over time the meritocratic upper class will more and more resemble a hereditary aristocracy. If this class absorbs the left-wing views that prevail in the universities, then once it is in power it will resemble Djilas's arrogant Communist bureaucracy. The emotional charge of conservative attacks on organizations, such as the Corporation for Public Broadcasting and the arts endowments, and on more amorphous institutions, such as the news media, Hollywood, and the Washington establishment, comes from the idea that they are made up of superior-feeling, fancily educated, isolated New Class members who want to force their own cultural mores on ordinary folks toward whom they feel contempt.

Populism usually arises from a general discontent that precedes the specific identification of the villain. In the case of American populism today, the source of the discontent is the stagnation in incomes over the last generation and the loosening of traditional mores regarding sexual behavior, family structure, and authority. Men with no higher education, especially, have dramatically lost ground over the last two decades. People feel that things are out of control, socially and economically; the idea that it's the fault of the meritocratic elite is an easy sell.

Unlike most populist ideas, the notion that an elite group whose defining quality is intelligence runs the country is naturally attractive to intellectuals. It tends to be accepted unquestioningly by people who write books about the state of the society. The pampered liberal meritocrat has become the contemporary equivalent of the Organization Man of the 1950s, the symbol of the age for purposes of middlebrow handwringing. A typical capsule description of the type comes from an article about the "overclass" by Michael Lind:

> the closer you get to the centers of American politics and society, the more everyone begins to look the same.... the people who run big business bear a

> remarkable resemblance to the people who run big labor, who in turn might be mistaken for the people in charge of the media and the universities. They are the same people ... most of the members of the American elites went to one of a dozen Ivy League colleges or top state universities.... They talk the same. They walk the same.[4]

The unmatched pair of Christopher Lasch's *The Revolt of the Elites*[5] and William Henry's *In Defense of Elitism*[6] makes another good demonstration of how broad the consensus is around the idea that a meritocratic elite runs America, because while disagreeing about nearly everything they describe the group under discussion almost identically.

Henry positions himself as the lone defender of the elite, which conveniently obscures the familiarity of what his book's real purpose is, to make fun of political correctness. He writes as a lovable curmudgeon who is willing to publish what others dare say only in the club dining room, for example, "The dominant mood of contemporary American culture is the self-celebration of the peasantry." Despite his stance of blustery unconventionalism, though, he defines the elite, almost unthinkingly, as being made up of those people who were smart and studious enough to go to the best schools. "Intelligence varies genetically and ... intelligence by and large determines economic success," he says, precisely echoing Michael Young.

Lasch's elite is made up of the same people, "the new aristocracy of brains" or "the knowledge class," only he hates them. In Lasch's previous book, *The True and Only Heaven*, he presented himself as the champion of a provincial lower middle-class culture in which the highest good is community, not ambition.[7] Although Lasch grew up in Omaha, his own life story would not seem to indicate that he would have wound up identifying so powerfully with people on the outside. His father was a successful journalist and his father-in-law a famous academic (Henry Steele Commager). Lasch himself went to Harvard and became a tenured professor and prominent intellectual. He provided a clue about the source of his rebellion against his own subculture in this passage about his children in *The True and Only Heaven*:

> Our failure to educate them for success was the one way in which we did not fail them—our one unambiguous success. Not that this was deliberate either; it was only gradually that it became clear to me that none of my children, having been raised not for upward mobility but for honest work, could reasonably hope for any conventional kind of success.

That his own flesh and blood were being somehow cast out of the elite would have at the very least added some emotional *oomph* to whatever reservations Lasch was feeling about the elite to begin with.

Always in rebellion against mainstream liberal thought, Lasch by the time of *The True and Only Heaven* had settled into a comfortable self-identification with a "petty bourgeoisie" of "small proprietors, artisans, tradesmen, and farmers," which he admired for "its moral realism, its understanding that everything has its price, its respect for limits, its skepticism about progress." In *The Revolt of the Elites*, published posthumously, he hasn't changed this position at all, but he directs his main energies toward damning the people he dislikes rather than praising the ones he admires. Also, although all of Lasch's books have a patchwork quality, *The Revolt of the Elites* especially does: it is a slim collection of reworked reviews, and it careens almost wildly from subject to subject. This, along with the strength of his emotional animus against his subjects—he might have called the book *The Revolting Elites*—means that there is no place where he patiently lays out a theory of who these people are and where they came from.

It's easy, however, to extract a picture of them from the book. They have a "growing insularity"; they inhabit "an artificial world"; they spend too much time talking, exercising, and going to restaurants, they are excessively mobile"—at home only in transit, en route to a high-level conference, to the grand opening of a new franchise, to an international film festival, or to an undiscovered resort"; and they partake of a sob-sisterish brand of social-issue liberalism:

> a liberalism obsessed with the rights of women and minorities, with gay rights and unlimited abortion rights, with the allegedly epidemic spread of child abuse and sexual harrassment, with the need for regulations against offensive speech, and with curricular reforms designed to end the cultural hegemony of "dead white European males."

Lasch understands American life as a grand struggle between these people and the petty bourgeoisie: everything that helps the former group hurts the latter. For example, "feminism's appeal to the professional and managerial class" is simply that it "provides the indispensable basis of their prosperous, glamorous, gaudy, sometimes indecently lavish way of life." By implication, there's nothing in it for the unlavish lower middle class. The elite gets richer and ordinary people lose ground economically. The elite migrates and ordinary people remain in their deteriorating neighborhoods. The elite promotes cosmopolitanism and ordinary people feel their steady, traditional, provincial civic life slipping away.

Origin of Species

The idea that America is run by a sophisticated, brainy, culturally liberal group for whom education was the route to success is quite new. Most twentieth-century depictions of the elite treat it as being made up of conservative, vulgar, rich businessmen (and their heirs) whose salient characteristic is ruthless aggressiveness, not academic intelligence. Even as recent a novel about the elite as Tom Wolfe's *The Bonfire of the Vanities* hasn't got a trace of the idea of a meritocratic New Class sitting atop the society.[8] (Remember Larry Kramer, the smart liberal lawyer with the too-small apartment?) So where did it come from so suddenly?

The origin of the perception of an all-powerful meritocratic elite can't be precisely pinned down—it's part Gingrich, part Robert Reich's influential book about the importance of "symbolic analysts," *The Work of Nations*, part the ascension of our super-meritocratic President and First Lady, William Jefferson and Hillary Rodham Clinton, to the White House.[9] The most coherent theory of the rise of the meritocratic elite can be found not in Lasch (he's too scattered) or in Henry (he's too assumptive) but in Richard Herrnstein and Charles Murray's *The Bell Curve*, which, in addition to the material on racial differences in IQ that got most of the attention when it was published, contains a long, not much commented on section delineating the rise of the meritocratic elite. Because it's difficult to evaluate a phenomenon that hasn't been precisely defined, the Herrnstein–Murray argument is worth laying out as a prelude to discussing the work of Lasch and other attackers of the elite.

Until the midpoint of the twentieth century, Herrnstein and Murray say, people of high intelligence were scattered more or less randomly through the social structure. But during the 1950s, the highly intelligent embarked on an "invisible migration" to top universities and thence to positions of power, and so a distinct "cognitive elite" began to emerge. The reason this happened was that American society had become so complex that it could no longer operate without a high-IQ leadership: "A true cognitive elite requires a technological society." The new elite's rise represents a dramatic change: "Cognitive stratification as a central social process is something genuinely new under the sun." Today, with the cognitive elite controlling so much of the economic apparatus—and no wonder, because "intelligence is fundamentally related to productivity"—we are seeing take shape "an unprecedented coalition of the smart and the rich."

To people who have themselves been through the kind of cognitive-elite life trajectory that Herrnstein and Murray describe—straight As, high SAT scores, admission to a selective college and graduate school, and off to the races—their story has an almost overwhelming face validity. It fits Herrnstein and Murray personally (both came from obscure backgrounds to Harvard during its meritocratizing period). In the old days, the Ivy League colleges and the institutions into which they fed—upper academia, Wall Street, big law firms, research hospitals, the Foreign Service—all had a distinctly clubby, Episcopalian cast; they were filled with the kind of person Calvin Trillin (Yale '57) derisively refers to as Baxter Thatcher Hatcher. Now they're run by people who feel they deserve their high positions instead of having inherited them who, as they were constantly reminded on the way up, have earned the top jobs by being brighter and more hardworking than anyone else. If you're at Cravath, Swaine, and Moore, or McKinsey & Company, surrounded by people with similar personal histories to your own, the temptation to accept the cognitive elite theory at face value is well-nigh irresistible.

But the theory has two major shortcomings: the idea of all preceding elites' having been noncognitive is impossible to prove, and the degree to which the entire upper income stratum is now made up of people with high IQs and prestigious educations is greatly exaggerated. These aren't just dry conceptual problems; they have distinct political consequences.

The idea that intelligence used to be randomly distributed across the class structure is inherently problematic for Herrnstein and Murray. They contend that what IQ tests measure is general intelligence, or *g* in the impressive-sounding shorthand: the most important human talent, which would have been found at above-average levels even among "the chief ministers in Cheops's Egypt." But if you take this argument too far, you wind up saying that there was always a cognitive elite and that previous aristocracies were meritocracies engaging in "assortative mating." How could the cognitive elite have failed to emerge earlier if *g* is so crucial? Herrnstein and Murray's answer is that only a technological society can coax a true cognitive elite into existence—but society has been becoming steadily more technological for centuries, so shouldn't the cognitive elite have been taking form gradually over time, rather than overnight in the 1950s?

A more concrete difficulty in establishing the emergence of the cognitive elite is that reliable statistics on the distribution of intelligence before 1950 are hard to come by. The first mass-administered mental tests were the Army Alpha and Beta for military recruits during World War I, and, although Herrnstein and Murray don't say so, the results actually proved the opposite of their point, having shown that officers were markedly more intelligent than enlisted men. In *The Bell Curve* there are wonderfully persuasive charts that show those high in IQ beginning to "bunch up" in elite colleges during the 1950s, but the data underlying them are weak. For example, Herrnstein and Murray's figures on the (relatively low) average IQ scores at Ivy League schools in 1930, when pursued through the footnotes, turn out to come from the first administration of the Scholastic Aptitude Test to 8,040 students on June 23, 1926, and then the conversion of the scores to an IQ scale. But the takers were not actually students at Ivy League colleges; they were a self-selected group of high school students thinking of applying to Ivy League colleges. What Herrnstein and Murray report as the average IQ of Radcliffe College students is actually the average IQ of 233 high school girls who told the test administrators they'd like their scores sent to Radcliffe College.

The best statistical support for the idea of a scattered pre–World War II cognitive elite comes from the Pennsylvania Study, a major research effort conducted during the 1920s and 1930s by the Carnegie Foundation for the Advancement of Teaching. As ambitious as the Pennsylvania Study was, it still was not universally accepted in testing circles because its authors, William S. Learned and Ben D. Wood, were, all through the period when they were doing their research, active proselytizers for the use of educational admissions tests to separate the cognitive wheat and chaff. Carl Brigham, the author of the SAT, wrote to James Bryant Conant, the president of Harvard, in 1938:

> Dr. Learned and I long ago agreed to disagree on the results of this Pennsylvania Study.... He finally agreed that his conclusions could not be justified from his data, yet he insisted that the conclusions were good propaganda for the educational world and he has continued to preach them ... Dr. Learned feels that certain desirable social results may be obtained even though the methods of obtaining them are wrong.

An alternate explanation for the emergence of the cognitive elite would be that the key variable was not so much the personnel demands of a technological society as it was the development and promotion of mental tests. Probably before there were mental tests, the intelligent were overrepresented at the upper end of the social structure. The advent of testing rather than economic change is what made the fit tighter in certain professions. As soon as mental tests are used as a screen for entry into a field, that field will, by ironclad tautology, be dominated by people who score high on mental tests. With typical assurance, Herrnstein and Murray report that "only people from a fairly narrow range of cognitive ability can become lawyers," but to the extent that's true, it's because only people who get above average scores on the Law School Aptitude Test are allowed to become lawyers, not because you have to be intellectually gifted to draw up wills and deeds of sale.

The way to test this explanation would be to figure out whether high test scorers dominate fields that don't require high test scores as a condition of entry (such as business and the military) to the same extent that they dominate fields that do require high test scores (such as law and medicine). Herrnstein and Murray, of course, insist that the key issue is the trait the test scores measure, *g*, not the score's credential value. So, they say, high scores are predictive not just of admission to selective schools but of performance in a wide range of jobs. Their evidence here is susceptible to the same criticism Carl Brigham made of the Pennsylvania Study. Although they do a lot of drumrolling about the "scholarly consensus" to which "the top experts on testing and cognitive ability" now subscribe with "near unanimity," the people whose work Herrnstein and Murray usually cite in support of the omni-predictive power of IQ tests are known in the field as unusually zealous believers in *g*: Malcolm Ree and James Earles, military psychometricians in San Antonio; John Hunter of Michigan State University; Frank Schmidt of the University of Iowa; and Arthur Jensen of the University of California at

Berkeley. Herrnstein and Murray's section on "The Link Between Cognitive Ability and Job Performance," which is heavily based on studies by Ree and Earle, turns out to be mainly about the (unsurprising) link between scores on the Armed Forces Qualifying Test and grades in military training schools.

Herrnstein and Murray assert that business, especially the corporate world, fits their model: formerly run by people who weren't very bright, today controlled by the cognitive elite. Their evidence is pretty light. As at other key junctures in their argument, they depart from their statistics-wielding men-in-white-coats stance and simply present a bold assertion as true. (My favorite of these is, "Intermarriage among people in the top few percentiles of intelligence may be increasing far more rapidly than suspected"—no footnote.) About the business elite, they say, "Both common sense and circumstantial evidence suggest that people who rise to the upper echelons of large businesses tend to have high IQs and that this tendency has increased during the course of this century." To back this up, they cite a 1976 article in *Fortune* saying that 40% of the chief executive officers of the top 500 corporations had a background "in finance or law, fields of study that are highly screened for intelligence." Ergo the grip of the cognitive elite on the boardroom is tightening. A much more recent and comprehensive source is *Forbes* annual survey of the chief executives of the 800 largest corporations, which lists their higher education credentials. By now most of these are people who went to college in the 1950s and 1960s, that is, during what Herrnstein and Murray identify as the meritocratic era. The *Forbes* survey supports *The Bell Curve* to some extent but presents a much more nuanced picture. The college that provided the most CEOs is cognitively screened Harvard, with 23, and Cornell and Princeton are tied for second, with 18 each. It must be said, though, that only a couple of the CEOs from these schools have ethnic-sounding names, and that some of them leap out as belonging to the boss' son rather than the cognitive-elite category, such as John Hess, of Amerada Hess, Harvard '75. On the other hand, schools with much less cognitive screening are also important suppliers of CEOs, such as the University of Wisconsin (13 CEOs, 69% of applicants accepted), Purdue (10 CEOs, 90%), the University of Michigan (10 CEOs, 69%), and Vanderbilt (8 CEOs, 58%). Herrnstein and Murray say that because college graduates on average have high IQs (citing as proof an estimate from 23 years ago), and because most business executives have college degrees, "in the neighborhood of 70 or 80 per cent" of those in the ranks of management must have IQs of over 120. But of the 2,000 4-year colleges and universities in the United States, only 66 don't accept a majority of their applicants. The cognitive screening for a bachelor's degree is so much less severe than the financial screening as to make the proposition that college graduates are likely to come from comfortable backgrounds much easier to support than the proposition that they are likely to be highly intelligent.

Who Is Really Elite?

The sketchy estimates of the dimension of the new elite usually put it at about 20% of the population—the "fortunate fifth," as Reich called it. It's impossible to get to any number this big—about 50 million people—without going far outside the bounds of the elite as it's usually described. Just a cursory glance at the statistics demonstrates that the top fifth of the income distribution is made up mostly of people who didn't attend highly selective universities and aren't politically liberal. Conversely, most highly educated Americans are in an income bracket that doesn't fit the picture of a truly pampered class.

Every few years the Internal Revenue Service publishes a booklet called *Individual Income Tax Returns*; the most recent one is for 1992. The number of tax returns in 1992 that reported income of more than $100,000 was 3.8 million. It's hard to think of people making less than $100,000 as members of "the elite," but returns of $100,000 and above represent only 3% of the total. To get an elite of more than 10% of tax filers, you have to drop down to an income level of $60,000 a year. (Although the elite is routinely depicted as having exited the public school system, only 1% of American schoolchildren are enrolled in private nonsectarian schools costing more than $2,500 a year.) There are more than 23 million Americans with bachelor's degrees, and 12.7 million people, according to Herrnstein and Murray, in the top IQ decile. So the overwhelming majority of both groups must not belong to the over-$100,000 economic elite. In other words, there are a lot of highly educated and intelligent people who aren't especially affluent.

Also, as income rises above $100,000, the percentage of it derived from "salaries and wages" and "business or profession" steadily declines, and the percentage derived from long-term capital gains steadily increases, which hints at an upper end of the economic elite made up of inheritors, entrepreneurs, and financiers, not high-IQ professionals. In other words, perhaps there are also still a lot of rich people who don't have high IQs and prestigious degrees. Finally, a consistent finding of polls is that the more money you make, the more likely you are to vote Republican. So there are a lot of elite people who aren't liberal either. The one public-opinion finding that supports the idea of a liberal elite is that more people with graduate degrees vote Democratic than do people with only bachelor's degrees.

Clearly there is a group in the country that is affluent, highly educated, professional, and liberal, but the extent to which this group and "the elite" defined economically are the same has been wildly exaggerated by people who have spent their lives in the liberal/professional subgroup. Most books about the elite culture make it sound like a clonally enlarged version of the rarefied enclaves where the authors live: Manhattan, Cambridge, Palo Alto, and Washington, D.C. west of Rock Creek Park. It would be easier to make the case that the American elite is comprised mainly of Republican businessmen and their families, some of whom were put on the road to the top by fortuitous college and graduate school admissions and some not, and that places such as Buckhead, Irvine, North Dallas, and Rye are much more representative of the true American elite culture.

Even if you trimmed the meritocratic elite back from one fifth to, say, one hundredth of the population, there would still be a big problem with the cognitive elite theory, arising from the sociology of its members. Let's say that Herrnstein and Murray are right that the most selective colleges today efficiently "soak up" virtually all the American adolescents with the highest IQs and that these people are plainly the most able members of their age cohort. Ever since the meritocratic apparatus in the United States became mature, around 1970, the undergraduate atmosphere in Ivy League colleges has been one of intense anxiety about the perils of pursuing careers outside the professions of law, medicine, and MBA business (which generally leads to management consulting or investment banking, not, as Herrnstein and Murray imply, to corporate management). Because these fields are tightly screened on the basis of test scores and grades, they present the lowest risk career option to people who have very high test scores and grades. You enter them as part of an extremely limited pool of people, most of whom are virtually guaranteed high incomes.

In fields that aren't as tightly screened, from entrepreneurship to show business to corporate management, the odds are much longer for members of the cognitive elite than they are in the professions. So a stampede into the professions, driven by risk aversion, is a key cultural phenomenon for meritocratic Ivy League students, the subject of many hand-wringing commencement speeches (and, years later, of the students' middle-aged longueurs). It would even be possible to take the perverse position that people with high IQs may have actually become less powerful in recent decades because they are so firmly channelled into advisory roles in the professions. This is why the Ross Perots and David Geffens of the world don't find themselves competing with many anointed members of the cognitive elite during their rise into the economic elite.

In *In an Age of Experts*, Steven Brint sets out to create a more precise taxonomy of the professional classes than is available in the books that sweepingly assert the presence of a unitary new elite.[10] Although Brint attempts to prove his points via somewhat woolly efforts to quantify, such as enlisting a squadron of research assistants to code articles from 16 publications numerically on each of 100 variables, most of what he says has the simple ring of common sense. His main theory is that the professions are internally differentiated in ways that mirror wider divisions in the society. Professionals in "business services" (corporate lawyers, for example) and "applied science" (engineers) tend to be affluent and conservative. Professionals in "human services" (social workers, schoolteachers) are liberal but can't plausibly be presented as part of an economic elite. Professionals in "consultative" jobs are more liberal than those in "command-oriented" jobs. To the extent that there is a liberal strain common to all professionals, it would be found in their positions on social issues such as abortion, not economic issues. To the extent that there is a "new class," it would be found in western Europe, where there is a tradition of elite university graduates becoming career government officials, not in America.

It may be too much to hope that the prevailing intellectual view of ordinary American life will ever be accurate and nonprojective; certainly, the voluminous warnings about "conformity" during the 1950s now seem quaint. But the cartoon version of the all-powerful meritocratic elite that has recently emerged has the special disadvantage of serving to distract the country from responding to its real problems. Income inequality is growing. The economic condition of much of the workforce is stagnant or deteriorating. Most Americans don't feel themselves to be part of a political and economic system that they can believe in and that serves their needs. To lay all this at the feet of a high-IQ upper class that is supposedly feasting on the corpse of the social compact doesn't much help anybody.

References

1. Herrnstein, R.J. and Murray, C. (1994), *The Bell Curve: Intelligence and Class Structure in American Life*, The Free Press, New York.
2. Djilas, M. (1957), *The New Class: An Analysis of the Communist System*. Praeger, New York.
3. Young, M.D. (1958), *The Rise of the Meritocracy*, Thames and Hudson, London.
4. Lind, M. (1995), "To Have and Have Not: Notes on the Progress of the American Class War," *Harper's Magazine*, 290, 35–39.
5. Lasch, C. (1955), *The Revolt of the Elites and the Betrayal of Democracy*, W.W. Norton, New York.
6. Henry, W.A. III (1994), *In Defense of Elitism*, Doubleday, New York.
7. Lasch, C. (1991), *The True and Only Heaven: Progress and its Critics*, W.W. Norton, New York.
8. Wolfe, T. (1987), *The Bonfire of the Vanities*, Farrar, Straus, Giroux, New York.
9. Reich, Robert B. (1991), *The Work of Nations: Preparing Ourselves for 21st Century Capitalism*, Alfred A. Knopf, New York.
10. Brint, S. (1994), *In an Age of Experts: The Changing Role of Professionals in Politics and Public Life* Princeton University Press, Princeton, NJ.

chapter fifteen

Science, Public Policy, and *The Bell Curve*

DANIEL P. RESNICK AND STEPHEN E. FIENBERG

In the final section of *The Bell Curve* (pp. 387–552), Herrnstein and Murray deal explicitly with the implications of their argument for public policy.[1] In so doing, they move with different degrees of success from the realms of humanist argument about inequality, through statistical analysis of quantitative social data, to public policy advocacy. We cannot take the measure of so ambitious a work without stepping back and answering some pointed questions. Where does it break new ground? Where does it rehearse old arguments? What are its public policy implications?

In the effort to move between scientific views of the nature and distribution of intelligence and imperatives for public policy, Herrnstein and Murray follow the trail opened up almost 130 years ago by the publication of Sir Francis Galton's *Hereditary Genius*.[2] Galton argued that there was a scientific basis for believing that differences in individual capacity, established at birth, were responsible for different life outcomes. These differences were heritable, and because society rewarded talent in many fields and across generations, there was a necessary inequality in the position of English families and classes. Through lectures and publications, the organization of the first eugenics society, and support for applied university research, he promoted policies that were consistent with his scientific views. *The Bell Curve*, which briefly acknowledges the work of Galton (pp. 1–2, 26, 284) and offers a lengthy section on dysgenesis (pp. 341–68), recognizes its own pedigree. But, as we explained in Chapter 1, Britain did not by itself give us a convincing scientific way to represent intelligence; Americans did.

Leading American psychologists Robert Yerkes, Henry Goddard, and Lewis Terman, all of whom were open hereditarians at least into the late 1920s, viewed the scores on the IQ tests they created and used as good indicators of native intelligence. Intelligence, they believed, was fixed at birth, heritable, and unequally distributed among the races of humankind. IQ scores became for generations of social researchers not simply an indirect indicator for intelligence but an expression of native endowment.

Social researchers left it to the measurement psychologists to improve the quality of their instrument, but used the IQ test without too much cavil, trying to see how well it predicted careers, wealth, crime, education, family size, and the measured intelligence of offspring. Murray and Herrnstein follow this well-worn path in the analysis of IQ and its social correlates in the National Longitudinal Survey of Youth. It is an exploration of the nature–nurture relationship.

Nature–nurture arguments have long been understood to bear on both science and policy. Early advocates of eugenics in Britain, from Galton through Karl Pearson and R.A. Fisher, were viewed as social progressives and they stressed the positive eugenics approaches that they believed fit with their analyses of available data. But later, and especially in the United States, defenders of heritable "natural ability" have been drawn toward and have contributed to conservative public policy. They have generally identified with calls for limited immigration, low taxes, self-reliance, voluntary giving, and a weak state. The supporters of a determining role for nurture have, in turn, asked for expansion of public schooling, public housing, health services, school lunches, wage floors, and income equalization, all requiring a larger role for the state in society. By arguing for intelligence as a fixed, heritable quantity, responsible for life outcomes, Herrnstein and Murray have allied themselves with the "nature" school. Their hostility to government and government programs, their defense of the privileges of a wealthy educated class (albeit accompanied by considerable hand-wringing), and their appeal to self-reliance make the identification complete.

The critique of their contribution to social science must then proceed on two levels—a scientific examination (genetic, psychological, and statistical) and an investigation of policy implications. For Herrnstein and Murray in *The Bell Curve*, scientific analysis and policy advocacy are clearly inter-related. They write

> Our analysis provides few clear and decisive solutions to the major domestic issue of the day. But, at the same time, there is no major domestic issue for which the news we bring is irrelevant. (p. 387)

Here we recap what we have learned about Herrnstein and Murray's scientific investigations, and then we consider the policy implications of this work.

Science: The Genetics–Intelligence Link

Genetics is an important part of the science that bears on intelligence, even though it receives little attention in *The Bell Curve* (pp. 292–315). There may be close to a hundred thousand genes in the human system. A substantial fraction of these genes may contribute to measured intelligence, but less than a handful have been characterized. Moreover, the relationship between genes, gene systems, and external environment is interactive, defying and challenging the nature–nurture formulation. Where Herrnstein and Murray see fixed divisions of genes and environment affecting measured intelligence, based on the nature–nurture paradigm, many geneticists see no such division and therefore considerable malleability in the development of intellectual capacity. Herrnstein and Murray claim no special competence in genetics, although assumptions about genetic structures and development clearly influence their arguments.

The authors of the chapters dealing with this topic in our volume differ strongly with Herrnstein and Murray about the extent to which genetics and environment matter. Environment is an all-encompassing, yet nebulous, force affecting IQ. In their examination of environment, our contributors recognize specific nutritional, maternal, biological, social, cultural, and economic factors.

Daniels, Devlin, and Roeder, in their genetic analysis of classic familial IQ studies, find that Herrnstein and Murray's claim of heritability rates for intelligence, in the neighborhood of 60% to 80%, is a gross exaggeration. Their analyses differentiate between two kinds of heritability: broad sense and narrow sense. Only the latter matters for Herrnstein and Murray's arguments about IQ's impact on the evolution of American society. In light of their narrow-sense estimate, Daniels and colleagues conclude that Herrnstein and Murray's evolutionary arguments have no basis in science.

Daniels and colleagues also find evidence to support potent environmental effects on IQ, namely, through the maternal (womb) environment and through the shared-family environment. Wahlsten, focusing on how environment impacts IQ, argues for the malleability of intelligence and the need to explore further the interaction between nature and nurture. Finally, Singer and Ryff, focusing on health outcomes, explain how alternative biological and psychosocial models can be used to explain effects that others have traditionally associated with genetic inheritance.

Although the analyses and assessments we report on clearly allow for some genetic component associated with intelligence, other things matter, and considerably so.

Science: Intelligence and the Measurement of IQ

Psychological measurement as a field has developed and refined the IQ test. What does it show to those who use it as a research tool? *The Bell Curve* sees intelligence as a fixed variable, established at birth and relatively unchanged through life. IQ tests are used as an indicator of this native intelligence. Herrnstein and Murray have treated general intelligence, *g*, as a given, rather than discussing its basis in factor analysis. Some, such as Stephen Gould in the new edition of *The Mismeasure of Man*, have argued that Herrnstein and Murray's reliance on factor analysis is the Achilles heel of their entire effort and have dismissed it accordingly.[3] For us, their approach raises more questions than it answers and calls for a deep internalist consideration of how *g* was identified and how its presence is confirmed.

John Carroll reviews our knowledge of the psychometric domain, the history of the search for generalized intelligence, and the relationships of more specialized capacities to a general factor. He finds a large consensus within the psychometric community about the existence of generalized intelligence and the ability of IQ tests to measure it. These findings, he argues, support the existence of a general intelligence factor, along with evidence of more specialized capacities.

Earl Hunt remains far more tentative about the existence of *g*, and he presents an overview of how intelligence is viewed in modern cognitive theory, pointing to the evidence for and arguments about multiple factors for intelligence and considering their potential implications.

In our reading of these contributions, we see that these psychologists (1) make no claim for heritability of IQ, (2) have differing judgments about the unidimensional features of intelligence, and (3) reach at most limited conclusions about the implications for success in life, or for social policy, of *g* or its multidimensional alternatives.

Science: Analyzing the Outcomes Data

The Bell Curve has been a "Pied Piper" for some segments of the social sciences. It deserves kudos for its way with words, drawing in readers who would otherwise be intimidated by numbers, whether or not they agree with Herrnstein and Murray's conclusions. This success, however, has its price. We have looked a lot harder at the numbers than the typical reader is invited to do, and we are disappointed. Even though a deep reading requires familiarity with the quantitative tools and the models of the social sciences, no readers are turned away. Lay readers have received special attention in this work, although we fear that these readers have been led down a primrose path. The book depends on a rhetorical coup—opening up a technical analysis to nonscientific readers with a narrative line that requires no attention to technical detail. How do they do it?

Murray and Herrnstein write that while there is "considerable technical detail" in their work, most of that technical apparatus is placed in footnotes and appendixes. Their prefatory note indicates that this detail is not essential, and they invite readers to follow their argument by paying attention to the flow of the narrative (p. xix): "We have designed *The Bell Curve* to be read at several levels."

> At the simplest level, it is only about thirty pages long. Each chapter except the Introduction and the final two chapters opens with a precis of the main findings and conclusions minus any evidence for them, written in an informal style free of technical terms. You can get a good idea of what we have to say by reading just those introductory essays.
>
> The next level is the main text. It is accessible to anyone who enjoys reading, for example, the science section of the news magazines. No special knowledge is assumed; everything you need to know to follow all of the discussion is contained within the book.

The invitation from Herrnstein and Murray to read without attention to technical details and without reference to any assumed special knowledge opened their work to a broad audience of nonscientists in the first round. Their response was often polemical, as the preface to *Intelligence, Genes, and Success* indicates, but rarely grounded in the technical details. A healthy pause, however, has permitted a second round of readers to get into the "technical terms," the footnotes, the appendixes, the relevant databases, and to place this work in the history of psychology, genetics, and social research. Most importantly, as required by "good science," it has pushed others to see if they can "replicate" *The Bell Curve*'s findings.

Such replication has centered on the authors' use of data extracted from a widely used social science database, the National Longitudinal Survey of Youth (NLSY). Murray and Herrnstein draw upon data from the NLSY in Chapters 5 through 13.* Analyses of that database, we have found, can support interpretations far different from the ones made by Murray and Herrnstein. What is the relationship of IQ to crime, educational attainment, employment, and income? What do the regression analyses show? Here is what the contributors to our volume, *Intelligence and Success*, have found.

John Cawley et al. have reanalyzed the NLSY data related to earnings, and they found *The Bell Curve's* argument that *g* predicts differences in wages to be overstated. Other components of ability are also associated with wage levels. But even when lumped with education and experience, ability of all kinds doesn't predict wage levels very well. And the difference in earnings of

---

**The Bell Curve*'s index underplays its importance, offering only nine citations. See, however, pp. 36, 49, 56, 98, 113, 118–120, 124, 131–133, 146-147, 152–153, 165, 174, 176, 182, 184, 187, 191, 194, 197, 203, 205, 213, 214, 218, 220, 226, 230, 235, 245–249, 253–254, 263–265, 273, 275, 277–280, 286, 289,290, 318–340, 347, 350–352, 354–355, 360, 367–368, 369–370, 375, 378–379, 386, 569–577, 590, 593.

demographic groups with the same cognitive levels, but of different race and gender, is considerable. Those differences undermine fundamentally the claim that the job market is meritocratic.

The Bell Curve's claims for the existence of a meritocratic wage structure and against affirmative action are also attacked by Alexander Cavallo et al., who find that race differences in earnings have been masked by Herrnstein and Murray's statistical procedures. Controlling for ability, Cavallo et al. find a sizeable gap in the earnings of black and white males. Education and experience differences may also be affected by racism, in ways that are less easily established. The labor market, in this analysis of NLSY data, is far from meritocratic.

For Winship and Korenman, *The Bell Curve* fails to deal adequately with the relationship between education, family background, and IQ. In their analysis of the NLSY data, every year of school beyond 8th grade is likely to increase IQ by 2 to 4 points. *The Bell Curve*'s statistical procedure is faulted for its treatment of the effects of both family background and years of education. IQ is not a totally freestanding variable. Instead, Winship and Korenman show IQ to be affected by both family background and education. Educational investments would thus appear to be a promising way to impact earnings.

The Bell Curve has revived a favored eugenics argument that IQ is a strong predictor of criminal behavior, among other significant life outcomes. Manolakes challenges Murray and Herrnstein's view that, controlling for socioeconomic background, people of low IQ are more likely to commit criminal actions. She faults their explanatory model for the inadequate specification of socioeconomic background variables and the unwillingness to allow for interactions among variables. Interactions between race and education are shown to be important. Contrary to what *The Bell Curve* argues, IQ and level of parental education have very different effects on criminality for black and white respondents in the NLSY survey. There appear to be no simple cognitive causes for crime.

Clark Glymour has argued, however, that problems of causal analysis in *The Bell Curve* run deeper than even these critiques suggest. The main statistical technique of *The Bell Curve* is regression, linear and logistic. Both multiple linear regression and multiple logistic regression are powerful tools that can yield causal information under very specified conditions, but statistical textbooks routinely caution against using regression as a method for inferring either the existence or strength of causes from nonexperimental data. In addition to arguing about the circularity of the causal interpretation of the factor analyses of intelligence, Glymour explains that this kind of regression modeling cannot be used to offer causal explanations under nonexperimental conditions with the types of variables that appear in this NLSY data. For example, he explains the problems in the scientific model created by variables that could have an effect on both IQ and whatever social outcome is being examined and are not present in the regression. He concludes: "A great deal of social science, including *The Bell Curve*, is in no position to make a case for the causal assumptions necessary to use regression reliably in causal inference."

Based on these and other assessments of Herrnstein and Murray's core regression analyses of the NLSY data, we are unpersuaded about the "dominant" role of IQ in determining life outcomes and conclude that propositions about the distribution of IQ have no demonstrated relevance for public policy. These analyses do not justify privileging one set of public measures over another.

Genetics, Race, and IQ

Only one chapter in *The Bell Curve* deals exclusively and explicitly with race, but as the book's preface acknowledges, "This book is about differences in intellectual capacity among people and groups and what these differences mean for America's future." Differences "among people," individual differences, quickly become differences of class and caste. Differences among "groups," in turn, get resolved to differences between races. The decision of Herrnstein and Murray to move from arguments about individual differences, where the data are abundant, to theories of race difference, where data are sparser and more contested in meaning, is puzzling.

The theory of cognitive castes, for example, does not require any racial arguments. By definition, cognitive castes emerge from the spread in the distribution of IQ among individuals, as well as the presumed genetic basis for this distribution and Herrnstein and Murray's assumption that assortative mating for IQ is an increasingly important feature of American society. Moreover, in a society presumed to be meritocratic, those with more capacity will gain greater income, social status, and professional position, thereby creating environmental differences among castes that reinforce the "genetic pressure" forming the castes.

But Herrnstein and Murray are hereditarians, who appear to have a theory about race as well as class. Like Galton, they hold that success in life is a measure of native endowment, and seem to believe that the denial of success to a population, over the long run, is evidence of the group's poorer native endowment. They even extend their hereditarian theories to include the cultural transmission of behaviors, values, or environments that affect IQ. "The correlation between parents and children is just that: a statistical tendency for these things to be passed down, despite society's attempt to change them, without any necessary genetic component" (p. 314). Cultural transmission of traits that influence IQ is certainly possible. But they present no evidence for cultural stability and persistence, which is what their argument requires, and we find such stability unlikely given the rapid rate that culture can and does evolve.

There is no diasagreement that median IQ scores show a 15-point gap between blacks and whites in America, but there is a lot of disagreement about what this means. Here are some of the questions that have been raised. If IQ scores correlate only in the 0.2 to 0.4 range with occupational success and income, how much importance can we assign to the cultural transmission of traits affecting the IQ of progeny? And how will this culture change in the near and distant future? What effect, in turn, will changes in culture have on the correlation of IQ with success in life? Moreover, how can we discount the influence of underlying environmental factors on both IQ scores and life outcomes?

Herrnstein and Murray acknowledge that there are many differences in the environments in which black and white children in America are likely to grow up and that these can affect the ability to learn and achieve. Nutrition, schooling, parents' educational level, wealth, and medical care are among the variables that can create very different environments for black and white populations. To make their argument effectively, therefore, the authors of *The Bell Curve* have to statistically control for these variables. We are not satisfied that they have done so adequately.

The analyses of the NLSY data reported in our volume indicate that Herrnstein and Murray have not controlled adequately for residence, parental income, and gender, and that they have been inattentive to their interactive effects. Other variables, such as nutrition, quality of education, and peer culture, are, for various reasons, excluded from the analysis. As a result, we and our contributors cannot accept their argument that success in America is largely blind to the effects of environment and individual agency.

Herrnstein and Murray note that there is some indication that the gap between black and white scores is closing, but they see little reason to expect this trend to continue. They do not claim this difference is entirely genetic, although many reviewers have accused them of doing so. Certainly the strong case that they make for genetic explanations throughout the book, and their tendency to discount contrary environmental explanations, might lead one to believe that this is what they think. It is surprising to us, however, that they might even consider a genetic explanation for racial difference in IQ, especially given the dearth of theory supporting this explanation and the wealth of evidence suggesting the difference could, at least in large part, be due to differences in environment.

Public Policy

Despite their access to a relatively new database, the themes that Herrnstein and Murray explore are not new, their quantification does not proceed at the needed level of sophistication, and their policy recommendations are more the product of a moral persuasion and libertarian ideas than they are of solid evidence. Here are four of their findings, each tied to a policy recommendation:

1. Everyone cannot learn. Efforts to improve the lot of average and below-average students have taken place through the systematic neglect of the cognitively gifted minority. Put public money into programs for the gifted.
2. Affirmative action in the workplace has been tried and found wanting because it produces large racial discrepancies in job performance and incurs costs in efficiency and fairness. End attention to equal opportunity in employment; and end affirmative action. Market forces will continue to reduce racism.
3. More inequality and social tension are inevitable. Prepare to invest more heavily in internal security: expect to spend more money for prisons and for the protection of wealthier neighborhoods.
4. Government has failed the citizen. Virtues such as respect and self-reliance can be cultivated primarily in the private sphere. Reduce the role of government.

Zigler and Styfco take up Herrnstein and Murray's charge that little can be done to raise the intelligence of those in the poorest areas of our cities and countryside, an argument made in 1969 by Jensen in the *Harvard Educational Review*.[4] They do not disagree that the popular Head Start programs, made possible by federal legislation in 1965, and that now enroll more than half the eligible preschool population, are failing to raise IQ scores. To the contrary, they point out that the program was never designed to do so. But they note the benefits in other areas, such as school completion, and they make a strong civic case for this investment.

Affirmative action in the American workplace and in higher education is currently under assault—in the media, in Congress, and in the courts. It also comes in for rough treatment in the final chapters of *The Bell Curve*. Commentators, politicians, scholars, and the public more generally object to affirmative action programs on a variety of grounds, ranging from philosophical to those reflecting a racial backlash. Herrnstein and Murray's argument is somewhat philosophical, but largely "pragmatic" in tone, following from the syllogism that we explained at the outset. After all, because IQ is largely genetically determined and resistant to change, and because IQ causes the inequities in educational and social outcomes we observe in American society, laws aimed at ameliorating the inequity must fail.

Herrnstein and Murray single out for attack the federal government's requirements for affirmative action in the workplace (e.g., the 80% rule for hiring and promotion) and the landmark 1971 Supreme Court decision in *Griggs v. Duke Power Co.*, which eliminated the use of educational or test requirements for applicants when the employer could not show that they were clearly job related. Their analysis of the legal and administrative history of the past 30 years is, at best, selective, and it omits the clear, convincing cases that reinforced the Court's strong stance in discrimination matters, such as *International Brotherhood of Teamsters v. U.S.* (see the related review in Ref. 5). Herrnstein and Murray believe that because IQ and related general ability tests essentially capture the essence of job requirements, the Court's actions and the attendant rules and regulations simply lead, at best, to gross inefficiencies without changing the realities of what is needed in the workplace. Where the rules and regulations "demand" changes in representation of minorities in the workforce, Herrnstein and Murray argue, they often produce perverse outcomes. They conclude (p. 508):

> Whatever their precise amounts, the benefits to productivity and to fairness of ending the antidiscrimination laws are substantial. But our largest reason for wanting to scrap job discrimination law is our belief that the system of affirmative action, in education and the workplace alike, is leaking a poison into the American soul.

In fact, we agree with Herrnstein and Murray about many of the problems in the federal government's attack on racial discrimination in the workplace, including the naive and potentially misleading uses of statistics, particularly regression analyses (e.g., see Ref. 5 and the discussion in Ref. 6). We too are against reverse discrimination. But we disagree profoundly with the notion that all of the law and practice that has moved America away from its discriminatory past is perverse and inequitable. Nor do we think that simply allowing employers to once again use tests to discriminate in hiring will serve the nation or its people well.

Conclusions

How should Americans respond to the inequality in their midst? Is it an unfortunate byproduct of the otherwise laudable emergence of the cognitive elite? Readers of the *The Bell Curve* are asked to celebrate the growth of a more competent America in which those with advanced degrees go to the best schools, intermarry, earn well, live together in enclaves, send their children to good preparatory schools, and manage the nation. We are asked to recognize the rightness and necessity of this kind of development because a complex America now requires the services of those with high IQs.

Nicholas Lemann points out in his contribution to our volume that populist America doesn't like elites, and that *The Bell Curve's* message requires some sugarcoating. The sweet covering given to this argument by Herrnstein and Murray is that the growth of privileged elites, supported by the unacknowledged tax, regulatory, and appropriation policies of government, provides competent staffing of our necessary public and private services; that efforts to broaden educational and social opportunity by public investment are doomed to failure, because of natural limits created by the distribution of IQ; and that great inequality is socially tolerable if the police offer adequate protection to those in wealthy segregated enclaves.

Why are police powers the preferred and favored public service of government? *The Bell Curve*'s vision of the future is galling in the paralysis it suggests, even if it is no more than a thought experiment. "In short, by *custodial state*," they write, "we have in mind a high-tech and more lavish version of the Indian reservation for some substantial minority of the nation's population, while the rest of America tries to go about its business"(p. 526). While professing displeasure at growing inequality, Herrnstein and Murray see no alternative.

A good part of the problem may be *The Bell Curve*'s nostalgia for Jeffersonian America, small towns, and self-reliance. A few statistics may provide an antidote to its conflation of past and present. Two-hundred years ago, our population was 1/100th its present size, our territory no more than ten percent of what it is now, and there was only one city with more than 50,000 people.* Today, a very densely urbanized and diverse America meets its public needs through the services of large numbers of employees in federal, state, county, and municipal government. They account for about fifteen percent of the civilian American workforce.†

Government is legitimate, large, and powerful; public opinion, as we have discovered from experience, will not tolerate shutdowns of public services. Solutions to our national needs require the participation of government, and it is utopian to think otherwise.

What kind of society would we choose to sustain if we recognized the public resources on which we can draw? Large numbers of Americans have asked for labeled and inspected foods, clean water and air, old-age assistance, impartial and competent courts, regulated utilities, public transportation, educational opportunity, and a host of other protections, privileges, and amenities that only government can help to provide. This doesn't sound like a call to dismantle government and let unregulated markets run our lives.

---

*In 1800, Boston had about 25,000 inhabitants, New York about 80,000, Philadelphia about 40,000, and Baltimore about 27,000. These were the largest agglomerations. See Donald B. Dodd (1993), *Historical Statistics of the States of the United States: Two Centuries of the Census, 1790–1990*. Greenwood Press, Westport, CT, pp. 443–468.

†For government employment, see U.S. Bureau of the Census (1996), Public Employment, Series GE, No. 1, cited in U.S. Bureau of the Census, *Statistical Abstract of the United States: 1996*, 116th edit., Washington, D.C., Table 502, p. 319. The information is also available on an Internet site. See http://www.census.gov/ftp/pub/govs/www/index.html. Total government employment at the end of 1993—civilian and military, federal, state, county, and local—was estimated at 18.8 million. The largest single categories were education, health and hospitals, defense, postal service, and police. One out of every seven members of the civilian labor force of about 130 million was a government employee.

And there is no agreement that America's social inequality has brought us the most competent leadership in the world. Even in areas where our markets are very competetive and relatively free of restraint, such as management training, insiders with some baselines for comparison do not see us as a cut above our competitors. A recent McKinsey Institute report, drawing on international experience in consulting for business, makes the following judgment: "There are no general inherent differences among management abilities or natural talents across the countries we are looking at.... Managers in the U.S. are not inherently 'better' or more talented than those in Western Europe or Japan, and vice versa."[7]

Although Herrnstein and Murray deplore the absence of moral virtue in society, they miss an open opportunity to encourage its growth. Michael Sandel[8] reminds us of less antagonistic moments in our past when investment in public parks, playgrounds, schools, and civic movements could become the subject of public debate, and government was more actively involved in the moral and educational development of citizens. The teaching of civic virtue cannot succeed when it is limited to public school classes in history and civics.

Herrnstein and Murray share the consensus among social researchers about growing inequality in America, but they stand apart in their explanation of why this is happening and what can be done about it. They also stand apart from most of the research community in the tight and directional ties they posit between the smarts that people are born with and the kind of lives that they lead. And they stand apart from most Americans in the depth of their hostility to government, although many Americans are also angry about the cost of programs that have failed. Because of their principled opposition to government, Herrnstein and Murray have denied Americans the support of public institutions in the struggle against rising inequality. Without government, it will be a very unequal struggle.

References

1. Herrnstein, R.J., and Murray, C. (1994), *The Bell Curve: Intelligence and Class Structure in American Life*, The Free Press, New York.
2. Galton, Francis (1869), *Hereditary Genius: An Inquiry into Its Laws and Consequences*, MacMillan, London.
3. Gould, S.J. (1996), *The Mismeasure of Man* (Revised and Expanded Edition), Norton, New York.
4. Jensen, A. (1969), "How Much Can We Boost IQ and Scholastic Achievement?" in *Environment, Heredity and Intelligence*, Reprint series No. 2, *Harvard Educational Review*, 1–123.
5. Fienberg, S.E. (1989), "The Evolving Role of Statistical Assessments as Evidence in the Courts," Report of the Panel on Statistical Assessments as Evidence in the Courts, National Academy Press, Washington, D.C.

6. Gray, M.W. (1993), "Statistical Arguments in the Courts Concerning Faculty Salaries" (with discussion), *Statistical Science*, 8, 144–179.
7. McKinsey Global Institute. (1994), *Employment Performance*, pp. 14–15, cited in Derek Bok (1996), *The State of the Nation. Government and the Quest for a Better Society*. Harvard University Press, Cambridge, MA, p. 424.
8. Sandel, Michael J. (1996), *Democracy's Discontent: America in Search of a Public Philosophy*, Belknap Press of Harvard University, Cambridge, MA.

CONTRIBUTOR BIOGRAPHIES

Terry Belke is Assistant Professor in the Department of Psychology, Mount Allison University. He did his graduate training in the experimental analysis of behavior under Richard Herrnstein and Gene Heyman at Harvard University. His research interests cover a broad range of topics, including activity anorexia, the role of conditioned reinforcement in gambling situations, and the pharmacological basis of the reinforcing properties of running.

John B. Carroll is William R. Kenan, Jr. Professor of Psychology Emeritus, University of North Carolina at Chapel Hill, where he served on the faculty from 1974 on, retiring in 1982. He was formerly a Senior Research Psychologist at Educational Testing Service in Princeton, NJ, and the Roy E. Larson Professor of Education at Harvard University. He is a founding member of the National Academy of Education, and a former president of the Psychometric Society and of the Division of Evaluation, Measurement and Statistics of the American Psychological Association. He has specialized in psychometrics and the psychology of language, concerning himself with measurements of cognitive abilities and achievements, particularly with those pertinent to the learning and use of the native and foreign languages, and with statistical models for identifying abilities and factor-analytic studies of the structure of human abilities.

Alexander Cavallo is a Ph.D. student in economics at the University of Chicago. He is currently a Research Fellow with the University of Chicago—Northwestern University Joint Center for Poverty Research. He is investigating the impact of race and gender on earnings and education, and the effects of changes in welfare programs on economic outcomes.

John Cawley is a Ph.D. student in economics at the University of Chicago. He is a National Institute on Aging Predoctoral Fellow, a Graduate Fellow of the University of Chicago–Northwestern University Joint Center for Poverty Research, and an In-Absentia Fellow of the American Institute for Economic Research. He graduated magna cum laude with a bachelor's degree in economics from Harvard University. His current research concerns the economics of aging, health, and longevity.

Karen Conneely is a Ph.D. student in economics and a National Science Foundation Fellow at Princeton University. Her current research interests focus on labor market decisions made early in the career, especially educational attainment and choice of occupation.

Michael Daniels is Visiting Assistant Professor in the Department of Statistics, Carnegie Mellon University. His main area of research is in the area of Bayesian biostatistics. Specifically, his research has focused thus far on various issues in hierarchical modeling with applications to both health services research and meta-analysis for evaluating surrogate markers in clinical trials.

Bernie Devlin is Program Director for the Computational Genetics Program in the Western Psychiatric Institute and Clinic and an Assistant Professor of Psychiatry in the Department of Psychiatry, University of Pittsburgh School of Medicine. He serves on the DNA Advisory Board to the Federal Bureau of Investigation Director, regarding standards for forensic DNA testing laboratories, and the National Forensic DNA Review Panel for the National Institute of Justice, regarding the performance of proficiency tests. He specializes in the development and application of methods to solve statistical problems arising in the field of genetics, including biometrical and population genetics, evolution, genomics, DNA forensic science, and genetic epidemiology.

Hazem El-Abbadi is a graduate student in the Harris School of Public Policy at the University of Chicago. He was formerly a graduate student in economics at the University of Chicago. His research interests include the effect of special interest groups on international trade policy, and the role of race in the allocation of public and government goods and services.

Stephen E. Fienberg is Maurice Falk University Professor of Statistics and Social Science, Carnegie Mellon University. He was formerly Dean of the College of Humanities and Social Sciences at Carnegie Mellon and Vice President for Academic Affairs at York University in Toronto. He is a fellow of the American Association for the Advancement of Science, the American Statistical Association, and the Institute of Mathematical Statistics, and currently serves as the President of the International Society for Bayesian Analysis. He has also served as Chair of the National Research Council Committee on National Statistics, editor of the *Journal of the American Statistical Association*, and was a founding co-editor of *Chance*. He has published extensively on statistical methods for the analysis of categorical data, sample surveys and randomized experiments, the use of statistics in public policy and the law, and the role of statistical methods in census taking.

Clark Glymour is Alumni University Professor of Philosophy, Carnegie Mellon University, and Valtz Family Professor of Philosophy, the University of California, San Diego. He is the author of *Theory and Evidence* (Princeton, 1980), *Thinking Things Through* (MIT, 1993), and a co-author of *Discovering Causal Structure: Artificial Intelligence, Philosophy of Science and Statistical Modeling* (Academic Press, 1987), *Philosophy of Science* (Prentice Hall, 1993), and *Causation, Prediction and Search* (Springer, 1993). This essay was completed while he was a Fellow of the Center for Advanced Study in the Behavioral Sciences, supported by the Andrew Mellon Foundation.

James J. Heckman is Henry Schultz Distinguished Service Professor, Department of Economics, University of Chicago. He has also served on the faculties of Columbia and Yale Universities. His research interests include econometric methods, statistical models for the longitudinal analysis of labor market data, and the analysis of data from surveys and social experiments. He is a recipient of the John Bates Clark Medal of the American Economic Association, a fellow of the Econometric Society and an elected member of the National Academy of Sciences and the American Academy of Arts and Sciences.

Randal Heeb is a Ph.D. student in economics at the University of Chicago. He holds a M.P.A. from the Kennedy School of Government at Harvard University and serves as the president of Policy Planning Associates, a consulting firm specializing in regulatory economics. His current research interests include the economics of regulation, antitrust policy, and government policy interventions.

Earl Hunt is Professor of Psychology and Adjunct Professor of Computer Science at the University of Washington. His scientific interests are individual differences in cognition and in the use of computers to augment human intelligence. His most recent activities have been the study of levels of intelligence in the workforce and in ways to raise levels of understanding of science and mathematics by high school and junior high school students. Hunt is a Fellow of the American Association for the Advancement of Science and the Society of Experimental Psychologists. His book on intelligence and the workforce, *Will We Be Smart Enough?* won the William James Prize of the American Psychological Association in 1996.

Sanders Korenman is Associate Professor in the School of Public Affairs, Baruch College, CUNY, and member of the doctoral faculty in Economics at the CUNY Graduate Center. He is a Research Associate of the National Bureau of Economic Research. He was formerly an Associate Professor at the University of Minnesota and an Assistant Professor at Princeton University. He has written about poverty and child health and development, the social and health consequences of teenage childbearing, and social policy.

Nicholas Lemann is the national correspondent of *The Atlantic Monthly*. His work also appears regularly in *The New Republic*, *The New York Review of Books*, *The New York Times Magazine*, and other publications. His most recent book is *The Promised Land* (1991), a history of African-American migration from the rural South to the urban North. His new book on meritocracy in the United States will be published by Alfred A. Knopf in 1998.

Lucinda A. Manolakes is a Ph.D. student in the Sociology Department, State University of New York at Stony Brook. She has recently earned her masters degree and is currently working toward a doctoral degree. Her interest in pedagogical techniques led her to copresent a paper at an Eastern Sociological Society conference regarding innovative methods for teaching difficult topics such as rape and domestic violence in the classroom setting. Her research interests cover a broad range of topics, including race relations, the mass media, and volunteer organizations.

Daniel P. Resnick is Professor of History, Carnegie Mellon University. His research deals with the relationship of historical thinking and experience to public policy development. Educational policy is his special area, with attention to comparative perspectives on assessment, achievement and standards. Over the past 15 years, he has written about the context for policy reform, the history of competency testing, new and old forms of assessment, historical efforts to advance mass literacy, programs for the gifted, and the benchmarking of student achievement in different countries. He has served as a consultant to public agencies and has worked to enlarge the place of historical thinking in public policy analysis.

Kathryn Roeder is Associate Professor of Statistics, Carnegie Mellon University. She was formerly an associate professor in the Department of Statistics, Yale University. She is the recipient of the National Science Foundation Young Investigator prize, a fellow of the American Statistical Association and the Institute of Mathematical Statistics, and serves as an associate editor of the *Journal of the American Statistical Association* and *Biometrics*. Her research has focused on developing statistical methodologies for latent class models, errors-in-variables, semiparametric inference, and graphical diagnostic techniques. She also has a strong research interest in applied problems, including statistical genetics, DNA forensic inference, and criminology.

Carol D. Ryff is Professor of Psychology and Director of the Institute on Aging, University of Wisconsin-Madison. She is a fellow of the American Psychological Association and the Gerontological Society of America. Since 1988 she has been a member of the MacArthur Foundation Research Network on Mid-Life Development. Her research centers on the study of psychological well-being, using a theory-driven, empirically based approach to assessment multiple dimensions of positive psychological functioning. Her descriptive studies have used survey samples to document sociodemographic correlates of well-being (i.e., how positive mental health varies by age, gender, social class, ethnic/minority status). Explanatory studies have focused on individuals' life experiences and their interpretations of them to account for variations in well-being. Longitudinal investigations of mid-life development and old age explore processes of resilience and vulnerability via the cumulation of adversity and advantage.

Burton Singer is Professor of Demography and Public Affairs in the Office of Population Research, Princeton University. He was formerly a Professor at Yale University and Columbia University. His research interests include the epidemiology of tropical diseases, the demography and economics of aging, and the interrelationship between social stratification, psychosocial experiences, and their physiological sequelae. He served as Chair of the Steering Committee for Socio-economic Research in the World Health Organization Tropical Disease Research Program and Chair of the National Research Council Committee on National Statistics. He is a Fellow of the American Statistical Association and the American Association for the Advancement of Science, and an elected member of the National Academy of Sciences.

Sally J. Styfco is Associate Director of the Head Start Unit at the Bush Center in Child Development and Social Policy at Yale University and Research Associate in the Yale Child Study Center and Psychology Department. She is a policy analyst whose work has influenced policy directions for the Family and Medical Leave Act, Title I/Chapter 1 of the Elementary and Secondary Education Act, Early Head Start, and expansion and quality improvements in the national Head Start program. She writes extensively about early intervention services for children and families in poverty, comprehensive educational needs of at-risk youth, delinquency prevention, and the importance of quality childcare to the healthy development of all children and to the effectiveness of welfare reform.

Edward Vytlacil is a Ph.D. student in economics and a C.V. Starr Fellow at the University of Chicago. His research interests include dynamic discrete choice econometrics and empirical labor economics.

Douglas Wahlsten is Professor of Psychology and Adjunct Professor of Neuroscience at the University of Alberta. He formerly was Professor of Psychology at the University of Waterloo. His laboratory research focusses on hereditary defects of brain structure in mice, especially absence of the corpus callosum that normally connects the cerebral hemispheres. He has also written numerous articles on the theoretical and methodological foundations of behavioral genetics, pointing out the reductionist fallacies in additive models of development and the low power of common analytical methods to detect heredity–environment interactions. He has been invited to prepare the next chapter on genetics of brain and behavior for the *Annual Review of Psychology* and is writing a comprehensive treatise on neurobehavioral genetics.

Christopher Winship is Professor of Sociology at Harvard University. He was formerly Professor of Sociology, Statistics, and Economics at Northwestern University. He is editor of Sociological Methodology and Research. He is currently chair of the American Sociological Assocation's Methodology Section. His research focuses on explanations for changes in the social and economic status of African Americans in this century, and structural models for qualitative data.

Edward Zigler is the Sterling Professor of Psychology, Director of the Bush Center in Child Development and Social Policy, and head of the Psychology Section of the Child Study Center at Yale University. He was one of the planners of Head Start and was the federal official responsible for the program from 1970 to 1972, when he served as the first director of the U.S. Office of Child Development (now the Administration on Children, Youth and Families) and Chief of the U.S. Children's Bureau. Recently he served on the Advisory Committee on Head Start Quality and Expansion and on the planning committee for the Early Head Start program for families and children ages zero to three. He is also founder of the School of the 21st Century, which adds childcare and family service components to public schools, and cofounder of the CoZi model, which combines the 21st Century School model with James Comer's School Development Program. He is the author, coauthor, and editor of hundreds of scholarly publications and has conducted extensive investigations on topics related to child development, early intervention, psychopathology, and mental retardation.

BIBLIOGRAPHY

Herrnstein and Murray's *The Bell Curve* set off an avalanche of response, much of it unfavorable, including book reviews in the public press, more extensive reviews in periodicals, technical critiques in professional journals, and reanalyses of data. The book even generated several entire volumes of criticism. Much of this material complements that in the present volume. To pave the way for readers who would like to access this commentary, we believe that the panorama of public reception for *The Bell Curve* is readily discernible from the almost 200 publications contained herein.

Adler, J. (1994), "Beyond *The Bell Curve*: Forget Intelligence, What Matters in Our Society Is Looks," *Newsweek*, 124(19), Nov 7 1994, 56.

Aizawa, K. (1995), "The Gap Between Science and Policy in *The Bell Curve*," *American Behavioral Scientist*, 39(1), Sept-Oct 1995, 84–98.

Alland, A., Jr., Blakey, M.L., Brace, C.L., Goodman, A.H., Molnar, S., Rushton, P.J., Sarich, V.M., and Smedley, A., (1996), "Reviews: *The Bell Curve*," *Current Anthropology*, 37(Suppl), February 1996, s151–s181.

Allen, N.R., Jr. (1995), "Humanism and Intelligence: A Critique of *The Bell Curve* [Viewpoints]," *Free Inquiry*, 15(2), Spring 1995, 60–62.

Allman, W.F. (1994), "Why IQ Isn't Destiny," *US News and World Report*, 117(16), Oct 24 1994, 73–80.

Andrade, K. (1995), "Head Trips: Debunking *The Bell Curve* and the New Wave of Bio-Determinists," *LA Village View*, Dec-Jan 5 1995, 30–32.

Andrews, L.B., and Nelkin, D. (1996), "*The Bell Curve*: A Statement," *Science*, 271, Jan 5 1996, 13–14.

Barone, M., Shipman, P., van den Haag, E., Neuhaus, R.J., Genovese, E.D., Wilson, J.Q., Jensen, A.R., Glazer, N., Lomasky, L.E., Young, M., Berger, B., Loury, G.C., Novak, M., and Seligman, D. (1994), "*The Bell Curve*: A Symposium," *National Review*, 46(3), Dec 5 1994, 32–61.

Barrow, R. (1995), "Keep Them Bells a-Tolling," *The Alberta Journal of Educational Research*, 61(3), September 1995, 289–296.

Bateson, D.J. (1995), "How Can *The Bell Curve* Be Taken Seriously?" *The Alberta Journal of Educational Research*, 61(3), September 1995, 271–273.

Beardsley, T. (1995), "For Whom *The Bell Curve* Really Tolls," *Scientific American*, 272(1), Jan 1995, 14–17.

Belke, T.W. (1995), "A Synopsis of Herrnstein and Murray's *The Bell Curve: Intelligence and Class Structure in American Life* (New York, Free Press, 1994)," *The Alberta Journal of Educational Research*, 61(3), September 1995, 238–256.

Bernstein, J. (1995), "Welfare Bashing Redux: The Return of Charles Murray," *The Humanist*, 55(1), 22–25.

Blackman, R. (1996), "Psych Prof Speaks on Bell Curve," *The Peak*, 91(2), Sept 11 1996, Internet Resources.

Blinkhorn, S., "Willow, titwillow, titwillow!" *Nature*, 372 (1994), 417–419.

Block, N. "Race, Genes, and IQ," *Boston Book Review*, Internet Resources.

Bouchard, T., Jr. (1995), "Breaking the Last Taboo," *Contemporary Psychology*, 40(5), May 1995, Internet Resources.

Boyd, R. (1995), "Genetic Racism," *Denver Post*, Oct. 23, 1994, 21A.

Braden, J. (1995), "For Whom the Bell Tolls: Why *The Bell Curve* Is Important for School Psychologists," *School Psychology Review*, 24 Jan 1, 1995, 27.

Brewer, R.M. (1995), "Knowledge Construction and Racist 'Science'," *American Behavioral Scientist*, 39(1), Sept-Oct 1995, 62–74.

Brimelow, P. (1994), "Disadvantaging the Advantaged," *Forbes*, Nov. 21, 1994, 52–57.

Brimelow, P. (1994), "For Whom the Bell Tolls," *Forbes*, 154(10), Oct 24, 1994, 153–163.

Browne, M.W. (1994), "What Is Intelligence and Who Has It?" *New York Times Book Review*, Oct. 16, 1994, 3.

Bruning, F. (1994), "Book Reviews: *The Bell Curve*," *Maclean's*, 107(47), Nov 21 1994, 13.

Caldwell, C. (1995a), "Alarm Bell," *Scientific American*, Jan 1995, 62–67.

Caldwell, C. (1995b), "Book Reviews: *The Bell Curve*," *American Spectator*, 28(1), Jan 1995, 62–68.

Callahan, S. (1995), "Book Reviews: *The Bell Curve*," *Commonweal*, 133(3), Feb 10 1995, 14–18.

Carey, J. (1994), "Book Reviews: *The Bell Curve*," *Business Week*, (3397), Nov 7 1994, 16–18.

Case, J. (1995), "Is *The Bell Curve* Statistically Sound?" *Society for Industrial and Applied Mathematics (SIAM) News*, 28(1), Jan 1995, pp. 6 and 12.

Chidely, J. (1994), "The Brain Strain: Attacking Liberal Notions of Racial Equality," *Maclean's*, 107(48), Nov 28 1994, 72–74.

Chinyelou, M. (1995), *Debunking the Bell Curve and Scientific Racism*, Vol 1, New York: Mustard Seed Press.

Cohen, R. (1994), "Social Racism," *Pittsburgh Post Gazette*, Oct 20 1994.

Colson, J. (1994), "It's Just a Curve Ball," *The Aspen Times*, Oct 29, 1994, Internet Resources.

Connors, J.B. (1995), "Two Tails Are Better Than One: The Logic of *The Bell Curve*," *The Alberta Journal of Educational Research*, 61(3), September 1995, 342–348.

Conrad, C.A. (1995), "Race, Earnings, and Intelligence," *Black Enterprise*, 25(8), March 1995, 26.

Coulter, R.P. (1995), "And Academic Sexism Too: A Comment on *The Bell Curve*," *The Alberta Journal of Educational Research*, 61(3), September 1995, 308–311.

Cowley, G. (1994), "Testing the Science of Intelligence," *Newsweek*, Oct 1994, 56–60.

Crowell, O.L. (1995), "Book Reviews: *The Bell Curve*," 32(9), May 1995, 1527.

Csikszentmihalyi, M. (1995), "Scales of Inequality," *Educational Leadership*, April 1995, 75–76.

Culbertson, J.C., "Three Informal Fallacies in *The Bell Curve*," Internet Resources.

DeAngelis, T. (1995), "Psychologists Question Findings of Bell Curve," *American Psychological Association Monitor*, Oct. 1995.

DeParle, J. (1994), "Daring Research or Social Science Pornography," *New York Times Magazine*, Oct 9 1994, 48–80.

Devlin, B, Fienberg, S.E., Roeder, K., and Resnick, D. (1995), "Wringing *The Bell Curve*: A Cautionary Tale About the Relationships Among Race, Genes, and IQ," *Chance*, 8(3), 27–36.

Devlin, B., Fienberg, S.E., Resnick, D., and Roeder, K., (1995), "Galton Redux: Eugenics, Intelligence, Race, and Society," *Journal of the American Statistical Association*, 90, 1483–1488.

Dickens, W.T., Kane, T.J., and Schultze, C.L. (1995), "Does *The Bell Curve* Ring True?" *Brookings Review*, 13(3), Summer 1995, 18–23.

Dorfman, D.D. (1995), "Soft Science with a Neoconservative Agenda," *Contemporary Psychology*, May 1995, 40, 5.

Douglas, S. (1994), "Dumbing Down in America," *The Progressive*, 58(12), Dec 1994, 21.

Easterbrook, G. (1994), "The Case Against *The Bell Curve*," *Washington Monthly*, 26(12), Dec 1994, 17–26.

Ehrenrich, B. (1995), "Planet of the White Guys," *Time*, Mar 13 1995, 114.

Epling, W.F., (1995), *The Alberta Journal of Educational Research*, 61(3), September 1995.

Fancher, R.E. (1995), "*The Bell Curve* on Separated Twins," *The Alberta Journal of Educational Research*, 61(3), September 1995, 265–270.

Farrell, W.C., Jr., Johnson, J.H., Jr., Sapp, M., and Jones, C.K. (1995), "*The Bell Curve*: Ringing in the Contract with America," *Educational Leadership*, 52(7), April 1995, 77–79.

Feingold, S. (1995), "For Whom *The Bell Curve* Tolls," *Congress Monthly*, 62(2), Mar/April 1995, 3–6.

Feuerstein, R., and Kozulin, A. (1995), "*The Bell Curve*: Getting the Facts Straight," *Educational Leadership*, 52(7), April 1995, 71–74.

Finn, C.E., Jr. (1995), "Book Review of *The Bell Curve*," *Commentary*, 99(1), Jan 1995, 76.

Fischel, J. (1995), "Strange 'Bell' Fellows (Race-Based Eugenics)," *Commonweal*, 122(3), Feb 10 1995, 16–18.

Fischer, C.S., Hout, M., Jankowski, M.S., Lucas, S.R., Swidler, A., and Voss, K. (1996), *Inequality by Design: Cracking the Bell Curve Myth*, Princeton, NJ: Princeton University Press.

Fraser, S., ed. (1995), *The Bell Curve Wars: Race, Intelligence, and the Future of America*, New York: Basic Books.

Frisby, C. (1995), "When Facts and Orthodoxy Collide: *The Bell Curve* and the Robustness Criterion," *School Psychology Review*, 24, Jan 1, 1995, 12.

Furedy, J.J. (1995), "Race Studies: Contemptuous but Legitimate Science," *The Toronto Star*, Dec 10 1995, K19 and K21.

Gardner, H. (1995), "Cracking Open the IQ Box," *The American Prospect*, (20), Winter 1995, 71–80.

Gartrell, J., and Marquez, S.A. (1995), "The Spurious Relationship Between IQ and Social Behavior: Ethnic Abuse, Gender Ignorance, and Confounded Education," *The Alberta Journal of Educational Research*, 61(3), September 1995, 277–282.

Genovese, E. (1995), "Living with Inequality," in Russell Jacoby and Naomi Glauberman (eds.), *The Bell Curve Debate: History, Documents, Opinions*, New York: Times Books, pp. 331-334.

Goddman, A.H. (1996), "Reviews: *The Bell Curve*," *Current Anthropology*, 37 (Suppl), Feb 1996, s161–s165.

Goldberger, A.S., and Manski, C.F. (1995), "Review Article: *The Bell Curve*," *Journal of Economic Literature*, 33, June 1995, 762–776.

Gould, S.J. (1994), "Curveball," *The New Yorker*, 70(39), Nov 28 1994, 139–149.

Gould, S.J. (1995), "Ghosts of Bell Curves Past," *Natural History*, 104(2), Feb 1995, 12–18.

Gould, S.J. (1996), *The Mismeasure of Man* (Revised and Expanded Edition), New York: Norton.

Hanson, F.A. (1995), "Testing, *The Bell Curve*, and the Social Construction of Intelligence," *Tikkun*, 10(1), Jan-Feb 1995, 22–28.

Harris, M.H. (1996), "Reviews: *The Bell Curve*," *Library Quarterly*, 66(1), Jan 1996, 89–92.

Hauser, R.M. (1995), "Symposium: *The Bell Curve*," *Contemporary Sociology*, 24(2), March 1995, 149–161.

Haynes, N. (1995), "How Skewed Is *The Bell Curve*?" *Journal of Black Psychology*, 21(3), Aug 1995, 275–293.

Hazlett, T.W. (1995), "Ding Dong," *Reason*, 26(8), Jan 1995, 65–66.

Heckman, J.J. (1995), "Cracked Bell," *Reason*, March 1995, 49–56.

Heckman, J.J. (1995), "Lessons from *The Bell Curve*," *Journal of Political Economy*, 5(103), Oct 1995, 1091–1121.

Herrnstein, R.J., and Murray, C. (1996), "The Eternal Triangle: Race, Class, and IQ," *Current Anthropology*, 37 (Suppl.), Feb 1996, s143–s151.

Hirsch, E.D., Jr. (1994), "Good Genes, Bad Schools," *New York Times*, Oct 29 1994.

Hitchens, C. (1994), "Minority Report," *Nation*, 259(18), Nov 28 1994, 640.

Holmes, S.A. (1994), "You're Smart If You Know What You Are," *New York Times*, Oct 23 1994.

Horwitz, L. (1995), "Review of *The Bell Curve* and *The Bell Curve* Debate," *Boston Book Review*, April 1995.

House, E., and Haug, C. (1995), "Riding *The Bell Curve*: A Review," *Education Evaluation and Policy Analysis*, 17(2), 263–272.

Huber, P. (1994), "Silicon on *The Bell Curve*," *Forbes*, 154(12), Nov 21 1994, 210.

Hudson, J.B., Jackson, J.J. Lewis, E., Duster, T., and Early, G. (1995), "The Black Scholar Symposium: A Critique of *The Bell Curve*," *The Black Scholar*, 25, 2–47.

Hunt, E. (1995), "The Role of Intelligence in Modern Society," *American Scientist*, 83, July/August 1995, 356–368.

Irvin, N., III. (1994), "*The Bell Curve* (Book on African-American Intelligence)," *Black Issues in Higher Education*, 11(18), Nov 3 1994, 108.

Jacoby, R., and Glauberman, N., eds. (1995), *The Bell Curve Debate: History, Documents, and Opinion*, New York: Random House.

Johnson, G. (1994), "Learning Just How Little Is Known About the Brain," *New York Times*, Oct 23 1994.

Jones, B.L., and Collins, M.E. (1995), "Differing Views on the State of Equity and Excellence in Schools: A Review," *Equity and Excellence in Education*, 28(1), Apr 1995, 69–75.

Kagan, J. (1996), "The Misleading Abstractions of Social Scientists," *The Chronicle of Higher Education*, 42(18), Jan 12 1996, A52.

Kamin, L.J. (1995), "Behind the Curve," *Scientific American*, 272(2), Feb 1995, 99–103.

Karagiannis, A. (1995), "*The Bell Curve*, A Review," *McGill Journal of Education*, 30(1), 111–116.

Kardon, S. (1996), "Review: *The Bell Curve*," *Social Work*, Jan 1996 41(1), 116–117.

Karlgaard, R. (1994), "IQ and IT (Intelligence and American Business Success)," *Forbes*, 154(13), Dec 5 1994, S9.

Kass, L. (1995), "Intelligence and the Social Scientist," *The Public Interest*, Summer 1995, 64.

Kaus, M., Gates, H.L., Jr., Ramos, D., Lippman, W., Star, A., Peretz, M., Crouch, S., Loury, G., Hacker, A., Rosen, J., Lane, C., Nisbett, R., Glazer, N., Pearson, H., Wolfe, A., Judis, J.B., Hulbert, A., Kennedy, R., Wieseltier, L., Lind, M., Murray, C., and Heckman, R.J. (1994), "Race, Genes, and IQ," *New Republic*, 211(18), Oct 31 1994, 5–37.

Kavanaugh, M. (1995), "*The Bell Curve*: Rising Fears, Fallen Dreams," *The Alberta Journal of Educational Research*, 61(3), September 1995, 297–301.

Kincheloe, J.L., Steinberg, S.R., and Gresson, A.D., III (eds.) (1996), *Measured Lies: The Bell Curve Examined*, New York: St. Martin's Press.

Kirby, J.R. (1995), "Intelligence and Social Policy," *The Alberta Journal of Educational Research*, 61(3), September 1995, 322–334.

Kissinger, M. (1994), "Explosive Book on Race Has Roots Here," *Milwaukee Journal*, Oct 23 1994, pp. 1 and 22.

Kohn, M. (1995), "Science and Race Matters," (reprinted from the *Independent* on Sunday, Sept 10, 1995), *World Press Review*, 42(12), Dec 1995, 48.

Kranzler, J. (1995), "Commentary on Some of the Theoretical Support for *The Bell Curve*," *School Psychology Review*, 24 Jan 1, 1995, 36.

Krishnan, P. (1995), "*The Bell Curve*: Some Statistical Concerns," *The Alberta Journal of Educational Research*, 61(3), September 1995, 274–276.

Krull, C.D., and Pierce, W.D. (1995), "IQ Testing in America: A Victim of Its Own Success," *The Alberta Journal of Educational Research*, 61(3), September 1995, 349–354.

Kupermintz, H. (1996), "*The Bell Curve*: Corrected for Skew," Education Policy Analysis Archives, 4(20), http://olam.ed.asu.edu/epaa/v4n20.html.

Lacayo, R. (1994), "For Whom the Bell Curves," *Time*, 144(17), Oct 24 1994, 66–67.

Lane, C. (1994), "The Tainted Sources of *The Bell Curve*," *The New York Review of Books*, 41(20), Dec 1 1994, 14–19.

Lemann, N. (1996), "A Cartoon Elite," *The Atlantic Monthly*, November, 109-116.

Leo, J. (1994), "Return to the IQ Wars," *US News and World Report*, 117(16), Oct 24 1994, 24.

Lerner, B. (1995), "Aim Higher," *National Review*, June 3, 1995, 56.

Levin, M. (1995), "For Whom *The Bell Curve* Tolls," *Globe and Mail*, Aug 17 1995.

Lieber, M.D. (1994), "An Anthropological Look at Race and Intelligence," *Chicago Tribune*, Oct 23 1994.

Lieberman, L. (1995), "Herrnstein and Murray, Inc," *American Behavioral Scientist*, 39(1), Sept-Oct 1995, 25–35.

Livingstone, D.W. (1995), "For Whom *The Bell Curve* Tolls," *The Alberta Journal of Educational Research*, 61(3), September 1995, 335–341.

MacPherson, E.D. (1995), "*The Time Machine*: The Next Generation," *The Alberta Journal of Educational Research*, 61(3), September 1995, 304–307.

Massey, D.S. (1995), "Review Essay: *The Bell Curve*," *American Journal of Sociology*, 101(3), Nov 1995, 747–754.

McInerney, J.D. (1996), "Why Biological Literacy Matters: A Review of Commentaries Related to *The Bell Curve*," *Quarterly Review of Biology*, 71(1), Mar 1996, 81–97.

McMinn, L.G. (1994), "For Whom The Bell Curves," *Christianity Today*, 38(14), Dec 12 1994, 19.

Miele, F. (1995), "An Interview with the Author of *The Bell Curve*, Charles Murray," *Skeptic*, 3(2), 34–41.

Miller, A. (1994), "Professors of Hate," *Rolling Stone*, Oct 20 1995, 106–118.

Miner, B. (1995), "Who Is Backing *The Bell Curve*?" *Educational Leadership*, 52(7), April 1995, 80–81.

Molnar, A. (1995), "*The Bell Curve*: For Whom It Tolls," *Educational Leadership*, 52(7), April 1995, 69–70.

Morgenthau, T. (Oct 1994), "IQ: Is It Destiny?," *Newsweek*, 124(17), 53–55.

Morin, R. (1995), "An Army from Academe Tries to Straighten Out *The Bell Curve*," *The Washington Post*, Jan 16 1995, A3.

Morrow, L. (1994), "The Cure for Racism," *Time*, 144(23), Dec 5 1994.

Murphey, D.D. (1995), "Rethinking the American Dream: Reactions of the Media to *The Bell Curve*," *Journal of Social, Political, and Economic Studies*, 20(1), Spring 1995, 93–131.

Murray, C. (1995), "*The Bell Curve* and Its Critics," *Commentary*, 99(5), May 1995, 23–31.

Naureckas, J. (1995), "Racism Resurgent: How Media Let *The Bell Curve*'s Pseudo-Science Define the Agenda on Race," *Extra*, Jan/Feb 1995, 20.

Neisser, U., Boodoo, G., Bouchard, T.J., Jr., Boykin, A.W., Brody, N., Ceci, S., Halpern, D.F., Loehlin, R.J.C.P., Sternberg, R.J., and Urbina, S. (1996), "Intelligence: Known and Unknowns," *The American Psychologist*, 51(2), Feb 1996, 77–103.

Newby, R.G. (1995), "*The Bell Curve*: Laying Bare the Resurgence of Scientific Racism," *American Behavioral Scientist*, 39(1), Sept/Oct 1995, 6.

Nielsen, F. (1995), "Book Review: *The Bell Curve*," *Social Forces*, 74(1), Sept 1995, 337–343.

Norwood, J.L. (1994), "Comments on *The Bell Curve*," National Urban League and Howard University Media Symposium, Washington D.C., Dec 12 1994.

Nunley, M. (1995), "*The Bell Curve*: Too Smooth to Be True," *American Behavioral Scientist*, 39(1), Sept-Oct 1995, 74–84.

Passell, P. (1994), "Bell Curve Critics Say Early IQ Isn't Destiny," *New York Times*, Nov 9 1994, B10.

Pattullo, E.L. (1996), "On *The Bell Curve*," *Society*, 33(3), Mar/April 1996, 86–88.

Perkins, D.N. (1995), *Outsmarting IQ: The Emerging Science of Learnable Intelligence*, New York: The Free Press.

Poole, D.A. (1995), "The Two Bell Curves," *American Behavioral Scientist*, 39(1), Sept/Oct 1995, 35–43.

Presston, S.H. (1995), "Book Reviews: *The Bell Curve*," *Population and Development Review*, 21(3), Sept 1995, 675–678.

Quaye, R. (1995), "The Assault on the Human Spirit: *The Bell Curve*," *The Black Scholar*, 25, 41–43.

Raspberry, W. (1994), "A Stink Bomb," *Pittsburgh Post-Gazette*, Oct 20 1994, B3.

Reed, A., Jr. (1994), "Herrnstein and Murray: *The Bell Curve: Intelligence and Class Structure in American Life*," *Nation*, 259(18), Nov 28 1994, 654–662.

Reed, A., Jr. (1994), "Intellectual Brown Shirts," *The Progressive*, 58(12), Dec 1994, 15–18.

Reese, W.A., II. (1996), "The Shaped Bell Curve and the Social Sciences," *Social Science Journal*, 33(1), Jan 1996, 113–119.

Reiland, R.R. (1995), "Charles Murray and Albert Einstein," *The Humanist*, 55(2), Mar-Apr 1995, 3–5.

Reviere, R. (1995), " 'Slow Through the Dark': A Commentary on *The Bell Curve*," *The Alberta Journal of Educational Research*, 61(3), September 1995, 312–315.

Richard, H.W., and Washington, M. (1996), "Critique of the Article 'How Skewed Is *The Bell Curve*'," *Journal of Black Psychology*, 21(3), Aug 1995, 293–297.

Ridgeway, J. (1994), "Behind *The Bell Curve*: Tracing the Roots of Charles Murray's Race Management," *Village Voice*, Nov 15 1994.

Robinson, D.N. (1994), "Book Reviews: *The Bell Curve*," *Business and Society Review*, (91), Fall 1994, 69–73.

Robitaille, D.F., and Robeck, E.C. (1995), "Distorted Vision: Education as Seen Through *The Bell Curve*," *The Alberta Journal of Educational Research*, 61(3), September 1995, 367–376.

Rosenthal, S.J. (1995), "The Pioneer Fund: Financier of Fascist Research," *American Behavioral Scientist*, 39(1), Sept-Oct 1995, 44–62.

Ryan, A. (1994), "Apocalypse Now," *New York Review of Books*, Nov 17 1994, 7–11.

Safire, W. (1994), "Of IQ and Genes," *New York Times*, Nov 19 1994, A19.

Scarr, S. (1994), "Book Reviews: *The Bell Curve*," *Issues in Science and Technology*, 11(2), Winter 1994, 82–86.

Schleifer, M. (1995), "Should We Change our Views About Early Childhood Education?" *The Alberta Journal of Educational Research*, 61(3), September 1995, 355–359.

Schultze, C.L., Dickens, W.T., and Kane, T.J. (1995), *Does "The Bell Curve" Ring True?*, Washington, D.C.: Brookings Institution.

Seabrook, J. (1996), "All in the Genes," *New Yorker*, Feb 12 1996, 79–81.

Shell, E.R. (1995), "Book Reviews: *The Bell Curve*," *Technology Review*, 98(4), May-June 1995, 75–79.

Siegel, L. (1995a), "For Whom *The Bell Curves*: The New Assault on Egalitarianism," *Tikkun*, 10(1), Jan-Feb 1995, 27–33.

Siegel, L.S. (1995b), "Does the IQ God Exist?," *The Alberta Journal of Educational Research*, 61(3), September 1995, 283–288.

Siegel, R. (1994), "Interview with Charles Murray," *National Public Radio*, Oct 10, 1994.

Singham, M. (1995), "Race and Intelligence: What Are the Issues?" *Phi Delta Kappan*, 77(4), Dec 1995, 271–279.

Snell, L. (1995), *Chance News*, 11 Jan. 1995 to 1 Feb 1995, Internet Resources.

Sowell, T. (1995), "Ethnicity and IQ," *American Spectator*, 28, Feb 1, 1995, 32.

Sperling, S. (1994), "Beating a Dead Monkey," *Nation*, 259(18), Nov 28, 1994, 663–665.

Sternberg, R.J. (1995), "For Whom *The Bell Curve* Tolls: A Review of *The Bell Curve*," *Psychological Science*, 6(5), Sept 1995, 257-261.

Sternberg, R.J. (1996), "The School Bell and *The Bell Curve*: Why They Don't Mix," *NASSP Bulletin*, 80(577), Feb 1996, 46–56.

Sternberg, R.J. (1996), "For Whom Does *The Bell Curve* Toll?" (lecture excerpt), *The Vocational Educational Journal*, 71(5), May 1996, 62.

Sternberg, R.J., Callahan, C., Burns, D., Gubbins, E.J., Purcell, J., Reis, S.M., Renzulli, J.S., and Westberg, K. (1995), "Book Reviews: *The Bell Curve*," *Gifted Child Quarterly*, 39(3), Summer 1995, 177–180.

Subramanian, S. (1995), "The Story in Our Genes," *Time*, 145(3), Jan 16 1995.

Sullivan, A. (1996), "Closed Minds in the Land of the Free," *Sunday Times*, April 28 1996.

Taylor, H. (1995), "Review Symposium of *The Bell Curve*," *Contemporary Sociology*, 24, 153–158.

Traub, J. (1995), "Norman Podhoretz: *Leaving It*," *The New Yorker*, Feb 6 1995, 30.

Van Brunschot, E.G., and Brannigan, A. (1995), "IQ and Crime: Dull Behavior and/or Misspecified Theory?" *The Alberta Journal of Educational Research*, 61(3), September 1995, 316–321.

Wahlsten, D. (1995), "Increasing the Raw Intelligence of a Nation is Constrained by Ignorance, Not Its Citizens' Genes," *The Alberta Journal of Educational Research*, 61(3), September 1995, 257–264.

Walters, R. (1995), "The Impact of the 'Bell Curve' Ideology on African American Public Policy," *American Behavioral Scientist*, 39(1), 98–109.

Wangler, D.G. (1995), "Is *The Bell Curve* a Ringer?" *The Alberta Journal of Educational Research*, 61(3), September 1995, 360–367.

Wattenberg, B. (1994), "A Conversation With Charles Murray," *Think Tank*, Oct 14 1994.

Weisberg, J. (1994), "Who, Me? Prejudiced?" *New York*, 27(41), Oct 17 1994, 26–28.

Whatley, W. (1995), "Wanted: Some Black Long-Distance Runners—The Message of *The Bell Curve*," *The Black Scholar*, 25, 44–46.

Winship, C. (1994), "Lessons Beyond *The Bell Curve*," *New York Times*, Nov 15 1994.

Wooldridge, A. (1995), "Bell Curve Liberals," *New Republic*, Feb 27 1995, 22–24.

Wright, R. (1994), "For Whom *The Bell Curves*," *Time Magazine*, Oct 24, 1994, 66–67.

Wright, R. (1994), "Bell, Book, and Scandal (Intelligence Testing)," *Economist*, 333(7895), Dec 24 1994, 69–72.

Wright, R. (1994), "What If Intelligence Is Inheritable?" *Nature*, 371(6499), Oct 20 1994, 637.

Wright, R. (1994), "How Clever Is Charles Murray?" *Economist*, 333(7786), Oct 22 1994, 29–30.

Wright, R. (1994), "How Intelligence Findings Are Seen Poses the Danger," *The Daily Beacon*, 67(40), Oct 19 1994, 4.

Wright, R. (1994), "Inequity Quotient," *Nation*, 259(15), Nov 7 1994, 516.

Wright, R. (1994), "IQ: A Hard Look at a Controversial New Book on Race, Class and Success," *Newsweek*, Oct 24, 1994, 53–62.

Wright, R. (1994), "Mainstream Science on Intelligence," *Wall Street Journal*, Dec 13 1994.

Wright, R. (1995), "Dumb Bell," *New Republic*, 212(1), Jan 2 1995, 6.

Wright, R. (1995), "The Herrnstein and Murray Book: A Controversy (interview with Education Testing Service Pres. Nancy Cole)," *Journal of Teacher Education*, 46(1), Jan-Feb 1995, 7–10.

Wright, R. (1995), "Racial Differences Need Not Reflect Genetic Differences," *Brookings Review*, 13(3), Summer 1995, 23.
Wright, R. (1995), "Skeptic Magazine Interview with Robert Sternberg on *The Bell Curve*," *Skeptic*, 3(3), 72–80.
Wright, R. (1994), "The Bell Curve Agenda," *The New York Times*, Oct 24 1994, Editorial.
Wright, R. (1994), "Scholars Defend *The Bell Curve*," *Science* 266, Dec 16 1994, 181.
Wright, R. (1995), "Intelligence: Knowns and Unknowns," Report of a task force established by the Board of Scientific Affairs of the Amerian Psychological Association, Aug 7 1995.
Wright, R. (1996), "Inequality by Design: Cracking *The Bell Curve* Myth," *The Economist*, 340(7974), July 13 1996, 89–90.
Wright, R. "Task Force Releases Report in Response to *The Bell Curve*," Internet Resources.

AUTHOR INDEX

SUBJECT INDEX